Lecture Notes in Physics

Edited by J. Ehlers, München, K. Hepp, Zürich,
H. A. Weidenmüller, Heidelberg, and J. Zittartz, Köln
Managing Editor: W. Beiglböck, Heidelberg

48

Interplanetary Dust and Zodiacal Light

Proceedings of the IAU-Colloquium No. 31,
Heidelberg, June 10–13, 1975

Edited by H. Elsässer and H. Fechtig

Springer-Verlag
Berlin · Heidelberg · New York 1976

Editorial Board
J. S. Dohnanyi, Holmdel, U.S.A.
E. Grün, Heidelberg, BRD
B. A. Lindblad, Lund, Sweden
H. Link, Heidelberg, BRD
J. Rahe, Bamberg, BRD

Editors
Prof. H. Elsässer
Max-Planck-Institut für Astronomie
Königstuhl
6900 Heidelberg/BRD

Prof. H. Fechting
Max-Planck-Institut für Kernphysik
Saupfercheckweg
6900 Heidelberg/BRD

Library of Congress Cataloging in Publication Data

```
IAU Colloquium No. 31 on Interplanetary Dust and
Zodiacal Light, Heidelberg, 1975.
     Interplanetary dust and zodiacal light.

     (Lecure notes in physics ; 48)
     Sponsored by the IAU and COSPAR.
     Bibliography: p.
     Includes index.
     1. Cosmic dust--Congresses. 2. Zodiacal light--
Congresses. I. Elsässer, Hans, 1929-
II. Fechtig, H., 1929-      III. International
Astronomical Union. IV. International Council of
Scientific Unions. Committee on Space Research.
V. Title. VI. Series.
QB791.I15 1975       523.2        76-2597
```

ISBN 3-540-07615-8 Springer-Verlag Berlin Heidelberg New York
ISBN 0-387-07615-8 Springer-Verlag New York Heidelberg Berlin

This work is subject to copyright. All rights are reserved, whether the whole or part of the material is concerned, specifically those of translation, reprinting, re-use of illustrations, broadcasting, reproduction by photocopying machine or similar means, and storage in data banks.
Under § 54 of the German Copyright Law where copies are made for other than private use, a fee is payable to the publisher, the amount of the fee to be determined by agreement with the publisher.
© by Springer-Verlag Berlin · Heidelberg 1976
Printed in Germany
Printing and binding: Beltz Offsetdruck, Hemsbach/Bergstr.

Introduction

Research on particulate matter in interplanetary space has achieved considerable progress in recent years. Space experiments for the registration of interplanetary dust particles as well as for observations of the zodiacal light were performed and are at present still in operation on the Pioneer Jupiter and Helios solar probes. New data appeared which have been eagerly awaited for years.

The IAU Colloquium No. 31 on "Interplanetary Dust and Zodiacal Light" which took place in Heidelberg from June 10 until June 13, 1975, attempted to bring together all groups active in this field in order to render a thorough discussion of the new results. The response in all parts of the world and the large attendance demonstrated the justification for this meeting. It was sponsored by the IAU and by COSPAR. Three Commissions of the IAU, No. 15 (Physical Study of Comets, Minor Planets and Meteorites), No. 21 (Light of the Night Sky) and No. 22 (Meteors and Interplanetary Dust) supported the colloquium and were represented by their presidents or vice-presidents; on behalf of COSPAR the Panel 3. C on Cosmic Dust was responsible.

The Scientific Organizing Committee of the colloquium consisted of A.H. Delsemme, R. Dumont, H. Elsässer (Chairman), H. Fechtig, C.L. Hemenway, R.C. Jennison, B.A. Lindblad, R.E. McCrosky, V. Vanysek, J.L. Weinberg.

The local organisation was in the hands of H. Fechtig, E. Grün and H. Link. The excellent management of the conference office by Miss T. Filsinger and Miss E. Siepmann is gratefully acknowledged.

We are grateful to the Deutsche Forschungsgemeinschaft and the Max-Planck-Gesellschaft for financial support.

    H. Elsässer   H. Fechtig

1. L. Kohoutek, 2. P. Proisy, 3. Mrs. Proisy, 4. K.W. Michel,
5. D.E. Brownlee, 6. A. Mujica, 7. A. Llebaria, 8. W. Kokott,
9. Mrs. Gehrels, 10. C. Leinert, 11. T. Gehrels, 12. J.G. Delcourt,
13. S. Temesvary, 14. R. Bloch, 15. N. Pailer, 16. J.L. Weinberg,
17. Z. Ceplecha, 18. J.G. Sparrow, 19. W. Kempe, 20. G. Eichhorn,
21. E. Grün, 22. B. Donn, 23. D.A. Morrison, 24. B. Marsden,
25. B.K. Dalmann, 26. W. Gentner, 27. H.J. Völk, 28. M.S. Hanner,
29. F. Link, 30. R. Robley, 31. J.W. Rhee, 32. J. Trulsen,
33. Z. Sekanina, 34. H.J. Staude, 35. J.M. Alvarez, 36. E. Pitz,
37. T. Nishimura, 38. S. Drapatz, 39. H. Elsässer, 40. P. Szkody,
41. J.A.M. McDonnell, 42. E. Schneider, 43. V. Vanysek, 44. J. Rahe,
45. F.E. Roach, 46. J.R. Roach, 47. D.A. Tomandl, 48. S. Röser,
49. V. Stähle, 50. O.E. Berg, 51. J. Kissel, 52. R. Soberman,
53. J. Hartung, 54. H. Wolf, 55. J.C. Mandeville, 56. R. Dumont,
57. B. König, 58. J.E. Blamont, 59. A. Levasseur, 60. G.B. Burnett,
61. H.U. Keller, 62. J. Rosinski, 63. ? , 64. D.S. Hallgren,

65. L.W. Banderman, 66. D.W. Hughes, 67. C.F. Lillie, 68. H. Lee,
69. R.A. Howard, 70. R.D. Wolstencroft, 71. G. Morfill,
72. R.H. Munro, 73. G. Schwehm, 74. S. Hayakawa, 75. K.D. Schmidt,
76. H. Link, 77. M. Kröger, 78. R.H. Giese, 79. G. Braun,
80. P.W. Blum, 81. H. Tanabe, 82. M. Alexander, 83. A.H. Delsemme,
84. H. Fechtig, 85. F.L. Whipple, 86. J.S. Dohnanyi,
87. C.L. Hemenway, 88. P.M. Millman, 89. R. Wlochowicz

Other participants not appearing on the photo:

J. Appenzeller, Ch. Bertaux, R. Bien, A. Frey, D. Hablik,
E.G. Igenbergs, Ph.L. Lamy, D. Lemke, B.A. Lindblad, H. Mach,
K. Nagel, G. Neukum, K. Nock, C.L. Ross, M. Schmidt, Th. Schmidt,
K.-H. Schneider, H. Scholl, P. Strittmatter, C. Thum, K. Weiss,
R. Zerull

TABLE OF CONTENTS

1 Zodiacal Light

1.1 Observations from Space   1

   1.1.1   Space Observations of the Zodiacal Light
Invited Paper
J.L. Weinberg   3

   1.1.2   Helios Zodiacal Light Experiment
E. Pitz, C. Leinert, H. Link, and N. Salm   19

   1.1.3   Preliminary Results of the Helios A Zodiacal Light Experiment
H. Link, C. Leinert, E. Pitz, and N. Salm   24

   1.1.4   Pioneer 10 Observations of Zodiacal Light Brightness Near the Ecliptic: Changes with Heliocentric Distance
M.S. Hanner, J.G. Sparrow, J.L. Weinberg, and D.E. Beeson   29

   1.1.5   Star Counts in the Background Sky Observed from Pioneer 10
H. Tanabe, and K. Mori   36

   1.1.6   The $S_{10}(V)$ Unit of Surface Brightness
J.G. Sparrow, and J.L. Weinberg   41

   1.1.7   Polarization of the Zodiacal Light: First Results from Skylab
J.G. Sparrow, J.L. Weinberg, and R.C. Hahn   45

   1.1.8   Photometry of the Zodiacal Light with the Balloon-Borne Telescope THISBE
A. Frey, W. Hofmann, D. Lemke, and C. Thum   52

   1.1.9   OSO-5 Zodiacal Light Measurements
G.B. Burnett   53

   1.1.10   Evidence for Scattering Particles in Meteor Streams
A.C. Levasseur, and J. Blamont   58

   1.1.11   The Ultraviolet Scattering Efficiency of Interplanetary Dust Grains
Ch.F. Lillie   63

   1.1.12   Summary of Observations of the Solar Corona/Inner Zodiacal Light from Apollo 15, 16, and 17
C.L. Ross   64

   1.1.13   A Temporal Study of the Radiance of the F-Corona Close to the Sun
R.H. Munro   65

   1.1.14   Measurements of the F-Corona from Daily OSO-7 Observations
R.A. Howard, and M.J. Koomen   66

1.1.15	The Thermal Emission of the Dust Corona during the Eclipse of June 30, 1973 P. Lena, Y. Viala, D. Hall, and A. Soufflot	67
1.1.16	The Color Characteristics of the Earth-Moon Libration Clouds J.R. Roach	68
1.1.17	A Search for Forward Scattering of Sunlight from the Lunar Libration Clouds C.L. Ross	73
1.1.18	Presentation of Zodiacal Light Instrument Aboard the D2B Astronomical Satellite M. Maucherat, and P. Cruvellier	74
1.1.19	Visible and UV Photometry of the Gegenschein and the Milky Way A. Llebaria	78

1.2 Groundbased Observations 83

1.2.1	Ground-Based Observations of the Zodiacal Light Invited Paper R. Dumont	85
1.2.2	Polarimetry of the Zodiacal Light and Milky Way from Hawaii R.D. Wolstencroft, and L.W. Bandermann	101
1.2.3	Scattering in the Earth's Atmosphere: Calculations for Milky Way and Zodiacal Light as Extended Sources H.J. Staude	106
1.2.4	Scattering Layer of Interplanetary Dust in the Upper Atmosphere F. Link	107

1.3 Models and Interpretation 113

1.3.1	Some Formulae to Interpret Zodiacal Light Photopolarimetric Data in the Ecliptic from Ground or Space R. Dumont	115
1.3.2	Discussion of the Rocket Photometry of the Zodiacal Light C. Leinert, H. Link, and E. Pitz	120
1.3.3	Consequences of the Inclination of the Zodiacal Cloud on the Ecliptic R. Robley	121
1.3.4	Method for the Determination of the Intensity of Scattered Sunlight per Unit-Volume of the Interplanetary Medium A. Mujica, and F. Sánchez	122
1.3.5	On the Visibility of the Libration Clouds S. Röser	124
1.3.6	Scattering Functions of Dielectric and Absorbing Irregular Particles R. Zerull	130

1.3.7	The Compatibility of Recent Micrometeoroid Flux Curves with Observations and Models of the Zodiacal Light R.H. Giese, and E. Grün	135

2 In Situ Measurements of Interplanetary Dust

2.1 Measurements from Satellites and Space Probes — 141

2.1.1	In-Situ Records of Interplanetary Dust Particles - Methods and Results Invited Paper H. Fechtig	143
2.1.2	Preliminary Results of Micrometeoroid Experiment on Board Helios A E. Grün, J. Kissel, H. Fechtig, P. Gammelin, and H.-J. Hoffmann	159
2.1.3	Composition of Impact-Plasma Measured by a Helios-Micrometeoroid-Detector B.-K. Dalmann, E. Grün, and J. Kissel	164
2.1.4	Orbital Elements of Dust Particles Intercepted by Pioneers 8 and 9 H. Wolf, J.W. Rhee, and O.E. Berg	165
2.1.5	Flux of Hyperbolic Micrometeoroids J.S. Dohnanyi	170
2.1.6	The Cosmic Dust Environment at Earth, Jupiter and Interplanetary Space: Results from Langley Experiments on MTS, Pioneer 10, and Pioneer 11 J.M. Alvarez	181
2.1.7	Dust in the Outer Solar System - Review of Early Results from Pioneers 10 and 11 R. Soberman, J.M. Alvarez, and J.L. Weinberg	182
2.1.8	Sources of Interplanetary Dust: Asteroids Invited Paper J.S. Dohnanyi	187

2.2 Lunar Studies and Simulation Experiments — 207

2.2.1	Lunar Microcraters and Interplanetary Dust Fluxes Invited Paper J.B. Hartung	209
2.2.2	The Size Frequency Distribution and Rate of Production of Microcraters D.A. Morrison, and E. Zinner	227
2.2.3	The Long Term Population of Interplanetary Micrometeoroids G. Poupeau, R.M. Walker, and E. Zinner	232
2.2.4	Lunar Soil Movement Registered by the Apollo 17 Cosmic Dust Experiment O.E. Berg, H. Wolf, and J. Rhee	233

2.2.5	Electrostatic Disruption of Lunar Dust Particles J.W. Rhee	238
2.2.6	Microcraters Produced by Oblique Incidence of Projectiles V. Stähle, K. Nagel, and E. Schneider	241
2.2.7	Measurements of Impact Ejecta Parameters in Crater Simulation Experiments E. Schneider	242
2.2.8	Impact Light Flash Studies: Temperature, Ejecta, Vaporization G. Eichhorn	243

2.3 Particle Collection Experiments and Their Interpretation — 249

2.3.1	Submicron Particles from the Sun Invited Paper C.L. Hemenway	251
2.3.2	Analysis of Impact Craters from the S-149 Skylab Experiment D.S. Hallgren, and C.L. Hemenway	270
2.3.3	Micrometeorite Impact Craters on Skylab Experiment S-149 K. Nagel, H. Fechtig, E. Schneider, and G. Neukum	275
2.3.4	Extraterrestrial Particles in the Stratosphere D.E. Brownlee, D. Tomandl, and P.W. Hodge	279
2.3.5	Magellan Collections of Large Cosmic Dust Particles D.S. Hallgren, C.L. Hemenway, and R. Wlochowicz	284
2.3.6	Specific Sources of Extraterrestrial Particles J. Rosinski	289
2.3.7	Near-Earth Fragmentation of Cosmic Dust H. Fechtig, and C.L. Hemenway	290

Cometary Dust — 297

3.1	Dust in Comets and Interplanetary Matter Invited Paper V. Vanýsek	299
3.2	The Production Rate of Dust by Comets A.H. Delsemme	314
3.3	Can Short Period Comets Maintain the Zodiacal Cloud? S. Röser	319
3.4	Optical Properties of Cometary Dust S. Hayakawa, T. Matsumoto, and T. Ono	323
3.5	The Dust Coma of Comets K.W. Michel, and T. Nishimura	328

3.6	Dust Emission from Comet Kohoutek (1973f) at Large Distances from the Sun E. Grün, J. Kissel, and H.-J. Hoffmann		334
3.7	Predicted Favorable Visibility Conditions for Anomalous Tails of Comets Z. Sekanina		339
3.8	Study of the Anti-Tail of Comet Kohoutek from an Observation on 17 January 1974 Ph.L. Lamy, and S. Koutchmy		343
3.9	Condensation Processes at High Temperature Clouds B. Donn		345
3.10	Mariner Mission to Encke 1980 C.M. Yeates, K.T. Nock, and R.L. Newburn		346

4 Meteors and Their Relation to Interplanetary Dust — 356

4.1	Meteors and Interplanetary Dust Invited Paper P.M. Millman	359
4.2	Meteoroid Densities B.A. Lindblad	373
4.3	Possible Evidence of Meteoroid Fragmentation in Interplanetary Space from Grouping of Particles in Meteor Streams V. Porubčan	379
4.4	The Heliocentric Distribution of the Meteor Bodies at the Vicinity of the Earth's Orbit V.V. Andreev, O.I. Belkovich, and V.S. Tokhtas'ev	383
4.5	Fireballs as an Atmospheric Source of Meteoritic Dust Z. Ceplecha	385
4.6	Interplanetary Dust in the Vicinity of the Earth G.M. Teptin	389
4.7	Meteor Radar Rates and the Solar Cycle B.A. Lindblad	390
4.8	Evolution and Detectability of Interplanetary Dust Streams L. Kresák	391
4.9	On the Structure of Hyperbolic Interplanetary Dust Streams L. Kresák, and E.M. Pittich	396
4.10	Expected Distribution of Some Orbital Elements of Interstellar Particles in the Solar System O.I. Belkovich, and I.N. Potapov	400

5 Dynamics and Evolution — 401

5.1	Sources of Interplanetary Dust Invited Paper F.L. Whipple	403

5.2	Dynamics of Interplanetary Dust and Related Topics Invited Paper J. Trulsen	416
5.3	Modeling of the Orbital Evolution of Vaporizing Dust Particles Near the Sun Z. Sekanina	434
5.4	Orbital Evolution of Circum-Solar Dust Grains Ph. Lamy	437
5.5	Temperature Distribution and Lifetime of Interplanetary Ice Grains Ph. Lamy, and M.F. Jousselme	443
5.6	Radial Distribution of Meteoric Particles in Interplanetary Space J.W. Rhee	448
5.7	Rotational Bursting of Interplanetary Dust Particles S.J. Paddack, and J.W. Rhee	453
5.8	Lunar Ejecta in Heliocentric Space W.M. Alexander, and M.A. Richards	458
5.9	Radiation Pressure on Interplanetary Dust Particles G. Schwehm	459
5.10	Are Interplanetary Grains Crystalline? S. Drapatz, and K.W. Michel	464
5.11	A Technique for Measuring the Interstellar Component of Cosmic Dust D.A. Tomandl	469

6 Concluding Summaries   473

6.1	The Zodiacal Light H. Elsässer	475
6.2	In Situ Measurements of Dust O.E. Berg	478
6.3	Can Comets be the Only Source of Interplanetary Dust? A.H. Delsemme	481
6.4	Meteors Z. Ceplecha	485
6.5	Final Remarks F.L. Whipple	489

Authors Index   494

1 ZODIACAL LIGHT

1.1 Observations from Space

1.1.1   SPACE OBSERVATIONS OF THE ZODIACAL LIGHT

J. L. WEINBERG
Space Astronomy Laboratory, State University of New York at Albany

Abstract. A listing and discussion are given of balloon, rocket, satellite, and space probe observations of the zodiacal light. The paucity of space observations in several critical areas (e.g., in the ultraviolet and near the sun) is noted and suggestions are made for experiments to meet these needs.

Figure 1 shows the zodiacal light as it typically appears at a low, northern latitude site in the mid-January morning sky. This photograph was taken in January 1967, several weeks before the first international symposium on interplanetary dust and the zodiacal light. At that time there were relatively few space observations, and most of our knowledge of the zodiacal light was derived from ground based observations. In recent years there has been an increasing number of space experiments, and it is most appropriate to have a second meeting at this time and in this place, where interplanetary dust and zodiacal light have been studied intensively for a number of years and where several important space experiments have recently been performed:

1. The first observations of the inner zodiacal light during the day and outside of eclipse (Leinert, et al. 1974a);
2. Multicolor observations of the zodiacal light from a balloon (Frey, et al. 1974; Hofmann, et al. 1973);
3. A zodiacal light experiment (Leinert, et al. 1974b) and a micrometeoroid detection experiment (Dietzel, et al. 1973) on Helios A, the first probe of the inner solar system; and
4. Near-earth and interplanetary measurements of micrometeoroids from HEOS 2 (Hoffmann, et al. 1975).

From Ground to Space. As the results of different ground observers came into better agreement and more opportunities arose for space observations, attempts were made to verify earlier zodiacal light results and to make observations not able to be made from the ground. One reason for going into space is illustrated in Figure 2 which shows the F-corona/inner zodiacal light as photographed from lunar orbit during the Apollo 15 mission (Mercer, et al. 1973b). This region is masked by twilight and is not observable from the ground (the apex of the cone in the Apollo photograph is approximately 20 degrees from the sun; i.e., it would be behind the outcropping of rocks shown in Figure 1). Having gone "from ground to space", it is appropriate for the author to enumerate the reasons for going (and, subsequently, to indicate what may be found when

Fig. 1. Photograph of the morning zodiacal light by P. B. Hutchison; Mt. Haleakala, Hawaii, 12 January 1967. 100 sec exposure on Kodak Tri-X. The outcropping of rock at the base of the zodiacal light cone is approximately 20 degrees from the sun.

Fig. 2. Apollo 15 photograph of the F-corona/inner zodiacal light from lunar orbit using a 70mm Hasselblad electric camera and Kodak 2485 film. Mercury, at 28 degrees from the sun, and Regulus are seen above the apex of the light cone.

one gets there):
1. To observe near the sun. Zodiacal light observations must avoid twilight, and corrections for airglow continuum and scattered light are uncertain closer than 10 degrees to the horizon. Therefore, observations from the ground are essentially restricted to elongations greater than 30 degrees.
2. To observe in the ultraviolet. Observations are difficult even in the blue and near ultraviolet. Atmospheric absorption is high, and there are relatively few windows that are free of airglow line and band emission.
3. To observe in the near infrared and parts of the infrared that are obscured by atmospheric emission and absorption.
4. To measure the change in zodiacal light with heliocentric distance in and out of the ecliptic and, thereby, to derive information on the spatial distribution of interplanetary dust.
5. To perform combined micrometeoroid/zodiacal light experiments.
6. To avoid uncertainties arising from airglow line and continuum emission, atmospheric scattering and extinction, and light pollution.
7. To increase the useable observing time by avoiding bad weather, the moon, and the personal equation.
8. To avoid confusion from the changing celestial aspect inherent in ground observations.

Although many of these difficulties are removed in space observations, a new, in many ways more difficult, set appears:
1. Limitations on location, weight, volume, power, and data rate.
2. Instrument design, test, and delivery are often based on system delivery schedules rather than on engineering or scientific readiness, resulting in insufficient time or access to the instrument for testing and calibration prior to launch.
3. Launch can abort or the instrument can fail, with little or no chance for repair.
4. Scheduling difficulties or degradation of data can result from spacecraft maneuvers, changes in experiment or mission priorities, contamination (spacecraft corona or particulate deposition), competition for astronaut time, interior vehicle lighting, etc.
5. Inadequate vehicle or instrument pointing information.
6. Effects of vehicle mechanical or electrical limitations or failures on instrument operation.
7. Environmental effects such as temperature variations associated with cyclic day-night-day operation, radiation effects during passage through the South Atlantic anomaly, ultraviolet degradation of optical components, high or low temperature operation, etc.
8. Stray light from direct or indirect solar radiation (Leinert and Klüppelberg 1974).
9. There are often long delays in getting correct data tapes.
10. In many cases, insufficient funds or time are provided for data analysis.

This represents a formidable set of obstacles, and without extensive engineering and scientific assistance one is certain to have several years of lower scientific productivity. In spite of these difficulties, a number of experimenters have survived and a great deal of new information from space experiments is now available on the observational characteristics of the zodiacal light.

To discuss every observation of zodiacal light from space in detail would be outside the scope of this paper and probably of limited value. A listing of these observations can, however, help to show what experiments have been performed and which areas require additional study. Therefore, we present in the next section an annotated listing of space observations of the zodiacal light and follow that with a discussion of selected results and with suggestions for additional experiments.

Table 1

An Annotated Listing of Space Observations of the Zodiacal Light

Institution	Date of Observation	Type of Observation or Detector	Wavelength Coverage	Polarization*	Selected References	Notes
A. Balloons.						
Univ. of Minnesota	1962-1965	photographic, PMT's	broadband: blue, visual	x	Gillett (1966); Gillett, et al. (1964)	1
Univ. of New Mexico	1965, 1966	PMT	broadband: 4000, 4350, 5500A	x	Regener and Vande Noord (1967); VandeNoord (1970)	2
MPI - Astronomy, Heidelberg and Heidelberg Observ.	1970	PMT	medium band: 5500	x	Gabsdil (1971)	3i
	1972	PbS	broadband: 2.4μ	x	Hofmann, et al. (1973)	3ii
	1972	PMT's	medium band: 3500, 5000, 7100, 8200	x	Frey, et al. (1974)	3iii

Notes:
1. The first space observations of the zodiacal light.
2. Observations of the inner zodiacal light, 25 to 50 degrees elongation, during solar minimum. No correlation was found with solar activity or lunar age, although the evening zodiacal light was found to be 30% brighter than the morning zodiacal light.
3. These observations were performed with the balloon-borne telescope THISBE. No deviation from solar color was found between 3500A and 2.4μ. Polarization observations at 5000A (iii), from 80 to 180 degrees elongation in the ecliptic, confirmed the existence of negative polarization** at large elongations.

---
* x indicates that polarization observations were made, although not necessarily at all of the indicated wavelengths.
** Electric vector parallel to the scattering plane.

Table 1, continued

An Annotated Listing of Space Observations of the Zodiacal Light

Institution	Date of Observation	Type of Detector	Wavelength Coverage	Polarization*	Selected References	Notes
B. Rockets.						
Kitt Peak National Observatory	Sept 1964	PMT	medium band: 4500, 7030A	x	Wolstencroft and Rose (1967)	1
University of Wisconsin	Sept 1964	PMT	2200 to 2900 and 4170; 7 colors		Lillie (1968, 1972)	2
Tokyo Astronomical Observatory	July 1965	PMT	medium band: 4300, 5300, 6000		Tanabe and Huruhata (1967)	3
Nagoya University	Jan 1969	PbS	broad band: .52, 1.23, 1.57, 2.16 μ		Hayakawa, et al. (1970); Nishimura (1973)	4
Cornell University	Dec 1970	copper doped germanium	5-16 μ 12-14 μ 16-23 μ		Soifer, et al. (1971)	5
MPI - Astronomy, Heidelberg and Heidelberg Observ.	July 1971	PMT	broad band: 4680, 4755, 5915	x	Leinert, et al. (1974a)	6
Los Alamos Scientific Laboratory	July 1972 eclipse	PMT	medium band: 3500, 4500, 5500, 6500	x	Sandford, et al. (1973)	7

* x indicates that polarization observations were made, although not necessarily at all of the indicated wavelengths.

Table 1, continued

An Annotated Listing of Space Observations of the Zodiacal Light

Institution	Mission or Spacecraft	Date of Observation	Type of Observation or Detector	Polarization*	Selected References	Notes
C. Satellites. (1) Manned						
University of Minnesota	Gemini 5	Aug 1965	photographic, Kodak Tri-X		Ney and Huch (1965)	1
University of Minnesota	Gemini 9	June 1966	photographic, Kodak Tri-X		Ney (1966)	
NASA and Dudley Observatory	Apollo 14	Jan 1971	photographic, Kodak 2485		Dunkelman, et al. (1971)	2
Dudley Observatory, NASA, High Altitude Observatory	Apollo 15	July-Aug 1971	photographic, Kodak 2485		Mercer, et al. (1973b)	2
High Altitude Observ. and NASA	Apollo 16	April 1972	photographic, Kodak 2485		MacQueen, et al. (1973b)	2
NASA and Dudley Observatory	Apollo 16	April 1972	photographic, Kodak 2485		Dunkelman, et al. (1972)	2
Dudley Observatory and NASA	Apollo 17	Dec 1972	photographic; broad band - blue and red	x	Mercer, et al. (1973a)	2i
High Altitude Observ. and NASA	Apollo 17	Dec 1972	photographic, Kodak 2485		MacQueen, et al. (1973a)	2
Johns Hopkins University	Apollo 17	Dec 1972	ultraviolet spectrometer		Fastie, et al. (1973)	3

* x indicates that polarization observations were made, although not necessarily at all of the indicated wavelengths.

Table 1, continued

An Annotated Listing of Space Observations of the Zodiacal Light

Institution	Mission or Spacecraft	Date of Observation	Detector	Wavelength Coverage	Polarization*	Selected References	Notes
C. Satellites.							
(1) Manned, continued							
State Univ. of NY at Albany and Dudley Observatory	Skylab	June, Aug 1973	PMT	narrow to medium band: 4000 to 8200A, 10 colors	x	Weinberg, et al. (1975)	4
NASA, State Univ. of NY at Albany, Dudley Observatory	Skylab	Aug 1973	Kodak 2485 film	visual		Kessler, Zook, Mercer, and Weinberg (unpub.)	5
(2) Unmanned							
University of Minnesota	OSO-B2	Feb 1965– Nov 1965	PMT	broad band: blue, visual	x	Sparrow and Ney (1968)	1
California Institute of Technology	Surveyor 6 Surveyor 7	Nov 1967 Jan 1968	camera/ vidicon	visual		Bohlin (1971)	2
University of Wisconsin	OAO-2	Dec 1968–	PMT	medium band: 1050 to 4250, 12 colors		Lillie (1972)	3
University of Minnesota	OSO-5	Jan 1969–	PMT	broad band: 4180, 5410, V, R, 6820	x	Burnett, et al. (1972)	4
Rutgers University	OSO-6	Aug 1969– Jan 1972	PMT	medium band: 4000, 5000, 6100	x	Roach, et al. (1972)	5
Service d'Aeronomie du CNRS	D2A (Tournesol)	Apr 1971– June 1973	PMT	narrow band: 6530		Levasseur and Blamont (1973a,b)	6

* x indicates that polarization observations were made, although not necessarily at all of the indicated wavelengths.

Table 1, continued

An Annotated Listing of Space Observations of the Zodiacal Light

Institution	Mission or Spacecraft	Date of Observation	Detector	Wavelength Coverage	Polarization*	Selected References	Notes
D. Space Probes.							
Soviet unmanned interplanetary station, Venus-3		Nov-Dec 1965	nitrous oxide filled Geiger counters	1050-1340A 1225-1340		Kurt and Syunyaev (1967)	1
University of Arizona, State Univ. of NY at Albany, Dudley Observatory	Pioneer 10 Pioneer 11	Mar 1972- Apr 1973-	dual-channel Channeltrons	broad band: blue, red	x	Weinberg, et al. (1973)	2
General Electric, Philadelphia	Pioneer 10 Pioneer 11	Mar 1972- Apr 1973-	PMT's	broadband, visual		Zook and Soberman (1974)	3
MPI - Astronomy, Heidelberg and Heidelberg Observatory	Helios A	Dec 1974-	PMT's	UBV	x	Leinert, et al. (1974b)	4

* x indicates that polarization observations were made, although not necessarily at all of the indicated wavelengths.

Notes to Table 1B:
1. The first observations of the brightness and polarization of the zodiacal light from above the scattering atmosphere. Negative polarization was found at large elongations, and circular polarization was found between 40 and 65 degrees elongation near the ecliptic. The zodiacal light was found to be bluer than the sun at elongations beyond 110 degrees and 15 per cent brighter at the south ecliptic pole than at the north ecliptic pole. The absolute brightnesses were substantially higher than those found in most ground observations.
2. First ultraviolet observations of the zodiacal light. The brightness of zodiacal light was found to increase sharply below 2500A.
3. First rocket observations of the inner zodiacal light, 15 to 41 degrees elongation. Brightnesses were found to increase sharply at small elongations (toward the horizon), suggesting that the observations were undercorrected for airglow emission at these elongations.
4. First infrared observations of the zodiacal light. Except for a weak enhancement at 1.6μ, the zodiacal light was found to have the color of the sun from $.52\mu$ to 2.16μ.
5. Thermal emission was detected in the zodiacal light at 160 degrees elongation near the ecliptic.
6. The first observations of the inner zodiacal light during the day and outside of eclipse. Observations were made in 360 degree circles around the sun at elongations 15, 21, and 30 degrees. The zodiacal light was found to be slightly red compared to the sun, and the minimum to maximum brightness ratios for each circle were found to be approximately .32, in agreement with results at 90 degrees elongation (Sparrow and Ney 1972; Dumont and Sanchez 1973). The degree of polarization in the plane of symmetry was found to be higher than expected from ground based observations, and it showed a large decrease away from this plane.
7. Observations were made at 12 elongations, 0 to 30 degrees, east and west of the eclipsed sun. The data were degraded by scattered light and by a trajectory error, but brightness observations west of the sun were less affected and suggest a reddening at very small elongations.

Notes to Table 1C(1):
1. The first photographs of zodiacal light and Gegenschein from a manned spacecraft.
2. The Apollo photography of the F-corona and zodiacal light was carried out from Command Modules while in the "double umbra"; i.e., that dark region free from sunlight and earthshine. Cameras used: Maurer 16mm, Nikon 35mm, Hasselblad 70mm. During Apollo 17, the first measurements were made of color and polarization (2i). Preliminary results showed the zodiacal light within 15 degrees of the sun near the ecliptic to be brighter in the red than in the blue and to be less bright in the red for all other regions.
3. Twice during lunar orbit the lunar atmosphere ultraviolet spectrometer was used to measure the inner zodiacal light. Although a preliminary analysis is said here to generally support the OAO-2 observations (i.e., enhanced brightness at short wavelengths), Lillie, in a private communication to C. Leinert (Leinert 1975), reports no excess brightness in the zodiacal light at 1470A.
4. This experiment is discussed later in this paper.
5. Photographs were taken of a region of sky containing the Gegenschein using equipment of Skylab experiment T025 (externally occulted coronagraph with the occulter rotated out of position and a 35mm Nikon camera).

Notes to Table 1C(2):
1. Observations of brightness and polarization at 90 degrees elongation for ecliptic latitudes from 50 degrees to the north pole. No changes greater than 10 per cent were detected in the polarized brightness, including observations during periods of strong magnetic index, K_p.
2. Inadequate calibration precluded polarimetry and absolute photometry, although some relative photometry of the F-corona/inner zodiacal light was accomplished by normalizing the Surveyor data to published mean values for the corona.
3. Fixed position, all-sky observations made during satellite night. The zodiacal light was found to be redder than the sun for wavelengths above 2500A with a sharp increase in brightness below 2500A.

Notes to Table 1C(2), continued:
4. Long-term multicolor observations of brightness and polarization at 90 degrees elongation for all ecliptic latitudes. No difference was found between zodiacal light brightnesses at northern and southern ecliptic latitudes, and no changes were found with solar activity, lunar phase, the orbital plane of Comet Encke, or an annual cycle (Burnett, 1976).
5. Three-color observations of brightness and polarization at 5-degree intervals of elongation from 180 to ± 10 degrees. Observations were only possible during satellite day, and stray light has limited analysis to observations near the antisolar point. Large, short-term changes were observed in the direction and amount of polarization (Roach, et al. 1974) and in the brightness in this region, the suggestion being made that the latter changes are a result of scattering by dust in the "general region of the earth-moon system" (Roach, et al. 1973).
6. Observations of brightness at 90 degrees elongation for all ecliptic latitudes. Observed short and long period changes were ascribed to local meteor streams and the changing position of the earth with respect to the invariant plane, respectively. Burnett, et al. (1974) attribute at least part of the short term changes to scattered moonlight.

Notes to Table 1D:
1. 2×10^{-12} ergs/cm^2 sec ster A is given as an upper limit for the intensity of zodiacal light at 1300A.
2. This experiment is discussed later in this paper.
3. Brightness measurements in the zodiacal light mode of the Asteroid Meteoroid Detector have been made periodically from both spacecraft since launch (see, also, Soberman, et al., 1976).
4. The first observations of the zodiacal light inside the earth's orbit (perihelion near 0.3 AU). Measurements of brightness and polarization with three photometers, fixed at 75, 60, and 0 (toward the south ecliptic pole) degrees with respect to the spacecraft spin axis (Pitz, et al., 1976, Link et al., 1976).

Table 1 illustrates the enormous growth in space observations of the zodiacal light in just the past decade - and their diversity. At the same time, there have been relatively few ground based observations of the zodiacal light, especially during the past five years. In spite of this space activity, several important areas have still received little or no study: polarization measurements in the ultraviolet and infrared, observations near the sun, and all-sky observations (color and polarization).

Selected Results and a Look at the Future

1. Observations from Skylab. Multicolor observations of sky brightness and polarization over large regions of the antisolar hemisphere were obtained with a photoelectric polarimeter during Skylab missions SL-2 and SL-3 (Weinberg, et al. 1975). The original plan to also observe the solar hemisphere, to within 15 degrees of the sun, had to be abandoned following the loss of use of a solar-pointing scientific airlock after launch. The instrument was extended outside the spacecraft and could be operated in fixed-position or sky scanning modes. Table 2 summarizes observing programs performed with the photometer during the Skylab missions (see, also, Figure 1 in Sparrow, et al.; 1976). The availbility of only limited amounts of valid data has restricted analysis to parts of fixed position (Mode 1) programs. Preliminary results have been derived on the spacecraft corona (there were levels of sky brightness in

daylight only five per cent above those at night) and on the polarized brightness of
the zodiacal light (Sparrow, et al.; 1976). Much of the data have just
been made available (September-October 1975) in useable form, and it will now be
possible to derive results for the scanning programs and to complete the analysis of
fixed position data.

Table 2

Observations with the Skylab Photometer

Mission	Mission day	1973 date	Mode	Program*
SL-2	19	June 12	4	sky map (7 colors, 5300-8200)
	19	12	1	I north celestial pole
	19	12	1	II south ecliptic pole
	19	12	1	III contamination, 2 parts: night/day, day/night
	22	15	4	sky map (9 colors, 4000-7100)
	22	15	1	IV north galactic pole
	22	15	2	V elevation scans
	22	15	4**	sky map (4000, 4760)
	23	16	4**	sky map (7 colors, 5080-7100)
	23	16	3	azimuth scans (5 colors, 4000-5577)
	23	16	3**	azimuth scans
	24	17	3**	azimuth scans
SL-3	5	August 1	4**	sky map (4000)
	6	2	4**	sky map (4760)
	7	3	1	Gegenschein (5 colors, 4000-5577)
	7	3	1	contamination, night/day
	7	3	1	night/day scan, gravity-oriented
	7	3	1	Gegenschein
	7	3	2**	elevation scans
	8	4	2**	elevation scans

*10-color observations, unless otherwise indicated.
**Program started one day and ended the following day.

2. Observations from Pioneers 10 and 11. Imaging photopolarimeters on the Pioneer 10
and 11 probes are being used by our group to map sky brightness and polarization in the
blue and red from heliocentric distances beyond 1 AU. The combination of telescope
stepping and spacecraft roll produces sky data for the entire sky between 28 and 170
degrees (near the antisun direction) from the spacecraft spin axis in a sky map or grid
of observations consisting of 78 data rolls and 64 effective fields of view per roll per
color. Since the spin axis moves slowly on the celestial sphere, most of the sky within
approximately 30 degrees of the sun is eventually covered with a resolution better than

the 1.8 degree roll-to-roll (cone angle) separation of fields of view in a single sky map. Figure 3 shows the trajectories for Pioneers 10 and 11 and the spacecraft positions from where sky maps were made. Measurements are continuing with both spacecraft.

Major results to date: The Gegenschein was observed when the spacecraft antisun direction was 3.4 degrees from the earth antisun direction and when the spacecraft was 1.011 AU from the sun (Weinberg, et al. 1973). Brightnesses in the ecliptic at elongations greater than 90 degrees show no change (i.e., negligible zodiacal light) beyond 3.3 AU (Hanner, et al. 1974). Observations from beyond the asteroid belt are being used to derive a map of the background starlight in two colors with high spatial resolution (Weinberg, et al. 1974). Observations from 1 to 3 AU are being used to derive

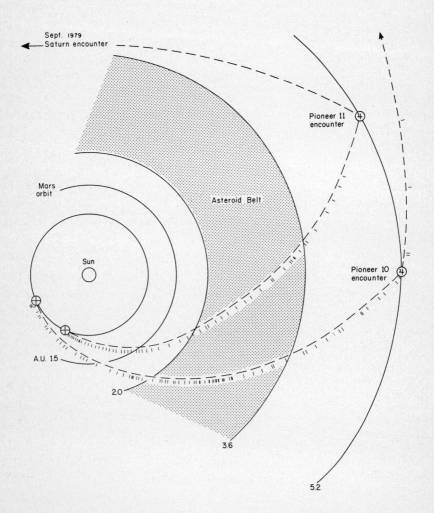

Fig. 3. Trajectories and celestial aspect for Pioneers 10 and 11. Tick marks along the trajectories indicate spacecraft positions from where sky maps were made with the imaging photopolarimeters.

the change in zodiacal light brightness with heliocentric distance (Hanner, et al.; 1976).

3. Wavelength Coverage. Multicolor brightness observations by Peterson (1967), Nishimura (1973), Hofmann, et al. (1973), Frey, et al. (1974), and others show the zodiacal light to have the color of the sun from the near ultraviolet out to 2.4μ. Similarly, Sparrow, et al. (1976) find the polarized brightness of the zodiacal light to have the solar color in the visible. Therefore, the degree of polarization is not wavelength dependent in the visible. Additional information on the sizes of the particles would be available from polarization observations in the ultraviolet, especially if the degree of polarization was found to change with wavelength in that region. No polarization observations have been made in the ultraviolet, although balloon observations of brightness and polarization are planned by A. Frey, rocket observations are planned by E. Pitz and by C. F. Lillie, and observations of brightness are planned from the French satellite D2B (Maucherat and Cruvellier; 1976; Llebaria, 1976). Limited ultraviolet-visible observations of the zodiacal light may be possible with a photopolarimeter during the cruise phase of the Mariner Jupiter-Saturn missions (Lillie 1974). Additional observations in the ultraviolet and infrared might disclose signatures or features (e.g., 2200A in the UV and 10μ in the IR) and, thereby, provide more direct information on the nature and sizes of the particles that give rise to the zodiacal light.

4. Spatial Coverage. As noted earlier, ground observations are restricted to elongations greater than 30 degrees, which means that they cannot see the effects of particles closer than 0.5 AU to the sun. Although extremely difficult to observe, brightness and polarization observations from the ultraviolet to the infrared in an annulus from 5 to 30 degrees from the sun probably contain more information on the nature of the particles than any other region. There are relatively few all-sky observations beyond 30 degrees elongation, even in the visible. The information-laden regions at large elongations, for example, are underobserved, and accurate observations of brightness, polarization, and color extending into the ultraviolet are not yet available.

Zodiacal light observations from Pioneer 10 and 11 and from Helios can provide information on the spatial distribution of interplanetary dust throughout the region from 0.3 to 3.3 AU (Hanner and Leinert 1972). Dumont (1972, 1973), Leinert (1975), and others have discussed inversion of the brightness integral to obtain information on the spatial distribution and scattering properties of the particles. It is generally assumed that the scattering properties do not depend on heliocentric distance, R, although this restriction is probably not realistic, especially in the inner solar system. If Pioneer 10 and 11 observations of polarization versus elongation (beyond 90 degrees) differ with R, it is likely that the scattering properties do change with heliocentric distance.

Concluding Remarks

Observations close to the sun can best provide information on the nature and sizes of the zodiacal dust, on possible solar storm effects, and on the vaporization zone; ultraviolet observations of brightness and polarization over a large range of elongations can provide information on the sizes and, perhaps, composition of the zodiacal dust;

more extensive observations in the ultraviolet and infrared may provide signatures or features characteristic of particular materials.

We now have preliminary results on the spatial extent of the zodiacal light and its appearance from the inner solar system to the asteroid belt and more consistent results on its brightness from the near ultraviolet to the near infrared. If we are to go beyond understanding its observational characteristics and derive information on the particles that give rise to the zodiacal light, we must have the aforementioned observations and further laboratory and theoretical studies of the properties of small and moderate size particles of different shapes.

Acknowledgements. This study received support from NASA contracts NAS2-7963 (Pioneer) and NAS8-30251 (Skylab). R. D. Mercer kindly provided the photograph shown in Figure 2 and additional information on Apollo Program studies of zodiacal light.

References.
Bohlin, J. D., 1971, Photometry of the Outer Solar Corona from Lunar-Based Observations, Solar Physics 18, 450-457.
Burnett, G. B., J. G. Sparrow, and E. P. Ney, 1972, OSO-5 Dim Light Monitor, Applied Optics 11, 2075-2081.
Burnett, G. B., J. G. Sparrow, and E. P. Ney, 1974, Is the Zodiacal Light Intensity Steady? Nature 249, 639-640.
Burnett, G.B., 1976, OSO-5 Zodiacal Light Measurements 1969-1975, this volume.
Dietzel, H., G. Eichhorn, H. Fechtig, E. Grün, H.-J. Hoffmann, and J. Kissel, 1973, The HEOS 2 and HELIOS Micrometeoroid Experiments, J. Phys. E: Sci. Instr. 6, 209-217.
Dumont, R., 1972, Intensité et polarisation de la lumière solaire diffusée par un volume isolé de matière interplanétaire, Compt. Rend. 275, 765-768.
Dumont, R., 1973, Phase Function and Polarization Curve of Interplanetary Scatterers from Zodiacal Light Photopolarimetry, Planetary Space Sci. 21, 2149-2155.
Dumont, R. and Sanchez-Martinez, 1973, Photométrie de la lumière zodiacale hors de l'écliptique en quadrature et en opposition avec le Soleil, Astron. and Astrophys. 22, 321-328.
Dunkelman, L., C.L. Wolff, R.D. Mercer, and S.A. Roosa, 1971, Gegenschein-Moulton Region Photography from Lunar Orbit, in Apollo 14 Prelim. Sci. Report, NASA SP-272, 249-252.
Dunkelman, L., C.L. Wolff, and R.D. Mercer, 1972, Gegenschein-Moulton Region Photography from Lunar Orbit, in Apollo 16 Prelim. Sci. Report, NASA SP-315, 16-1 to 16-4.
Fastie, W.G., P.D. Feldman, R.C. Henry, H.W. Moos, C.A. Barth, G.E. Thomas, C.F. Lillie, and T.M. Donahue, 1973, Ultraviolet Spectrometer Experiment, in Apollo 17 Prelim. Sci. Report, NASA SP-330, 23-1 to 23-10
Frey, A., W. Hofmann, D. Lemke, and C. Thum, 1974, Photometry of the Zodiacal Light with the Balloon-Borne Telescope THISBE, Astron. and Astrophys. 36, 447-454.
Gabsdil, W., 1971, Ph.D. Dissertation, Heidelberg University.
Gillett, F.C., 1966, Zodiacal Light and Interplanetary Dust, Ph.D. Dissertation, University of Minnesota.
Gillett, F.C., W.A. Stein, and E.P. Ney, 1964, Observations of the Solar Corona from the Limb of the Sun to the Zodiacal Light, July 20, 1963, Astrophys. J. 140, 295-305.

Hanner, M.S. and C. Leinert, 1972, The Zodiacal Light as seen from the Pioneer F/G and Helios Probes, in Space Research XII, (F.A. Bowhill, L.D. Jaffee, and M.J. Rycroft, eds.), 445-455, (Berlin: Akademie-Verlag).

Hanner, M.S., J.L. Weinberg, L.M. DeShields II, B.A. Green and G.N. Toller, 1974, Zodiacal Light and the Asteroid Belt: the View from Pioneer 10, J. Geophys. Res. 79, 3671-3675.

Hanner, M.S., J.G. Sparrow, J.L. Weinberg, and D.E. Beeson, 1976, Pioneer 10 Observations of Zodiacal Light brightness near the ecliptic: changes with heliocentric distance, this volume.

Hayakawa, S., T. Matsumoto, and T. Nishimura, 1970, Infrared Observation of Zodiacal Light, in Space Research X, (T.M. Donahue, P.A. Smith, and L. Thomas, eds.), 248-251, (Amsterdam: North-Holland Publ. Co.).

Hoffmann, H.-J., H. Fechtig, E. Grün, and J. Kissel, 1975, First Results of the Micrometeoroid Experiment S 215 on the HEOS 2 Satellite, Planetary Space Sci. 23, 215-224.

Hofmann, W., D. Lemke, C. Thum, and U. Fahrbach, 1973, Observations of the Zodiacal Light at 2.4 m, Nature Phys. Sci. 243, 140-141.

Kurt, V.G., and R.A. Syunyaev, 1967, Measurements of the Ultraviolet and X-ray Background outside the Earth's Atmosphere and their Role in the Study of Intergalactic Gas, Cosmic Res. 5, 496-512.

Leinert, C., 1975, Zodiacal Light - a Measure of the Interplanetary Environment, Space Science Rev., in press.

Leinert, C., and D. Klüppelberg, 1974, Stray Light Suppression in Optical Space Experiments, Applied Optics 13, 556-564.

Leinert, C., H. Link, and E. Pitz, 1974a, Rocket Photometry of the Inner Zodiacal Light, Astron. and Astrophys. 30, 411-422.

Leinert, C., H. Link, and E. Pitz, 1974b, Extraterrestrial Polarization of the Zodiacal Light: Rocket Measurements and the Helios Project, in Planets, Stars and Nebulae Studied with Photopolarimetry, (T. Gehrels, ed.), 766-767 (Tucson: The University of Arizona Press).

Levasseur, A.C., and J.E. Blamont, 1973a, Satellite Observations of Strong Balmer Alpha Atmospheric Emissions around the Magnet Equator, J. Geophys. Res. 78, 3881-3893.

Levasseur, A.C., and J.E. Blamont, 1973b, Satellite Observations of Intensity Variations of the Zodiacal Light, Nature 246, 26-28.

Lillie, C.F., 1968, An Empirical Determination of the Interstellar Radiation Field, Ph.D. Dissertation, University of Wisconsin.

Lillie, C.F., 1972, OAO-2 Observations of the Zodiacal Light, in The Scientific Results from the Orbiting Astronomical Observatory (OAO-2), NASA SP-310, 95-108.

Lillie, C.F., 1974, The Mariner Jupiter/Saturn Photopolarimeter Experiment, in The Rings of Saturn, NASA SP-343, 131-139.

Link, H., C. Leinert, E. Pitz, and N. Salm, 1976, Preliminary results of the Helios A Zodiacal Light experiment, this volume.

Llebaria, A., 1976, Visible and UV photometry of the Gegenschein and the milky way, this volume.

MacQueen, R.M., C.L. Ross, and R.E. Evans, 1973a, Solar Corona Photography, in Apollo 17 Prelim. Sci. Report, NASA SP-330, 34-4 to 34-6.

MacQueen, R.M., C.L. Ross, and T. Mattingly, 1973b, Observations from Space of the Solar Corona/Inner Zodiacal Light, Planetary Space Sci. 21, 2173-2179.

Maucherat, M., and P. Cruvellier, 1976, Presentation of Zodiacal Light instrument aboard the D2B astronomical satellite, this volume.

Mercer, R.D., L. Dunkelman, and R.E. Evans, 1973a, Zodiacal Light Photography, in Apollo 17 Prelim. Sci. Report, NASA SP-330, 34-1 to 34-4.

Mercer, R.D., L. Dunkelman, C.L. Ross, and A. Worden, 1973b, Lunar Orbital Photography of Astronomical Phenomena, in Space Research XIII, (M.J. Rycroft and S.K. Runcorn, eds.), 1025-1031, (Berlin: Akademie-Verlag).

Ney, E.P., 1966, Night-Sky Phenomena photographed from Gemini 9, Sky and Telescope 32, 276-277.

Ney, E.P., and W.F. Huch, 1965, Gemini V Experiments on Zodiacal Light and Gegenschein, Science 150, 53-56.

Nishimura, T., 1973, Infrared Spectrum of Zodiacal Light, Publ. Astron. Soc. Japan 25, 375-384.

Peterson, A.W., 1967, Multicolor Photometry of the Zodiacal Light, in Proceedings, Symposium on the Zodiacal Light and the Interplanetary Medium, NASA SP-150, (J.L. Weinberg, ed.), 23-31.

Pitz, E., C. Leinert, H. Link, and N. Salm, 1976, Helios Zodiacal Light Experiment, this volume.

Regener, V.H., and E.L. VandeNoord, 1967, Observations of the Zodiacal Light by means of Telemetry from Balloons, in Proceedings, Symposium on the Zodiacal Light and the Interplanetary Medium, NASA SP-150, (J.L. Weinberg, ed.), 45-47.

Roach, F.E., B. Carroll, L.H. Aller, and L. Smith, 1972, Surface Photometry of Celestial Sources from a Space Vehicle: Introduction and Observational Procedures, Proc. Nat. Acad. Sci. USA 69, 694-697.

Roach, F.E., B. Carroll, J.R. Roach, and L.H. Aller, 1973, A Photometric Perturbation of the Counterglow, Planetary Space Sci. 21, 1185-1189.

Roach, F.E., B. Carroll, L.H. Aller, and J.R. Roach, 1974, The Linear Polarization of the Counterglow Region, in Planets, Stars and Nebulae Studied with Photopolarimetry, (T. Gehrels, ed.), 794-803, (Tucson: The University of Arizona Press).

Sandford II, M.T., J.K. Theobald, and H.G. Horak, 1973, Observations of the F Corona and Inner Zodiacal Light during the 1972 July 10 total Solar Eclipse, Astrophys. J. 181, L15-L17.

Soberman, R.K., J.M. Alvarez, and J.L. Weinberg, 1976, Dust in the Outer Solar System - Review of early Results from Pioneer 10 and 11, this volume.

Soifer, B.T., J.R. Houck, and M. Harwit, 1971, Rocket-Infrared Observations of the Interplanetary Medium, Astrophys. J. 168, L73-L78.

Sparrow, J.G., and E.P. Ney, 1968, OSO-B2 Satellite Observations of Zodiacal Light, Astrophys. J. 154, 783-787.

Sparrow, J.G., and E.P. Ney, 1972, Observations of the Zodiacal Light from the Ecliptic to the Poles, Astrophys. J. 174, 705-716.

Sparrow, J.G., J.L. Weinberg, and R.L. Hahn, 1976, Polarization of the Zodiacal Light: first Results from Skylab, this volume.

Tanabe, H., and M. Huruhata, 1967, Rocket Observations of the Brightness of the Zodiacal Light, in Proceedings, Symposium on the Zodiacal Light and the Interplanetary Medium, NASA SP-150, (J.L. Weinberg, ed.), 37-40.

VandeNoord, E.L., 1970, Observations of the Zodiacal Light with a Balloon-Borne Telescope, Astrophys. J. 161, 309-316.

Weinberg, J.L., M.S. Hanner, D.E. Beeson, L.M. DeShields II, and B.A. Green, 1974, Background Starlight observed from Pioneer 10, J. Geophys. Res. 79, 3665-3670.

Weinberg, J.L., M.S. Hanner, H.M. Mann, P.B. Hutchison, and R. Fimmel, 1973, Observations of Zodiacal Light from the Pioneer 10 Asteroid-Jupiter Probe: Preliminary Results, in Space Research XIII, (M.J. Rycroft and S.K. Runcorn, eds.), 1187-1193, (Berlin: Akademie-Verlag).

Weinberg, J.L., J.G. Sparrow, and R.C. Hahn, 1975, The Skylab Ten Color Photoelectric Polarimeter, Space Science Instr. 1, 138-149.

Wolstencroft, R.D. and L.J. Rose, 1967, Observations of the Zodiacal Light from a Sounding Rocket, Astrophys. J. 147, 271-292.

Zook, H.A. and R.K. Soberman, 1974, The Radial Dependence of the Zodiacal Light, in Space Research XIV, (M.J. Rycroft and R.D. Reasenberg, eds.), 763-767, (Berlin: Akademie-Verlag).

HELIOS ZODIACAL LIGHT EXPERIMENT

Pitz,E., Leinert,C., Link,H. and Salm,N.
Max Planck Institut für Astronomie

D 6900 Heidelberg, Königstuhl, F.R.G.

The Z.L. experiment consists of 3 photometers which are mounted rigidly into the s/c with orientations of about 15°, 30° and 90° south of the s/c - XY - plane, which coincides in orbit with the ecliptic plane (see Fig.1). Helios is spinning with 1 Hz, and the integration time of the experiment is 513 revolutions. The 90° - photometer always looks to the south ecliptic pole and one

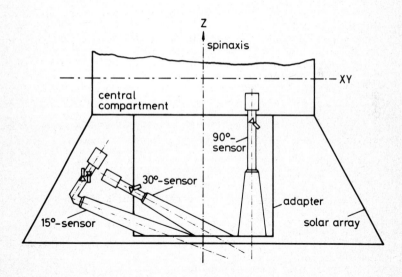

Fig.1: Schematic view of the southern s/c - cone with the mounting positions of the Z.L. - photometers.

revolution of the s/c is divided into 8 sectors to get information on the polarization of Z.L. The polarization is measured by a fixed polaroid foil within the photometer which is rotated by the s/c. In the other two photometers one revolution is split into 32 sectors with different angular resolution. Near the antisun where the gradient in Z.L. intensity is small, the sector length

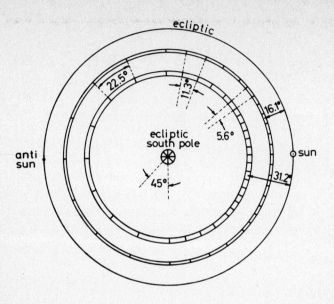

Fig.2: Sectoring of the photometer scans on the sky.

is 4 times the length near the sun (see Fig.2). In these 2 photometers the polarization is obtained by 3 differently oriented polarization foils moved by stepping motors. Intensity and polarization of Z.L. is measured in 3 different colors, which are near the international UBV - system (Ažusienis and Straižys 1969), the effective wavelength shifted by about 100 Å to the blue end.

To obtain the intensity of Z.L., we have to multiply the recorded counting rates by a calibration factor and we have then to subtract dark current, star background and possible stray light contribution.

Calibration of broad band photometers is not trivial and has been done on ground several times very carefully. The method is discribed by Leinert et.al. (1974). Inflight calibration by means of bright star crossing is used to check the ground based calibration. The stability of the sensors is monitored during the mission by means of internal lamps which are switched on from time to time to give a reference signal. In the upper curve of Fig.3 this signal is shown versus time after launch for the $30°$ sensor. The lower curve is the temperature near the sensor unit and it can be seen that the variation of the signal by $\pm$ 3% is mainly a temperature effect. The dark current of the sensor is about a factor 3 higher than on ground, due to cosmic radiation effects, but still less than

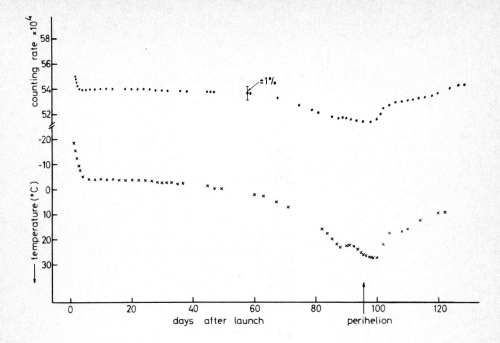

Fig.3: Signal of internal lamp and temperature of the 30° sensor versus time after launch.

1% of the average signal.

As an interplanetary probe is always exposed to sunlight which is 13 orders of magnitude brighter than the average Z.L., extreme care has to be taken to avoid unwanted stray light. Fig.4 shows the concept of stray light suppression. We call it a 4 step reduction system. For details see Leinert and Klüppelberg (1974).

Step 1: The photometer openings are mounted in the shadow of the s/c solar array.

Step 2: Reflected sunlight from the inner s/c walls is catched in the first baffle room of the extended baffle system associated with each photometer.

Step 3: Light from the first baffle room cannot reach the first optical element.

Step 4: Threefold attenuated stray light reaching the optical elements comes from outside the field of view and is therefore imaged onto the field stop.

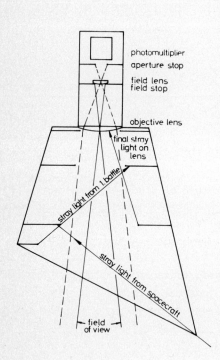

Fig.4: Principle of optical system and stray light suppression.

The worst condition with respect to stray light contamination is in the $15°$ - photometer observing the antisolar region. Fig.5 shows data of this region, measured during two different attitudes of the s/c. In the insert the position of the sun with respect to the southern s/c rim is shown. The "dashed" position means that no sunlight enters the s/c cone and therefore no stray light contribution is expected. In the "solid" position the inner s/c wall is partly illuminated and stray light may contaminate the data. Changing the attitude of the s/c has a measurable effect in the star background as can be seen in the milky way regions. The bright star α CMi has left the field of view after the attitude maneuver. Outside these regions no stray light effects can be seen and we conclude that stray light is less than 1% in our signal.

Star background correction will be the most difficult part of recuction and we hope to get some assistance from Pioneer 10 and 11 observations which are made now in the absence of Z.L.

Up to now, the experiment works perfectly and so far no anomalies can be reported.

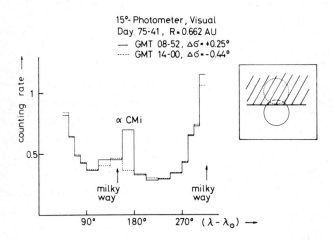

Fig.5: Antisolar data of the 15^0 - photometer during two different s/c attitudes.

Acknowledgments.

We wish to thank K. Mertens, B. Kunz and coworkers at DFVLR, Institut für Satellitenelektronik, for their basic work in developing the electronics and the test equipment. Dornier System with Project Managers E. Achtermann and R. Hartig was responsible for the technical realization of the instrument. The friendly cooperation and professional workmanship is gratefully acknowledged. This work has been supported by the Bundesministerium für Forschung und Technologie by grants RS 21 and WRS 0107 I.

References.

Ažusienis, A. and Straižys, V. 1969, Soviet Astronomy AJ, __13__, 316
Leinert, C., Link, H. and Pitz, E. 1974, Astron. & Astrophys., __30__, 411
Leinert, C. and Klüppelberg, D. 1974, Appl. Optics, __13__, 566

PRELIMINARY RESULTS OF THE HELIOS A ZODIACAL LIGHT EXPERIMENT

Link, H., Leinert, C., Pitz, E. and Salm, N.
Max-Planck-Institut für Astronomie

D 6900 Heidelberg-Königstuhl, F.R.G.

Helios A was launched on December 10, 1974 into a highly elliptical orbit with a perihelion of 0.31 A.U., which was reached on March 15, 1975. The zodiacal light experiment on Helios, described in the preceding paper, worked flawlessly and provided the first observations of the zodiacal light from inside the Earth's orbit. A typical example from the raw data of the 15^0 - photometer is shown in Fig.1. There is a strong intensity increase towards the sun and a remarkably flat intensity distribution at large elongations. The Milky Way is superimposed

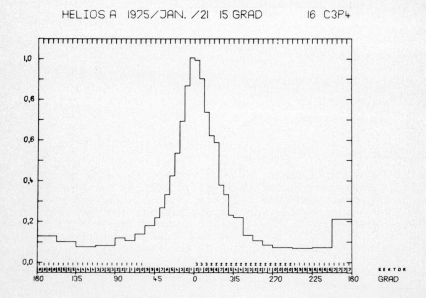

Fig.1: Raw data of the 15^0 - photometer, visual spectral band. The total observed intensity at $\beta = -16^0$ is given as a function of helioecliptic longitude $\lambda - \lambda_\odot$. The data have been normalized to the average intensity of the two subsolar sectors, no. 1 and no. 32. Heliocentric distance of Helios A is R = 0.84 A.U.

on the zodiacal light at longitudes 135° to 180° and 315° to 360°. Due to the
orbital motion of Helios the star background is being shifted with respect to
the zodiacal light, which will facilitate its separation from the total observed intensity. The peak at the right side of Fig.1 is due to the star α CMi,
which we intend to use for the calibration of the instrument in addition to
the ground calibrations. A preliminary evaluation showed less than 20% difference between the two calibrations. This is typical for other stars and for the
other photometers, too, and gives a safe upper limit for the accuracy of the
absolute calibration. Temperature effects are comparatively small and have not
been corrected so far.

Although for Helios the correction for star background is much less important
than for an outward bound space probe like Pioneer 10 or 11, it still remains
one of the main difficulties in the data reduction. Therefore we first evaluated the zodiacal light intensity integrated over the whole range of longitudes
scanned by the respective photometer. With this geometry the star background is
constant, except for the effect of a slow drift of the Helios spin axis by a
few tenth degrees, and the trends of zodiacal light brightness should show
clearly. Fig.2. shows the integrated intensity for the 15° - photometer, visual

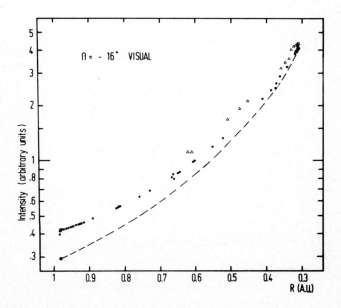

Fig.2: Integrated intensity observed in the 15° - photometer, visual band.
Points refer to measurements before, triangles to measurements after
perihelion. Intensity after star background correction is given by
dashed line.

band, as a function of heliocentric distance. There is a steady intensity increase by a factor of 10 from 1 A.U. to perihelion. Down to 18 solar radii, the closest approach of the line of sight, we therefore see no effect of a dust free zone. This is not surprising since interplanetary dust has been localized even at 3.4 $R_\odot$ by infrared measurements (Peterson 1967). On the way back from perihelion the observed total intensity was higher by 10% to 20%. Changes in the attitude of Helios and the inclination of the Helios orbital plane with respect to the symmetry plane of interplanetary dust both contribute to this effect. It may be noted that the data of Fig.2 show no sudden intensity changes as reported by Levasseur and Blamont (1975). If we subtract the star background as determined from the tables of Roach and Megill (1961) and the Catalogue of Bright Stars (Hoffleit 1964) the resulting integrated zodiacal light intensity is given by the broken line in Fig.2, with an increase from 1 A.U. to perihelion by a factor of 14. The data for the blue and ultraviolet bands of the 15° - photometer look almost the same as those of the visual band. The observed increase in zodiacal light intensity is nearly independent of wavelength.

The 30° - photometer leads to similar results (Fig.3). The total integrated intensity increases by a factor of 8. After subtraction of star background we

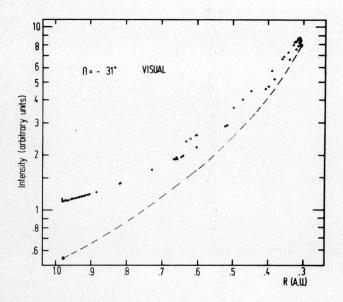

Fig.3: Integrated intensity observed in the 30° - photometer, visual band. Symbols are explained in Fig.2.

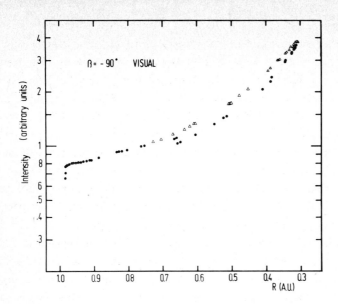

Fig.4: Integrated intensity observed in the 90° - photometer, visual band. Symbols are explained in Fig.2.

find an increase of zodiacal light intensity by a factor of 14 to 15 for the visual, blue and ultraviolet bands. Again higher intensities are observed on the way back from perihelion.

In the 90° - photometer the evaluation of the data (Fig.4) is more difficult because of the strong star background in the vicinity of the Large Magellanic Cloud. Reliable stars counts in this field around the South Ecliptic Pole would be of great value for us. A change in the attitude of Helios by 1°, as performed at a heliocentric distance of R = 0.7 A.U. already produced a change in the signal by 8%. On the basis of the preliminary attitude determination of Helios we were not able to define the star background with sufficient accuracy. However, we checked that the observed increase in total brightness was compatible with an increase of zodiacal light intensity by a factor 14.5, provided the zodiacal light intensity at the south ecliptic pole at 1 A.U. is between 53 and 63 S10 units (solar type stars of m_v = 10.0 per square degree), which is a realistic value. The increase of zodiacal light intensity seems to be nearly independent of the viewing direction.

We also find no important change in the polarization of the zodiacal light between 1 A.U. and perihelion. The degree of polarization for the sum of zodiacal light and star background near perihelion is 8% to 12% for the subsolar sectors of the 15°-and 30°-photometer and 14% to 16% for the 90° - photometer, which is consistent with earthbound observations.

Since the change in zodiacal light brightness is nearly independent of viewing direction, wavelength and direction of polarization, we may adopt the simple model that the radial dependence of the spatial density of interplanetary dust is given by a power law $n(r) \sim r^{-\nu}$ and the scattering function is constant over the volume considered. In this case the zodiacal light intensity varies with heliocentric distance R according to the power law $I_{zl} \sim R^{-\nu-1}$. From our measurement a value of $\nu \approx 1.3$ follows, which describes the spatial distribution between approximately 0.08 A.U. and 1.5 A.U. There is good agreement with the Pioneer space probe experiment of Hanner et al. (1976), which give $\nu \approx 1.5$ for $R > 1$ A.U. This spatial distribution is slightly steeper than the dependence $n(r) \sim 1/r$ resulting from the Poynting - Robertson - effect acting on particles in circular orbits, and it is quite different from the distribution of radio meteors, which according to Southworth and Sekanina (1974) show a minimum near 0.7 A.U.

Acknowledgement

We want to thank all members of the Helios team who by their continuing effort contributed to the success of the Helios mission and our experiment. This work was supported by the Bundesministerium für Forschung und Technologie with grant WRS 0107 I.

References

Hanner, M.S., Sparrow, J.G., Weinberg, J.L. and Beeson, D.E., 1976, this Volume.
Hoffleit, D., 1964, Catalogue of Bright Stars, 3rd Ed.
Levasseur, A.C. and Blamont, J.E., 1975, Space Res. XV, 295.
Peterson, A.W., 1967, Astrophys. J. Letters 148, L 37.
Roach, F.E. and Megill, R.R., 1961, Astrophys. J. 133, 128.
Southworth, R.B. and Sekanina, Z., 1973, NASA CR-2316, Washington.

1.1.4   PIONEER 10 OBSERVATIONS OF ZODIACAL LIGHT BRIGHTNESS
NEAR THE ECLIPTIC: CHANGES WITH HELIOCENTRIC DISTANCE

M. S. HANNER, J. G. SPARROW*, J. L. WEINBERG, AND D. E. BEESON
Space Astronomy Laboratory, State University of New York at Albany

Abstract.   Sky maps made by the Pioneer 10 Imaging Photopolarimeter (IPP) at sun-spacecraft distances from 1 to 3 AU have been analyzed to derive the brightness of the zodiacal light near the ecliptic at elongations greater than 90 degrees. The change in zodiacal light brightness with heliocentric distance is compared with models of the spatial distribution of the dust. Use of background starlight brightnesses derived from IPP measurements beyond the asteroid belt, where the zodiacal light is not detected, and, especially, use of a corrected calibration lead to considerably lower values for zodiacal light than those reported by us previously.

Introduction.   The Pioneer 10/11 zodiacal light experiment has provided the first opportunity to measure the zodiacal light from heliocentric distances greater than 1 AU. It is now complemented by the Helios zodiacal light experiment, which is observing the zodiacal light from positions between 0.3 and 1 AU. The imaging photopolarimeter (IPP) used for zodiacal light observations from Pioneers 10 and 11 is described by Pellicori, et al. (1973). In the sky mapping or zodiacal light mode (Weinberg et al., 1973), orthogonal brightness components are recorded in blue and red bandpasses (half-power wavelengths from 3950A to 4850A and 5900A to 6900A, respectively) as the IPP scans the sky with a 2.3 by 2.3 degree field of view.

Observations from a single heliocentric distance do not yield the particle spatial distribution directly, since the spatial distribution is multiplied by the unknown scattering function of the particles in the integration along the line of sight. While observations from a space probe still require integration along the line of sight, the change in zodiacal light brightness with heliocentric distance of the probe provides information on the large-scale spatial distribution of the dust (Hanner and Leinert, 1972).

The Pioneer 10/11 experiment has the additional advantage that the background starlight component can be observed directly from positions beyond the asteroid belt where the zodiacal light can no longer be seen with the IPP (Weinberg, et al., 1974). This background starlight is then subtracted from the total observed brightness to obtain the brightness of the zodiacal light. The method used to reduce Pioneer 10 brightness observations is described by Weinberg, et al.

---

*On leave from Aeronautical Research Laboratories, Melbourne, Victoria, Australia.

(1974) and was used to derive the zodiacal light brightness in the asteroid belt by Hanner, et al. (1974). In the present paper we extend that analysis by deriving the near-ecliptic zodiacal light brightnesses in the blue for Pioneer 10 distances from 1 to 2.4 AU.

Calibration. An empirical method was used with a number of stars to establish the relation between a star's blue (B) magnitude and the equivalent brightness it contributes over the IPP blue band (Weinberg, et al. 1974). The zero magnitude star calibration was converted to the response to Vega (B magnitude +0.04) and the absolute calibration was based on the flux of Vega (Hayes and Latham 1975) corrected for the effect of H_β, H_γ, and H_δ and integrated over the IPP blue bandpass. The conversion to $S_{10}(V)$* is based on Johnson's (1954) solar spectral irradiance integrated over the IPP blue bandpass and a solar visual magnitude of -26.73 (Stebbins and Kron, 1957). Details of the star calibration and the reasons for changing from the previously used diffuse source calibration are discussed by Weinberg, et al. (in preparation). if the Labs and Neckel (1970) solar irradiance is used, the brightnesses in $S_{10}(V)$ will be raised by 10% for the IPP blue band-pass. To correct for changes in instrument sensitivity, a relative calibration was derived for each sky map using star brightnesses.

The brightnesses given in earlier publications were based on a preflight absolute calibration of the IPP using a C^{14} diffuse source. Due to uncertainties in the absolute radiance of the C^{14} source and in the change of IPP sensitivity between laboratory calibration and flight, we believe the calibration using Vega to be more accurate (Weinberg, et al., in preparation). We estimate our final calibration accuracy to be ±8%, with the largest uncertainty in the effective field of view area. In the course of reanalyzing the calibration data, we also discovered a numerical error of a factor of 2 such that our previously published values are significantly too high:

1. Hanner and Weinberg, 1973. The Gegenschein brightnesses should be multiplied by 0.53. The relative brightness decrease between 1.01 and 1.25 AU remains approximately $1/R^2$. See discussion of Gegenschein in next section.

2. Hanner and Weinberg, 1974. The brightness levels and heliocentric dependence in that preliminary report are superceded by the results given in the present paper.

3. Weinberg, et al., 1974. The background starlight brightnesses should be multiplied by 0.57. This correction decreases the brightnesses in each of the reduction steps illustrated in Figure 3 and the brightness contour levels shown in Figure 4 of that paper.

4. Hanner, et al., 1974. The brightnesses should be multiplied by 0.57. The relative change in brightness with heliocentric distance is not affected.

Observations and Analysis. Data along the ecliptic at elongations $\epsilon > 90$ degrees were analyzed for 15 sky maps between 1.01 and 2.41 AU, following the method described by Weinberg, et. al. (1974) and Hanner, et al. (1974). The data presented here are restricted to the segment

---

*Equivalent number of 10th magnitude (V) stars of solar spectral type per square degree.

along the ecliptic between ecliptic longitudes $\lambda = 177$ and 230 degrees, where we have derived a composite of the background starlight from sky maps beyond the asteroid belt.

Representative data are shown in Figure 1 for two sky maps, Day 75/1972 and Day 104/1972. The upper data are brightnesses after subtraction of resolved stars but before subtraction of the background starlight. The lower plots show the zodiacal light component after subtraction

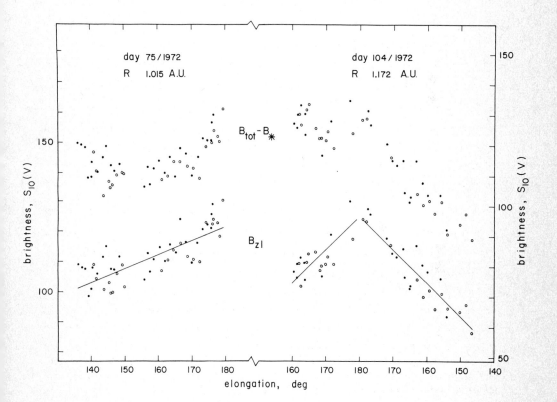

Fig. 1. Observed brightness vs. elongation in the blue spectral band. Filled circles are for $\beta > 0°$, open circles for $\beta < 0°$.

of the background starlight. Both Day 75 and 104 represent observations along the same section of the ecliptic. The lines represent least-squares linear fits to the data for the purpose of intercomparing the brightness versus elongation data from different maps and are not intended to fit the non-linear rise near the Gegenschein.

Although the spacecraft was still near 1 AU on Day 75, it was already 1.5×10^6 km below the ecliptic plane and 5.5×10^6 km below the invariant plane. Theoretical models of the zodiacal light brightness which would be observed from a spacecraft at 1.015 AU and 2 degrees below the plane of symmetry of the dust indicate that a brightness decrease of approximately 10 per cent compared to that observed at 1.00 AU in the symmetry plane can be expected.

It should be pointed out that the direction $\epsilon = 180$ degrees does not lie in the plane of symmetry of the dust distribution, or at ecliptic latitude $\beta = 0$ degrees, for a spacecraft position outside the plane of symmetry. As a result, the shape of the brightness distribution near $\epsilon = 180$ degrees is dependent on the off-ecliptic dust distribution as well as on the shape of the particle phase function. With the IPP scanning pattern and field of view we do not have measurements centered at exactly 180 degrees. Furthermore, since the effective field of view is elongated along the direction perpendicular to the ecliptic plane, any sharply peaked function will be smoothed out in our data numbers.

Figure 2 shows schematically the change with elongation and the level of the zodiacal light

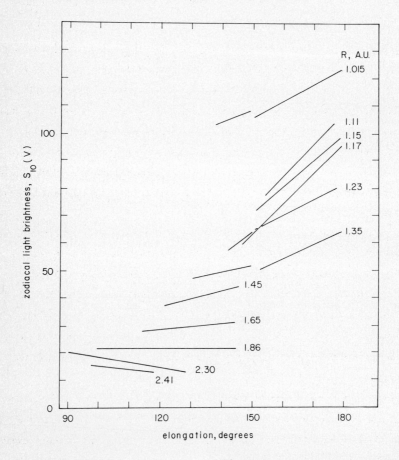

Fig. 2. Smoothed, linear-fit zodiacal light brightnesses in the blue spectral band. R = sun-spacecraft distance in A.U.

brightness for each of the sky maps. The lines represent a least-squares fit to the data, as illustrated in Figure 1. The line at R = 1.015 AU is the average for 5 sky maps made during the first ten days of the mission. The slope from $\epsilon = 150$ to 180 degrees is less steep near 1.015 AU than at R = 1.11 to 1.17 AU. The slopes at R = 1.23 and 1.35 AU are intermediate. We find no obvious reason for the difference in slopes, which is evident in the total brightness as well as in the zodiacal light component (see Figure 1). Since the same region of sky was scanned in all cases, the same resolved stars and background starlight were subtracted. Between R = 1.015 and 1.35 AU the position of the Gegenschein (ϵ = 180 degrees) as seen from the spacecraft moved from $\lambda = 177°$, $\beta = -0.5°$ to $\lambda = 228.6°$, $\beta = -1.9°$, with the result that the brightness plots at R = 1.015 and 1.35 AU are mirror images of each other in ϵ. At R = 1.11 to 1.17 AU, ϵ = 180 degrees was near the center of the plot and the slopes from ϵ = 150 to 180 degrees were identical for the two sides of the plot. Preliminary analysis of Pioneer 10 red data shows a slope on days 75 and 104 which is midway between the corresponding slopes in the blue. Pioneer 11 data will be similarly analyzed.

<u>Radial Dependence.</u>   The zodiacal light brightnesses at $130° \leq \epsilon \leq 145°$ were averaged for each sky map in order to derive the brightness variation with heliocentric distance. Figure 3 shows the averages, compared to the predicted brightness for various models of the dust distribution. 5 to 10 points are included in the averages, and the error bars represent the standard error of the mean at each solar distance. Points shown at heliocentric distances beyond 2.4 AU are data from Hanner, et al. (1974), corrected. As discussed in Hanner, et al. (1974), there is a drop in the zodiacal light brightness between 2.9 and 3.3 AU, indicating an abrupt decrease in the dust particle concentration near 3.3 AU. Beyond 3.3 AU no zodiacal light was detected at $\epsilon >$ 90 degrees.

Models of the zodiacal light brightness were computed for a power-law spatial distribution of the form $n(r) \propto r^{-\nu}$, assuming that the scattering properties of the dust remain constant between 1 and 3.3 AU. Scattering functions typical of silicate particles (Gegenschein) and iron particles (no Gegenschein) were used at several elongations in addition to ϵ = 135-145 degrees. Since neither scattering functions nor the exact position of the spacecraft affected the relative brightness decrease, the shape of the curves in Figure 3 is not strongly dependent on the model parameters, provided that the scattering properties of the particles do not vary strongly with heliocentric distance.

The upper pair of curves in Figure 3 shows the effect of a dust cutoff at the outer edge of the asteroid belt on the predicted zodiacal light brightness. The solid line gives the predicted brightness for ν = 1.5 and a dust distribution extended to infinity; in this case the brightness will vary as $I(R) \propto R^{-(\nu+1)}$. The dashed line represents the brightness variation for $\nu = 1.5$ at $r \leq 3.3$ AU and $n(r) = 0$ at $r > 3.3$ AU. The middle pair of curves compares ν = 1 and ν = 2, with a dust cutoff at 3.3 AU. From these examples and other, similar models, we conclude that a model with $\nu \approx 1$ and a cutoff near 3.3 AU gives the best fit to our observations for a single power-law distribution.

The lower curves give examples of the predicted brightness variation when an increased

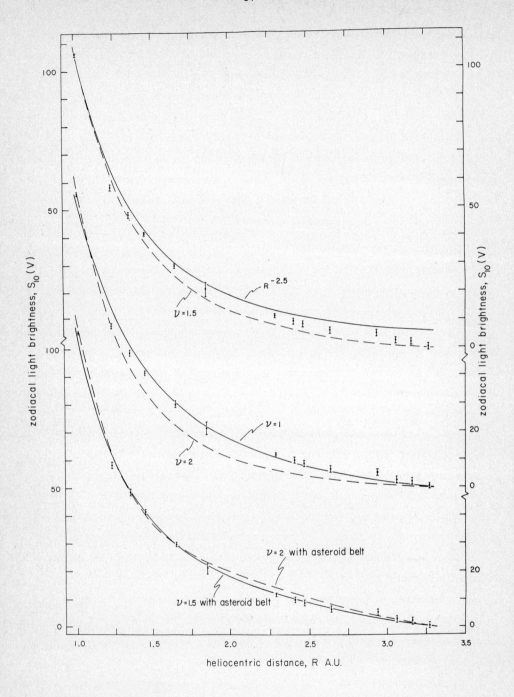

Fig. 3. Zodiacal light brightness vs. heliocentric distance. I = Pioneer 10 observations; ———, ----- are predicted models.

dust density in the asteroid belt is superimposed on a smooth power-law dust decrease. The composite dust distribution is of the form:

$$n(r) \propto n_0 r^{-\nu}, \qquad r < 2.3 \text{ AU}$$
$$n(r) \propto n_0 (r^{-\nu} + C), \qquad 2.3 \leq r \leq 3.3 \text{ AU}$$
$$n(r) = 0, \qquad r > 3.3 \text{ AU},$$

where n_0 is the interplanetary number density at 1 AU. The solid curve corresponds to $\nu = 1.5$, $C = 0.13$ and the dashed curve to $\nu = 2$, $C = 0.26$. Such a simplified 2-component model gives a particularly good fit for $\nu = 1.5$.

We conclude that, if the scattering properties of the dust particles (size, albedo) do not change significantly with heliocentric distance, then the spatial distribution can be satisfactorily represented by a smooth power law, $r^{-\nu}$, $\nu \approx 1$ or, equally well, by a 2-component model with $\nu \approx 1.5$ and increased dust in the asteroid belt. The observed brightnesses indicate a drop in the dust density near 3.3 AU, beyond which the zodiacal light is not detectable above the background starlight.

<u>Acknowledgements</u>. This research was supported by NASA contract NAS2-7963. Assistance was provided by Barbara Green, Helga Olender and Gary Toller.

References

Hanner, M. S. and C. Leinert, 1972, The Zodiacal Light as seen from the Pioneer F/G and Helios Probes, Space Research XII, 445-455.
Hanner, M. S. and J. L. Weinberg, 1973, Gegenschein Observations from Pioneer 10, Sky and Telescope, 45, 217-218.
Hanner, M. S. and J. L. Weinberg, 1974, Changes in Zodiacal Light with Heliocentric Distance: Preliminary Results from Pioneer 10, Space Research XIV, 769-771.
Hanner, M. S., J. L. Weinberg, L. M. DeShields II, B. A. Green, and G. N. Toller, 1974, Zodiacal Light and the Asteroid Belt: The View from Pioneer 10, J. Geophys. Res. 79, 3671-3675.
Hayes, D. S. and D. W. Latham, 1975, A Rediscussion of the Atmospheric Extinction and the Absolute Spectral Energy Distribution of Vega, Astrophys. J. 197, 593-601.
Johnson, F. S., 1954, The Solar Constant, J. Meteorol. 11, 431-439.
Labs, D. and H. Neckel, 1970, Transformation of the Absolute Solar Radiation Data into the International Practical Temperature Scale of 1968, Solar Phys. 15, 79-87.
Pellicori, S. F., E. E. Russell, and L. A. Watts, 1973, Pioneer Imaging Photopolarimeter Optical System, Appl. Opt. 12, 1246-1257.
Stebbins, J. and G. E. Kron, 1957, Six-Color Photometry of Stars. X. The Stellar Magnitude and Color Index of the Sun, Astrophys. J. 126, 266-280.
Weinberg, J. L., M. S. Hanner, H. M. Mann, P. B. Hutchison, and R. Fimmel, 1973, Observations of Zodiacal Light from the Pioneer 10 Asteroid-Jupiter Probe: Preliminary Results, Space Research XIII, 1187-1192.
Weinberg, J. L., M. S. Hanner, D. E. Beeson, L. M. DeShields II, and B. A. Green, 1974, Background Starlight Observed from Pioneer 10, J. Geophys. Res. 79, 3665-3670.

1.1.5

STAR COUNTS IN THE BACKGROUND SKY OBSERVED FROM PIONEER 10

H. Tanabe and K. Mori
Tokyo Astronomical Observatory
Mitaka, Tokyo, Japan

Abstract

Star counts in the sky region observed by Pioneer 10 at 4.64AU from the sun, where the contribution of zodiacal light is negligible, were made using the Palomar Sky Survey Atlas. Brightness of the integrated starlight derived from our star counts agrees, in general, with the Pioneer 10's observation.

1. Introduction

For several years, we have been measuring the photographic and red magnitudes of the stars in various regions on the Palomar Sky Survey Atlas with a star-counting instrument. The main purpose of our star counts is to estimate the integrated brightness and color of the star-fields for the elimination of the background light in the study of the zodiacal light.

Detailed explanation of our star-counting instrument and method of measurement have been given at the IAU-Colloquium No. 11 at Edinburgh in 1970 (Tanabe and Mori, 1971) and some results obtained by this instrument have been published (Tanabe, 1973).

In the meantime, in August 1973 the Asteroid-Jupiter probe Pioneer 10 observed the background sky brightness at a heliocentric distance of 4.64AU, where the contribution from the zodiacal light is negligible (Hanner et al., 1974), and a part of these results was reported by Weinberg et al. (1974).

Accordingly, we started to make star counts in the same sky region to obtain the brightness of the integrated starlight for comparison with the Pioneer 10's results and, if possible, isolation of the diffuse galactic light.

2. Measurements and Results

Fig. 1 shows a part of the sky observed by Pioneer 10. The fields of view of the Pioneer 10's telescope (FOV) are projected after a small correction following information from Weinberg (1975). Five small squares indicate the regions we measured. Each area of our regions is 0.871 square degrees, which corresponds to a square of side 5cm on a Palomar

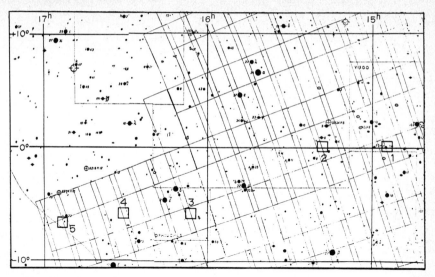

Fig.1. A part of the sky observed by Pioneer 10 with projection of the FOVs. Five small squares indicate the regions we measured.

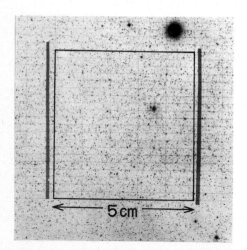

Fig.2. An example of measured regions on the Palomar Sky Survey Atlas.

Atlas print. This area may be too small in comparison with that of the FOV, but since measurements over the full area of the FOV require a long time by our manually operated instrument, we have selected those areas which are located in the central parts of the Palomar Atlas prints.

As an example of the regions we measured, a photograph of region No. 5 on the Palomar Atlas print is shown in Fig. 2. We have measured the diameters of 8694 star-images in this area and obtained the magnitude of each star with an experimental diameter-magnitude relation found from the standard stars. Besides the stars, about 5000 nebulae were counted.

Results of our star counts are shown in Table 1. The equatorial and galactic coordinates, total number of measured stars and brightnesses of integrated starlight in $S_{10}(p)$ and $S_{10}(r)$ units, the equivalent number of 10th-magnitude stars per square degree, obtained from blue (3500~5000A) and red (6200~6700A) photographs for each region are listed.

Table 1. Results of star counts ($m_p \geq 7.5$).

Region	α (1950) δ	l^{II} b^{II}	Stars	$S_{10}(p)$	$S_{10}(r)$
1	$14^h 54^m.7$ $+0°$ 4.7	356°.1 +49°.3	2049	23.6	89.2
2	15 18.7 +0 7.1	2.1 +44.9	3184	22.1	74.0
3	16 6.9 −5 47.4	5.9 +31.8	5327	15.0	77.6
4	16 30.9 −5 44.5	10.0 +27.0	5552	32.2	102.6
5	16 53.5 −6 37.3	12.6 +21.9	8694	35.7	200.0

3. Comparison with Pioneer 10's Results

Reported Pioneer 10's results are expressed in $S_{10}(v)$ units. These values were derived from observations in blue band (3900~5000A) on the basis of the solar color index. Accordingly, to compare our results with them, we obtained the brightness of $S_{10}(v)$ units using our results only from blue photographs with following processes.

As mentioned above, our measured area is so small that there is anxiety of being influenced by small-scale irregularities of the stellar distributions. For example, though a Pioneer's single FOV containes many stars of magnitude 7.5 or 8.0, distributions of these stars are so infrequent that they are not always present in our area.

To avoid this positional selection effect, therefore we used only the integrated value from 10th to the limiting magnitude in our data as a representative value of the integrated faint-star brightness for corresponding FOV, assuming that these faint-star distributions are rather uniform in a single FOV.

In order to obtain the total brightness of the stars fainter than magnitude 6 for each FOV, we counted all stars between photographic magnitudes 6 and 10 using the Star Catalog of the Smithsonian Astrophysical Observatory (1966) after conversion from visual to photographic magnitude with Seares' formula (1925) and added these to the above faint-star brightness in $S_{10}(p)$ units. The brightnesses of the nebulae (1.6~0.7 $S_{10}(p)$) were also added by estimating each magnitude.

Conversion from $S_{10}(p)$ to $S_{10}(v)$ units was based on the solar color index (Allen, 1973); $P - V = B - V - 0.11 = +0.54$.

Fig. 3 is a plot of our results on the figure in the paper of Weinberg et al. (1974). The upper full line shows Pioneer's observed brightness and the dashed line corresponds to the residual brightness after removal of the stars of blue magnitude ≤ 6. Dots indicate Roach and Megill's values (1961) derived from star counts data (van Rhijn, 1925).

Recently, Weinberg (1975) informed us of a corrected version of Pioneer 10's values, which involves more detailed calibration. These results are represented by the lower dashed line.

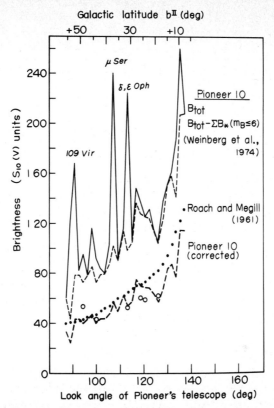

Fig.3. Comparisons of our results (open circles) with those of Pioneer 10 and Roach and Megill's values.

Our results for each corresponding FOV are plotted using open circles. Near the galactic latitude 25°, we obtained values for two adjacent FOVs from the result of our region NO. 4, because this region is on the boundary of two FOVs.

From this figure, it can be seen that our results, except region No. 1, agree well in general with Pioneer's observations, but they are slightly lower than Roach and Megill's values.

The discrepancy at region No. 1 must be due to a local effect of our measurement. Our star counts data show that the number of stars between photographic magnitudes 10 and 12 in this region is about 3/2 times those of next two adjacent regions; Nos. 2 and 3, whose galactic latitudes are lower than that of No. 1. The excess at region No. 1 is mainly due to the contribution from these stars.

Detailed study of the diffuse galactic light requires star counts in wider regions, comparable to the FOV.

4. Supplementary Remarks

For further analyses of the Pioneer 10's data or reductions of other zodiacal light observations, results of our star counts obtained in some basically important sky regions for the photometry are listed in Table 2. Some of them have been published already (Tanabe, 1973). Each measured area centered at the listed position is 3.48 square degrees, except the Celestial N. Pole, the area for which is 3.22 square degrees because of the halo due to the bright Polaris. They do not include the light from the nebulae.

Table 2. Integrated starlight ($m_p \geq 7.5$).

	Celestial N.Pole*	Ecliptic N. Pole*	Galactic N.Pole*	Galactic S.Pole	Equinox Vernal	Equinox Autumnal
α (1950)	$19^h 43^m.0$	$18^h 12^m.1$	$12^h 38^m.7$	$0^h 29^m.7$	$0^h 4^m.9$	$12^h 4^m.9$
δ	$+90°0'$	$+66°2'$	$+29°29'$	$-29°35'$	$+0°32'$	$-0°32'$
$S_{10}(p)$	22.0	29.0	17.7	21.0	17.6	15.0
$S_{10}(r)$	67.1	70.5	60.2	66.9	51.7	40.0

* Tanabe (1973)

We wish to thank Dr. J. L. Weinberg of the State University of New York at Albany, who kindly gave us detailed informations about Pioneer 10's observation.

References

Allen,C.W.(1973) "Astrophysical Quantities 3rd ed.", p.162, p.203, University of London, the Athlone Press.
Hanner,M.S., J.L.Weinberg, L.M.DeShields II, B.A.Green and G.N.Toller(1974) J. Geophys. Res., 79, 3671.
Roach,F.E. and L.R.Megill(1961) Astrophys. J., 133, 228.
Seares,F.H.(1925) Astrophys. J., 61, 114.
Smithsonian Astrophysical Observatory Star Catalog(1966) Smithsonian Institution, Washington, D.C.
Tanabe,H. and K.Mori(1971) Publ. Roy. Obs. Edinburgh, 8, 173.
Tanabe,H.(1973) "Papers on the Night Sky Light and Airglow Data at Chichijima", p.48, Tokyo Astronomical Observatory.
van Rhijn,P.J.(1925) Groningen Publ., No.43.
Weinberg.J.L., M.S.Hanner, D.E.Beeson, L.M.DeShields II and B.A.Green(1974) J. Geophys. Res., 79, 3665.
Weinberg,J.L.(1975) private communication.

THE $S_{10}(V)$ UNIT OF SURFACE BRIGHTNESS

J. G. SPARROW* AND J. L. WEINBERG
Space Astronomy Laboratory, State University of New York at Albany

<u>Abstract.</u> A source for some of the discrepancies among reported zodiacal light measurements is suggested, and a plea is made for more uniformity in the choice of units used to report zodiacal light results.

It is customary to express zodiacal light measurements in units of $S_{10}(V)$, the equivalent number of tenth magnitude (visual) stars of solar spectral type per square degree. As noted by one of its earliest proponents, F. E. Roach, this is a result of its historical use in studies of the background starlight and because the night sky brightness is conveniently four digits or less when expressed in these units. The conversion to these units is not always consistent from one author to another for a number of reasons:

(1)   The $S_{10}(V)$ unit relates to a comparison with stars of "solar type". Although the sun is now generally believed to be a G2V star (Morgan and Keenan, 1973), some authors have expressed their $S_{10}(V)$ units in terms of G0 stars (Weinberg, 1964; Tanabe, 1965; Huruhata, 1965). Roach and coworkers have at time expressed the $S_{10}(V)$ unit in terms of G0V stars (Roach et al., 1954) or of G2 stars (Roach, 1957). A color index B-V = .57 has been used by Weinberg (1967) and Lillie (1972), who obtained it from Roach and Smith (1964) who gave $S_{10}(VIS)/S_{10}(PHOT)$ = 1.69. An indication of the effect of these differences, when measurements are made at say 4400A, is seen by a comparison of the B-V color indices as given by Johnson (1966) viz: G0V = .59, G2V = .63. Croft et al. (1972) found the B-V color index of the sun to be +.631 and listed 7 other values given by earlier workers ranging from .62 to .68.

(2)   The apparent visual magnitude of the sun used by different authors contributes appreciably to the uncertainty in the $S_{10}(V)$ unit. Thus, Robley (1965) used -26.9, Dumont (1965) -26.72 and Weinberg (1964) -26.73, the last at an effective wavelength of 5300A. Leinert et al. (1974) used a value of -26.78 obtained from Allen (1963). In reviewing recent data on the solar spectrum, Code (1973) concludes that the sun is best represented by spectral type G2V, color index B-V = .63 and visual magnitude V = -26.74 (see, also, Allen 1973).

―――――――――――――
*On leave from Aeronautical Research Laboratories, Melbourne, Victoria, Australia

(3) Not all zodiacal light data are referenced to the same solar flux values. In order to preserve continuity with previous work from our group, we use the solar spectral irradiance values given by Johnson (1954); the more recent values of Labs and Neckel (1970) might be preferred as they are by Leinert et al. (1974) and Frey et al. (1974). Roach (1957), Dumont (1965), and Robley (1965) use the results of Abbot as published by Minnaert (1924). A comparison of the irradiance values from the three sources is given in Table 1.

TABLE 1

A Comparison of Values of the Solar Spectral Irradiance*

Wavelength A	Johnson (1954)	Labs and Neckel[i] (1970)	Minnaert[ii] (1924)
4000	.154	.138	.156
4500	.220	.197	.228
5000	.198	.193	.230
5500	.195	.185	.210
6000	.181	.175	.195
6500	.162	.156	.184
7000	.144	.143	.150
7500	.127	.127	.129
8000	.113	.115	.110

* in watts/cm$^2\mu$

[i] values obtained as average of two adjacent 100A means.
[ii] data taken from Roach (1957).

(4) Some authors using star calibrations convert from magnitudes to flux using the absolute flux given for Vega. Thus, Frey et al. (1974) chose the photometric measurements of Oke and Schild (1970), while we have preferred to use the more recent data from Hayes and Latham (1975), although the difference between these data sets as given in the paper by Hayes and Latham is generally less than 5%. It should be noted, however, that these flux values are given for the continuum of Vega, whereas the stellar magnitudes refer to the continuum plus lines. In the blue, the effect of the Balmer lines (H_β, H_γ, and H_δ) is to reduce the integrated continuum flux by about 6%.

(5) Earlier studies of zodiacal light were usually normalized to the International photographic or photovisual systems although the UBV system of Johnson and Morgan (1953) is now preferred. There is little difference between the effective wavelengths of the photovisual (λ_{eff} = 5427A for T = 10^4 K) and V systems (λ_{eff} = 5480A for T = 10^4 K; Allen 1955, 1973), and zodiacal light observers often use m_{pv} and m_v interchangeably. The effective wavelengths of the photographic (λ_{eff} = 4253A, T = 10^4 K) and blue (λ_{eff} = 4400A, T = 10^4 K) cause significant differences between zodiacal light brightnesses measured in the two systems.

(6) Finally, the question is asked: Is the $S_{10}(V)$ unit referred to the solar flux at the time of observation or to the mean solar output? Since the earth's heliocentric distance varies by 3.3% over the year, something is changing by 6.6% (assuming an r^{-1} dependence of the zodiacal dust). Is this the magnitude of the $S_{10}(V)$ unit or the brightness of the zodiacal light in $S_{10}(V)$? As yet our measurements are not accurate enough to show this change but one might hope that before long they will be.

Because of the above, we place great importance on the need to bring consistency into the use of the $S_{10}(V)$ unit, and we urge authors to specify the sources of the various factors that combine to determine the size of their $S_{10}(V)$ unit.

As zodiacal light studies become more precise (?), it is more important to have clarity and unanimity in the units used for reporting surface brightness or radiance. To this end we suggest that:

(1) the $S_{10}(V)$ unit be understood to represent 10th magnitude solar (G2V) stars per square degree at mean solar distance;

(2) the V refer to the visual color in the UBV system defined by Johnson and Morgan (1953);

(3) the apparent solar visual magnitude be taken as -26.73 and the B-V color index as .63;

(4) the solar spectral irradiance values of Labs and Neckel (1970) be used;

(5) when using Vega as a standard to obtain brightnesses in $S_{10}(V)$, +.04 be used as its magnitude at all wavelengths and the irradiance values of Hayes and Latham (1975) be used.

Finally, in quoting results, we suggest that the input parameters used to convert to $S_{10}(V)$ be quoted in a conspicuous manner in order to simplify intercomparison with other data in the literature.

Acknowledgement. This study received support from NASA contracts NAS8-30251 and NAS2-7963.

References.
Allen, C. W., 1973, Astrophysical Quantities, University of London, the Athlone Press; also, 1955 and 1963 editions.
Code, A. D., 1973, Ground Based and Extraterrestrial Observations of Stellar Flux, in Proceedings, IAU Symposium No. 54, Problems of Calibration of Absolute Magnitudes and Temperature of Stars, (B. Hauck and B. E. Westerlund, editors), 131-145, (Dordrecht-Holland: D. Reidel Publ. Co.).
Croft, S. K., D. H. McNamara and K. A. Feltz, Jr., 1972, The (B - V) and (U - B) Color Indices of the Sun, Publ. Astr. Soc. Pacific 84, 515-518.
Dumont, R., 1965, Séparation des composantes atmosphérique, interplanétaire et stellaire du ciel nocturne a 5000A. Application a la lumière zodiacale et du gegenschein, Ann. d' Astrophys. 28, 265-320.
Frey, A., W. Hofmann, P. Lemke and C. Thum, 1974, Photometry of the Zodiacal Light with the Balloon-Borne Telescope THISBE, Astron. and Astrophys. 36, 447-454.
Hayes, D. S. and D. W. Latham, 1975, A Rediscussion of the Atmospheric Extinction and the Absolute Spectral Energy Distribution of Vega, Astrophys. J. 197, 593-601.
Huruhata, M., 1965, Photoelectric Observations of the Photometric Axis of Zodiacal Light, Planet. Space Sci. 13, 237-241.

Johnson, H. L. and W. W. Morgan, 1953, Fundamental Stellar Photometry for Standards of Spectral Type on the Revised System of the Yerkes Spectral Atlas, Astrophys. J. 117, 313-352.

Johnson, F. S., 1954, The Solar Constant, J. Meteorol. 11, 431-439.

Johnson, H. L., 1966, Astronomical Measurements in the Infrared, In Annual Review of Astronomy and Astrophysics, Vol. 4, (L. Goldberg, editor), 193-206, (Palo Alto: Annual Reviews, Inc.).

Labs, D. and H. Neckel, 1970, Transformation of the Absolute Solar Radiation Data into the International Practical Temperature Scale of 1968, Solar Phys. 15, 79-87.

Leinert, C., H. Link and E. Pitz, 1974, Rocket Photometry of the Inner Zodiacal Light, Astron. and Astrophys. 30, 411-422.

Lillie, C. F., 1972, OAO-2 Observations of the Zodiacal Light, in The Scientific Results from the Orbiting Astronomical Observatory (OAO-2), (A. D. Code, editor), 95-108, NASA SP-310, (Washington: U.S. Government Printing Office).

Minnaert, M., 1924, Recent Data on Solar Radiation Converted into Absolute Measure, B.A.N. 2, 75-79.

Morgan, W. W. and P. C. Keenan, 1973, Spectral Classification, in Annual Review of Astronomy and Astrophysics, Vol. 11, (L. Goldberg, editor), 29-50, (Palo Alto: Annual Reviews, Inc.).

Oke, J. B. and R. E. Schild, 1970, The Absolute Spectral Energy Distribution of Alpha Lyrae, Astrophys. J. 161, 1015-1023.

Roach, F. E., 1957, Photometric Observation of the Airglow, IGY Instruction Manual, Part II, 115-138.

Roach, F. E., H. B. Pettit, E. Tandberg-Hanssen, and D. N. Davis, 1954, Observations of the Zodiacal Light, Astrophys. J. 119, 253-273.

Roach, F. E. and L. L. Smith, 1964, Absolute Photometry of the Light of the Night Sky, N.B.S. Tech. Note 214.

Robley, R., 1965, Lumière Zodiacale et Continuum Atmosphérique, Ann. de Geophys. 21, 505-513.

Tanabe, H., 1965, Photoelectric Observations of the Gegenschein, Publ. Astron. Soc. Japan 17, 339-366.

Weinberg, J. L., 1964, The Zodiacal Light at 5300A, Ann. d'Astrophys. 27, 718-738.

Weinberg, J. L., 1967, Summary Report II on Zodiacal Light, July 1967, Hawaii Institute of Geophysics Report HIG-67-13.

1.1.7
POLARIZATION OF THE ZODIACAL LIGHT: FIRST RESULTS FROM SKYLAB

J. G. SPARROW*, J. L. WEINBERG, AND R. C. HAHN
Space Astronomy Laboratory, State University of New York at Albany

Abstract. A brief description is given of the Skylab ten color photoelectric photometer and the programs of measurements made during Skylab missions SL-2 and SL-3. Results obtained on the polarized brightness of zodiacal light at five points on the antisolar hemisphere are discussed and compared with other published data for the north celestial pole, south ecliptic pole, at elongation 90 degrees on the ecliptic, and at two places near the north galactic pole.

Introduction. The Skylab ten color photoelectric polarimeter was designed to measure the surface brightness and direction and amount of polarization of the zodiacal light. The photometer and a 16mm camera were attached to an alt-azimuth mount at the end of a universal extension mechanism that could be deployed up to 5.5m out either the solar or antisolar scientific airlock of the Saturn workshop. Loss of the spacecraft meteoroid shield during launch removed the availability of the solar airlock for this experiment. Thus all observations with the photometer were made from the antisolar scientific airlock.

The Instrument. A brief description of the instrument will suffice here as a more detailed description is available elsewhere (Weinberg et al., 1975). The photometer consisted of a sunshield, a telescope cap with attached Pm 147-activated phosphor source for instrument performance monitoring, Fabry optics with a 6.35cm primary objective, two six position wheels each containing 5 interference filters and an open position, an HN32 rotating polaroid, a wheel with six field of view/neutral density filter combinations and a detector package.

The detector package contained an EMR 541E photomultiplier (with selected S-20 response), a high voltage supply, and an output voltage differential amplifier with time constant of approximately 2 msec. A thermoelectric cooler kept the photomultiplier photocathode at a temperature of approximately $-10^\circ C$; this temperature along with that of the filters and the cap source were monitored.

The photomultiplier output was sampled at 320 sps (nominal 2^8 bit words) while a similar channel was used to record the polaroid position pulse generated once per cycle as the HN32

---

*On leave from Aeronautical Research Laboratories, Melbourne, Victoria, Australia.

polaroid rotated at approximately 2 RPS. These two channels, together with other instrument data sampled less frequently, were stored by an onboard digital tape recorder and subsequently telemetered, or were telemetered to the ground in real time. The elevation and azimuth of the pointing direction of the photometer were recorded 20 and 10 times per second, respectively; however, in this paper we have used the right ascension and declination obtained from the camera photographs with an accuracy of better than $\pm 0.1°$.

Instrument Calibration. The calibration has been based on the signals measured during the scanning programs as the stars Vega, Arcturus and Spica traversed the field of view. Each of these stars was scanned at least twice at all ten wavelengths except 8200A, and the photometer star signal was related to the star's magnitude at each filter wavelength. The magnitude was obtained from a reciprocal wavelength curve fit to the Johnson, et al. (1966) and Mitchell and Johnson (1969) photometric data. Conversion to a calibration in units of $S_{10}(V)$* was obtained using the absolute flux of Vega given by Hayes and Latham (1975), the absolute solar flux given by Johnson (1954), an apparent solar visual magnitude of -26.73 (Stebbins and Kron, 1957) and the angular area of the field of view. We estimate the uncertainty of our calibration at each wavelength as not greater than: 8200A $\pm 20\%$; 4000A $^{+10}_{-20}\%$; 4760A $^{+6}_{-12}\%$; 6300A $\pm 10\%$ and $\pm 6\%$ at all other wavelengths. Further details of the Skylab photometer calibration are given in Sparrow et al. (1975).

Observational Programs. An automatic programmer was included to operate the photometer manually or in any of seven automatic modes:

    0 calibration
    1 fixed position
    2 vertical circle (scan in elevation h at fixed azimuth)
    3 almucantar (scan in azimuth at fixed elevation)
    4 limited-area sky map (almucantars separated 2.8 degrees in h)
    5 all-sky map (almucantars separated 5.6 degrees in h)
    6 stowage position return.

A description of the instrument is given by Weinberg, et al. (1975). Table 2 in Weinberg (1976) lists the observing programs which were performed with the photometer during the first (SL-2) and second (SL-3) missions. The 6 degree angular diameter field was used for most of the observations. Figure 1 illustrates the areas of the antisolar hemisphere for which useful data have been obtained from SL-2, including the five fixed pointing programs whose results are given here.

Data Analysis and Some Preliminary Results. The photomultiplier data are being analyzed by means of a least squares fit to an expression of the form PMT output = $A + Bt + C \sin(\omega t + D)$, where the frequency ω is determined from the polaroid position-pulse repetition rate. The

---

*the equivalent number of tenth magnitude (V) stars of solar spectral type per square degree (see preceding paper).

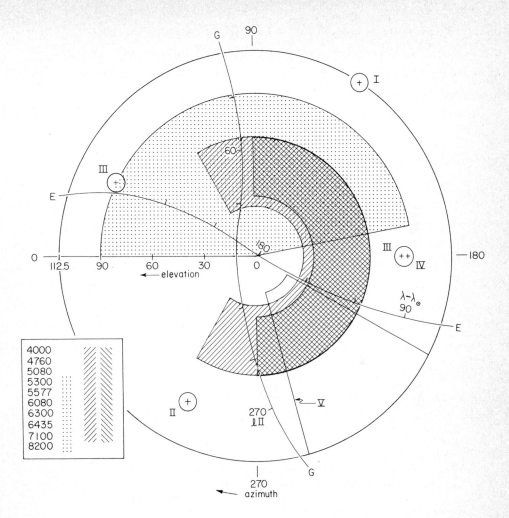

Fig. 1. Regions of the antisolar hemisphere for which useful data were obtained during SL-2. G and E show the positions of the galactic equator and ecliptic, respectively. l^{II} refers to System II galactic longitude in degrees and $\lambda - \lambda_\odot$ refers to the ecliptic longitude of the observed point with respect to the sun, in degrees. The inset shows the filters used with the sky-mapping programs.

phase angle D can be transposed to give the direction of the plane of polarization of the measured light. Comparison with that expected from zodiacal light gives some measure of confidence that additional components of polarized light are not contributing to the measured signal. This is particularly relevant since the SL-2 observations were taken when the moon was between 88% and 100% illuminated. For this paper, no analysis has been attempted during times when:

1. The spacecraft was in daylight;
2. The spacecraft was in the vicinity of the south Atlantic anomaly;
3. The photometer pointed below the spacecraft horizontal;

4. The photometer pointed within 60 degrees of the moon.

During these fixed pointing programs the photometer cycled through each of the 10 filters a number of times. Each color measurement took 10 seconds; between filters the shutter remained closed for 2 or 4 seconds to provide measurements of dark current. Unfortunately not all of the repetitions at each color, satisfactory according to the above four criteria, are presently suitable for analysis. This is caused by the introduction of an amplitude change compression factor (K3) during transmission of the data between ground stations prior to preparation of the experimenter tapes. This resulted in removal of all changes in photomultiplier output voltage that did not equal or exceed 80 mv (4 bits compared to full scale 256 bits) for some periods of data. Since the amplitude of the sinusoidal component was often less than this, some of the information on the polarized component of the zodiacal light was thereby lost from these periods of time. Regenerated experimenter tapes with no data compression (K1), expected to be available soon, will eliminate this problem.

For each of the fixed pointing programs, data without the amplitude change compression factor have been analyzed where available; for those periods where it was not available, we have used the K3 data as indicated in Table 1. It is anticipated that for these time periods, an

TABLE 1

Zodiacal Light Polarized Brightness

Central Wavelength Å	$\Delta\lambda$ Å	"north celestial pole" ϵ	β	"south ecliptic pole" ϵ	β	"90 deg elongation on ecliptic" ϵ	β	"north galactic pole" ϵ	β	"north galactic pole" ϵ	β
		$67.5°$	$66.0°$	$94.9°$	$-84.9°$	$89.0°$	$4.7°$	$96.3°$	$29.3°$	$93.6°$	$30.9°$
		$S_{10}(V)$		$S_{10}(V)$		$S_{10}(V)$		$S_{10}(V)$		$S_{10}(V)$	
4001	110.0	23.1		16.1		30.2		--		$20.5^{(2)}$	
4748	47.5	18.0		11.0		25.3		16.9		$16.1^{(2)}$	
5068	49.7	20.4		12.7		26.7		18.9		17.8	
5294	61.0	19.0		$14.2^{(1)}$		25.8		17.6		18.8	
5562	17.0	20.0		$13.2^{(1)}$		27.5		--		19.3	
6063	83.0	19.5		$12.4^{(1)}$		26.0		17.4		18.5	
6286	20.0	19.3		12.6		30.1		--		17.8	
6427	108.2	18.6		$9.2^{(1)}$		25.8		16.8		19.0	
7093	135.0	20.9		13.9		28.3		19.0		19.0	
8160	222.0	16.1		8.8		19.4		13.2		$14.4^{(2)}$	
Mean values		$19.5 \pm .2$		$12.4 \pm .2$		$26.5 \pm .3$		$17.1 \pm .3$		$18.1 \pm .2$	
Number of 10 second intervals/color		7 (mixed)		1 (K1)		6 (mostly K1)		1 (K3)		2 (K1)	

(1) 6 or 7 seconds of data only
(2) 10 seconds of data only

improvement in precision will be achieved when the new experimenter tapes (without the compression factor) are analyzed. During 4 of the 5 programs discussed here, the instrument pointing direction remained constant within $\pm 0.1°$ in right ascension and declination; during the final fixed pointing program, a spacecraft reorientation procedure was in progress and the pointing direction changed by about $5°$ during the course of the program. The photometer pointing direction is presently taken to be the same as that of the camera. The prelaunch boresite measurement coupled with the analysis of star crossings will enable any offset to be determined.

The results obtained from the five fixed pointing programs are shown in Table 1. It is apparent that the color of the polarized component of the zodiacal light is close to that of the sun.

TABLE 2

Zodiacal Light Polarized Brightness

Author(s)	Color	"90 deg elongation on ecliptic" $S_{10}(V)$	"north ecliptic pole" $S_{10}(V)$	"north celestial pole" $S_{10}(V)$
Behr and Siedentopf (1953)	5430A visual [1]	22[2]		
Elsässer (1958)[3]	visual	40		
Peterson (1961)	4355A blue [4]	33		
	5425A green [4]	46		
Beggs, et al. (1964)	blue	35[5]		
Weinberg (1964)	5300A	46		
Dumont (1965) Dumont and Sanchez (1966)	5000A	28[6]	9.7 - 12.6[7]	
Gillett (1966)	astronomical blue	28[8]		
Wolstencroft and Rose (1967)	7030A and astron. blue[8][9]		26.5	
Ingham and Jameson (1968)	5100A		10.7	
Sparrow and Ney (1968)	astronomical visual		10.5	
Weinberg, et al. (1968)	5080A			17[10]
Jameson (1970)	5100A	22	12	16[11]
Sparrow and Ney (1972)	4180A	30[8]	10.9[8]	
Divari and Krylova (1973)	5300A 5800A		14.3 20.6	
Frey, et al. (1974)	5000A	36[12]		
Dumont and Sanchez (1975)	4600A and 5020A	27.2		
Present study	10 colors	26.5	12.4	19.5
	ϵ, β	89, 4.7	94.9, 84.9	67.5, 66.0

Notes to Table 2:

(1) Authors give surface brightness at λ_{eff} = 5430A and degree of polarization at "visual".
(2) Obtained from product of surface brightness and degree of polarization.
(3) Values obtained from Elsässer (1963).
(4) Author gives surface brightness at λ_{eff} = 4355A and 5425A and degree of polarization at "blue" and "green".
(5) Authors apparently used different instruments for measurement of surface brightness and degree of polarization; the spectral response is not given for either instrument.
(6) Obtained from product of degree of polarization quoted in paper and surface brightness read from curve through combined easterly and westerly observations.
(7) Quoted in Dumont and Sanchez (1970).
(8) Calculated using solar color index.
(9) Authors averaged surface brightness in $S_{10}(V)$ from λ_{eff} = 7030A and astronomical blue; multiplying it by their interpolated degree of polarization at 7030A gave $26.5 S_{10}(V)$.
(10) Based on correcting the observed polarized brightness for atmospheric extinction and scattering and subtracting a polarized brightness of $4 S_{10}(V)$ which arises from atmospheric scattering (Dumont and Sanchez, 1975).
(11) Interpolated from table of polarized brightness as functions of $\lambda - \lambda_\odot, \beta$ to same celestial pole coordinates as Skylab observations.
(12) Value read from plot given in paper.

We believe the somewhat higher values in Table 1 at 4000A (15%) and the lower values at 8200A (25%) to be connected to the difficulties of calibration in the region of the convergence of the Balmer and of the Paschen lines. The 10% lower values at 4760A may be a consequence of our use of the solar irradiance values of Johnson (1954), which differ by about this amount from those given by Labs and Neckel (1970) at this wavelength.

The mean values given at the bottom of Table 1 have been transposed to Table 2 to compare with other published values for the polarized component of the zodiacal light on the ecliptic at $90°$ elongation ϵ, at the north celestial pole and at the ecliptic pole. No distinction has been made between measurements made at the north and south ecliptic poles.

New experimenter tapes with no compression would make it possible to obtain the polarized brightness of zodiacal light over much of the antisolar hemisphere. A search would be made for a polarized component associated with the L_4 and L_5 lunar libration regions. Studies are in progress of color differences in the polarization reversal at large elongations.

Acknowledgements. This research was supported by NASA contract NAS8-30251. Assistance was provided by Nancy Delaney, Robert Mercer, and Candy Toplansky.

References

Beggs, D. W., D. E. Blackwell, D. W. Dewhirst and R. D. Wolstencroft, 1964, Further Observations of the Zodiacal Light from a High Altitude Station and Investigation of the Interplanetary Plasma III. Photoelectric Measurements of Polarization, Mon. Not. R. Astron. Soc. 128, 419-430.

Behr, A. and H. Siedentopf, 1953, Untersuchungen über Zodiakallicht und Gegenschein nach lichtelektrischen Messungen auf dem Jungfraujoch, Zeits. für Astrophys. 32, 19-50.

Divari, N. B. and S. N. Krylova, 1973, On the Influence of the Atmospheric Dust on the Zodiacal Light Polarization, in Evolutionary and Physical Properties of Meteoroids, NASA SP-319, 275-280.

Dumont, R., 1965, Séparation des composantes atmosphérique, interplanétaire et stellaire du ciel nocturne a 5000A. Application a la photometrie de la lumière zodiacale et du gegenschein, Ann. d'Astrophys. 28, 265-320.

Dumont, R. and F. Sanchez Martinez, 1966, Polarisation du ciel nocturne et polarisation de la lumière zodiacale vers 5000A, sur l'ensemble de la sphere celeste, Ann. d'Astrophys. 29, 113-118.

Dumont, R. and F. Sanchez Martinez, 1970, Note on Results and Present Research Subjects at Observatorio del Teide (Tenerife) about Zodiacal Light and Green Atmospheric Continuum Photometry, in Trans. IAU. XIVB, (C. DeJager and A. Jappel, editors), 166, (Dordrecht-Holland: D. Reidel Publ. Co.).

Dumont, R. and F. Sanchez Martinez, 1975, Zodiacal Light Photopolarimetry II. Gradients along the Ecliptic and the Phase Functions of Interplanetary Dust, Astron. and Astrophys., in press.

Elsässer, H., 1958, Neue Helligkeits- und Polarisationsmessungen am Zodiakallicht und ihre Interpretation, Die Sterne. 34, 166-169.

Elsässer, H., 1963, The Zodiacal Light, Planet. Space Sci. 11, 1015-1033.

Frey, A., W. Hoffmann, D. Lemke and C.Thum, 1974, Photometry of the Zodiacal Light with the Balloon-Borne Telescope THISBE, Astron and Astrophys. 36, 447-454.

Gillett, F. C., 1966, Zodiacal Light and Interplanetary Dust, Thesis, University of Minnesota.

Hayes, D. S. and D. W. Latham, 1975, A Rediscussion of the Atmospheric Extinction and the Absolute Spectral-Energy Distribution of Vega, Astrophys. J., 197, 593-601.

Ingham, M. F. and R. F. Jameson, 1968, Observations of the Polarization of the Night Sky and a Model of the Zodiacal Cloud Normal to the Ecliptic Plane, Mon. Not. R. Astron. Soc. 140, 473-482.

Jameson, R. F., 1970, Observations and a Model of the Zodiacal Light, Mon. Not. R. Astron. Soc. 150, 207-213.

Johnson, F. S., 1954, The Solar Constant, J. Meteorol. 11, 431-438.

Johnson, H. L., R. I. Mitchell, B. Iriarte and W. Z. Wisniewski, 1966, UBVRIJKL Photometry of the Bright Stars, Comm. Lunar and Planetary Lab. No. 63, University of Arizona.

Labs, D. and H. Neckel, 1970, Transformation of the Absolute Solar Radiation Data into the International Practical Temperature Scale of 1968, Solar Phys. 15, 79-87.

Mitchell, R. I. and H. L. Johnson, 1969, Thirteen-Color Narrow-Band Photometry of One Thousand Bright Stars, Comm. Lunar and Planetary Lab. No. 32, University of Arizona.

Peterson, A. W., 1961, Three-Color Photometry of the Zodiacal Light, Astrophys. J. 133, 668-674.

Sparrow, J. G. and E. P. Ney, 1968, OSO-B2 Satellite Observations of Zodiacal Light, Astrophys. J. 154, 783-787.

Sparrow, J. G. and E. P. Ney, 1972, Observations of the Zodiacal Light from the Ecliptic to the Poles, Astrophys. J. 174, 705-716.

Sparrow, J. G., J. L. Weinberg, and R. C. Hahn, 1975, The 10-Color Gegenschein/Zodiacal Light Photometer, Applied Optics (Skylab Issue), in press.

Stebbins, J. and G. E. Kron, 1957, Six-Color Photometry of Stars. X. The Stellar Magnitude and Color Index of the Sun, Astrophys. J. 126, 266-280.

Weinberg, J. L., 1964, The Zodiacal Light at 5300A, Ann. d'Astrophys. 27, 718-738.

Weinberg, J. L., 1976, Space Observations of the Zodiacal Light, this Volume.

Weinberg, J. L., H. M. Mann and P. B. Hutchison, 1968, Polarization of the Nightglow: Line Versus Continuum, Planet. Space Sci. 16, 1291-1296.

Weinberg, J. L., J. G. Sparrow and R. C. Hahn, 1975, The Skylab Ten Color Photoelectric Polarimeter, Space Sci. Instrumentation, in press.

Wolstencroft, R. D. and L. J. Rose, 1967, Observations of the Zodiacal Light from a Sounding Rocket, Astrophys. J. 147, 271-292.

1.1.8   PHOTOMETRY OF THE ZODIACAL LIGHT WITH THE
         BALLOON - BORNE TELESCOPE THISBE

Frey, A., Hofmann, W., Lemke, D. and Thum, C.
Max Planck Institute für Astronomie

D-6900 Heidelberg-Königstuhl, F.R.G.

We report on new measurements extending the spectral range of our earlier photometry (Frey et al. 1974) to the near ultraviolet. The residual extinction caused by atmospheric ozone was found to be $0\overset{m}{.}25 \pm 0\overset{m}{.}13$ (2950 Å) and $0\overset{m}{.}36 \pm 0\overset{m}{.}13$ (2150 Å) at 41.5 km float altitude. Within the errors of 10-30% arising from calibration and the reduction procedure our measurements at 5000, 3450, and 2950 Å are compatible to a colour of the zodiacal light not different from that of the sun. Our result obtained at 2150 Å is an upper limit, since no reduction of airglow and integrated starlight has been done yet at that wavelength. This upper limit is 30% above a solar-like spectrum. This result is not in contradiction to the OAO-2 measurements (Lillie 1972). The strong intensity increase he found occurs at wavelengths below 2150 Å.

The airglow intensity found at 2950 Å is 0.77 R/Å corresponding to 1270 S_{10} - units (solar type stars of m_v = 10.0 per square degree). At 2150 Å we got an upper limit of 0.05 R/Å or 1730 S_{10}.

References

Frey,A., Hofmann,W., Lemke,D., Thum,C.: 1974, Astron & Astrophys. **36**, 447

Lillie,C.F.: 1972, in The Scientific Results from the Orbiting Astronomical Observatory OAO-2, NASA Sp.-310, 541

1.1.9   OSO-5 Zodiacal Light Measurements 1969-1975
George Burnett
University of Minnesota
Minneapolis, Minnesota 55455

Introduction:

The Minnesota zodiacal light experiment on board OSO-5 was reactivated in July of 1974, after having been off for 1-1/2 years and operated successfully through August 1975 at which time the satellite was again turned off. This gives another year's worth of data to compare with that taken by the same experiment from January 1969 to January 1973[1,2,3]. This paper reports on the seven year comparison.

Spacecraft and Experiment:

The OSO-5 spacecraft and the position of the Minnesota photometers is depicted in Figure 1. The spacecraft is spin stabilized at 0.5 RPS. The spin axis is held at 90° ecliptic elongation, and at varying ecliptic latitudes, usually within ± 25°. The average motion is 1° per day. The blue photometer, telescope #3, which recorded the zodiacal light measurements reported here, looks out along the anti-sail spin axis. It comprises a broadband system with an effective wavelength of 4180 Å, a 12° circular field of view and a fixed polaroid. A complete description is given in the literature.[4]

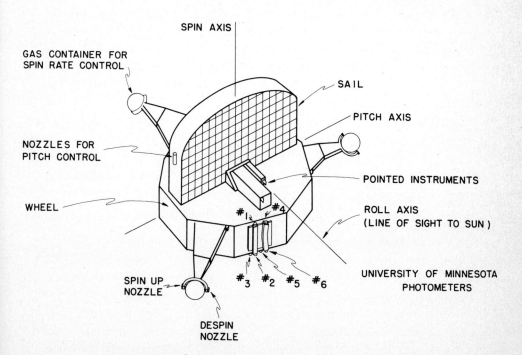

Figure 1. The OSO-5 spacecraft.

Calibration:

The calibration curves for telescope #3 are drawn in Figure 2. Curve B represents the change in sensitivity of the photomultiplier with time. This is measured by means of a light emitting photo diode turned on for five minutes each day. The mean curve through these readings is shown. Over the course of seven years, eight bright stars suitable for calibration came into the field of view. They are listed along the bottom. These eight stars are best fitted to the photo diode decay curve in a least squares sense to produce curve A, which gives the number of 10th magnitude blue (4180 Å) stars per count. The standard deviation of a point (SDP) here is 4%. This is consistent with the expected errors from temperature fluctuations in the photo diodes and the use of an effective wavelength with a broadband system to ascertain stellar magnitudes.

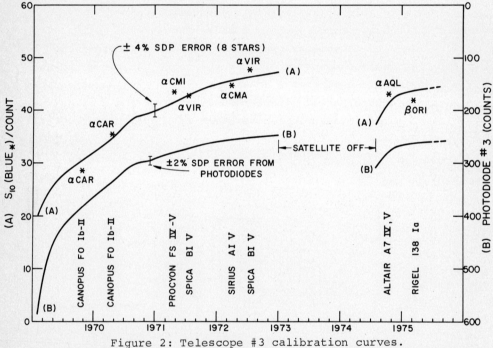

Figure 2: Telescope #3 calibration curves.

It is seen from the curves that a factor of two loss in gain took place over the seven years in a semi-exponential fashion with two points of interest. At turn-on in 1974 after 1-1/2 years rest, a 20% increase in gain took place. Thus some of the degradation was recoverable, related to the instrument being on, and possibly due to the high light levels encountered at sunset turn-on and sunrise turn-off. The small bump in the curve in mid 1970 corresponds to a period when substantial re-orientation took place in going from the South to North ecliptic pole. The experiment was off for two weeks during this time.

For comparison, the availability of a more complete set of calibration stars here has raised telescope #3's calibration by 10% above that used previously[1,2,3]. Also the B-V value of .65 used here to convert from B to V is substantially different from that used before[1].

Results:

A representative sample of the polarized amplitude data by year, from -25° to +25° ecliptic latitude is shown in Figure 3. The 1969 data are at E90° ecliptic elongation and the remaining data are at W90°. The standard deviation is ±2.5%. This is consistent with the error expected from the calibration curve, and that due to a ±0.5° uncertainty in the look direction of the spacecraft.

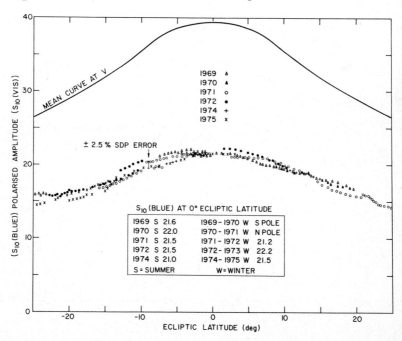

Figure 3: Telescope #3 polarized amplitude data by year.

Only data with a galactic latitude greater than 30° were used in this comparison. This resulted in the periods 15 February to 15 May and 15 August to 15 November not being used. Readings were also not used when the look direction was within 50° of the Moon, or when the look height was below 400 km. Throughout the remaining data periods at least two measurements per day were reduced. A complete description of the data reduction process is given in the literature.[1]

The data then fall into two groups -- summer and winter of each year. Two effects are at maximum at these times, the variation expected from the symmetry with respect to the invariant plane, and that due to the changing Earth-Sun distance. To the ±2.5% standard error, neither of these effects evidenced themselves in the data. This is illustrated in the box in Figure 3. The variations from month to month are as large as those seen between winter and summer, and between 1969 and 1975. No variation greater than ±2.5% is seen with time.

The mean curve is also plotted here at V using a solar color index of B-V equal to 0.65 from Allen.[5] The data to an accuracy of ±2.5% are symmetric with respect to the ecliptic plane, and E or W 90° ecliptic elongation. The mean values at 0° ecliptic latitude for the polarized component are: S_{10} Blue = 21.5 and S_{10} Visible = 39.5.

Figure 4 shows the zodiacal light brightness intensity year by year over the same period. Here the standard error is larger, ±6%, due to the problem of subtracting discrete and background starlight from the field. Again no variations with time are observed to ±6%. No significance in the double grouping of the data here is found. The low open circles to the right in 1971 occurred at a time when the

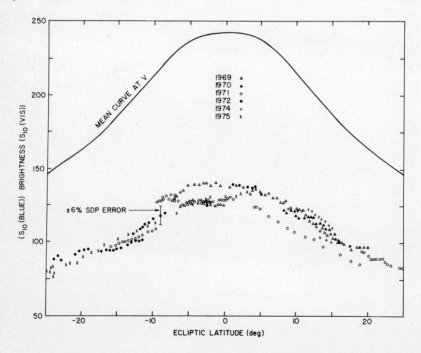

Figure 4. Telescope #3 brightness amplitude data by year.

spacecraft look direction was changing rapidly, resulting in an uncertainty in the look position of $\pm 2°$. The mean brightness values at $0°$ ecliptic latitude are: S_{10} Blue = 133 and S_{10} Visible = 243.

The above errors are all relative. They do not take systematic errors into account as they are not necessary to show changes. The systematic errors amount to an additional $\pm 10\%$.

Conclusions:

The results of the first four years have been reported on previously[1,2,3]. The 1974-1975 data reconfirm the earlier conclusion that the brightness and polarization of the zodiacal cloud varied by less than $\pm 10\%$, now over the seven year period from January 1969 through August 1975, as measured by a single well calibrated instrument.

No variations are found with the 11 year solar cycle, lunar phase, an annual period, or geomagnetic activity. Any or all such variations appear to be below the relative errors of $\pm 2.5\%$ for the polarized component and $\pm 6\%$ for the brightness intensity as measured at B at $90°$ elongation by the Minnesota OSO-5 zodiacal light experiment.

References:
1. Sparrow, J. G. and Ney, E. P., Ap.J. **174**, 705 (1972).
2. Sparrow, J. G. and Ney, E. P., Science **181**, 438 (1973).
3. Burnett, G. B., Sparrow, J. G. and Ney, E. P., Nature **249**, 639 (1974).
4. Burnett, G.B., Sparrow, J. G. and Ney, E. P., Appl. Optics **11**, 2075, (1972).
5. Allen, C. W., _Astrophysical Quantities_, The Athlone Press (1973).

1.1.10   EVIDENCE FOR SCATTERING PARTICLES IN METEOR STREAMS

Anny-Chantal LEVASSEUR - Jacques BLAMONT

Service d'Aéronomie du CNRS
BP n° 3, 91370 - Verrières-le-Buisson, France

I.- EVIDENCE FOR VARIATIONS IN THE ZODIACAL LIGHT INTENSITY

From observations made with a photometer placed on board D2A-Tournesol from April 1971 to June 1973 at $\varepsilon = \pm 90°$ for all ecliptic latitudes, we had deduced that the intensity of the zodiacal light is not constant, but shows fluctuations, some of which we correlated with cometary debris. The measurements were carried out by scanning the celestial sphere in a plane orthogonal to the Sun-Earth direction with a one minute period (LEVASSEUR et BLAMONT, 1973). BURNETT, SPARROW and NEY (1974) have pointed out that our data could have been contaminated by stray moonlight ; we have therefore refined our data analysis and conclude that indeed fluctuations of the intensity of the zodiacal light do exist, even when the Moon is much below the horizon.

The measurements taken into account are obtained over the night part of an orbit (solar depression angle < 25°) above the local horizontal plane ($\sim$ 600 km). Galactic latitude is greater than 30° and there is no planet, emissive nebula or bright star in the field of view. It may be remarked that the elimination of data with stars brighter than the 6th visual magnitude, instead of the 5.5th, has removed the erratic points from our previous curves giving the zodiacal light intensity as a function of ecliptic latitude at 90° elongation. We have studied the moonlight effect when the Moon is above the horizon for all the data obtained near the north ecliptic pole. We caracterize the Moon brightness by the usual arbitrary scale varying from 0 to 10, from new moon to full moon. The curves giving without any smoothing the intensity as a function of the (line of sight -Moon) angle showed no specular reflections and a least squares program could be used (fig. 1). The effect of the Moon is evident at the full moon where moonlight

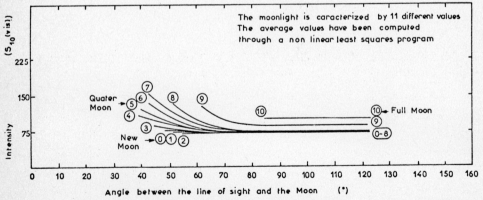

Figure 1 - MOONLIGHT EFFECT ON THE EXPERIMENT - The Moon is above the horizon and the zodiacal light intensity near the ecliptic pole is plotted as a function of the (line of sight-Moon) angle.

brightness is caracterized by 9 to 10 ; it may exist at the first or third quater for a (line of sight-Moon) angle smaller than 90°, as well as at the new moon for an angle smaller than 45°. Therefore, we have added to the data obtained when the Moon is below the horizon the data obtained when the Moon is above the horizon and when either E = 3, 4, 5, 6, 7, 8 with an angle greater than 90° or E = 0, 1, 2 with an angle greater than 45°. After this whole editing, it remains 41 445 data points to be compared to 60 440 data points previously taken into account (LEVASSEUR and BLAMONT, 1974).

For one orbit, as well as for one calendar day, the dispersion for the data obtained at the same latitude is smaller than 14 S_{10}(vis). The main sources of error are the evaluation of the stellar background contributing an average error of 5 S_{10}(vis), and the noise of the electronic system, increasing with the signal and contributing 9 S_{10} (vis) in the ecliptic plane and 5 S_{10}(vis) at the pole. When all the data points ($\sim$ 15) obtained on the same calendar day in the same direction are averaged (LEVASSEUR, 1975), the zodiacal light intensity is determined with a r.m.s. smaller than 10 S_{10}(vis) (fig.2)

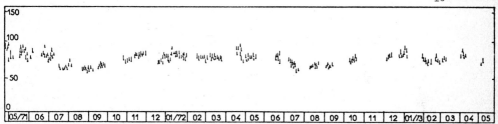

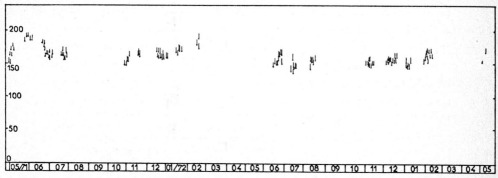

Figure 2 - INTENSITY (S_{10}(vis)) AS A FUNCTION OF TIME AT
    a- THE NORTH ECLIPTIC POLE
    b- THE ECLIPTIC PLANE $\varepsilon = \lambda - \lambda_\odot = -90°$

Two features may be noticed : An <u>annual variation</u> ($\sim$ 10%) due to the symmetry of the zodiacal cloud with respect to the invariant plane is obvious. BURNETT, SPARROW and NEY had thought that this variation might be explained by a $\sim$ 20 S_{10}(vis) per year degradation of the experiment. Anyway, we understand that such a variation, from 61 S_{10} (vis) in September to 75 S_{10}(vis) in March, has never been seen by OSO 5 (SPARROW and NEY, 1972), whose sensibility was therefore above 14 S_{10}(vis). <u>Short period fluctuations</u> are detected (up to 60 S_{10}(vis)). They present no monthly periodicity and may occur over a limited part of the sky ; they have consequently a much smaller probability of being detected by an experiment which is not scanning the celestial sphere. They are seen from one year to the next at the same time and in the same direction. Therefore they depend on the position of the Earth in its orbit and we interpret them as being due to local inhomogeneities of the zodiacal cloud.

II.- GEOMETRY OF THE OBSERVATIONS OF LOCAL INHOMOGENEITIES

The correlation in time and direction of the fluctuations with meteor streams, that is to say the identification of the detected inhomogeneities with local streams, has been discussed in previous papers (LEVASSEUR and BLAMONT, 1973 ; LEVASSEUR, 1974). We may nevertheless focus our attention on the two following points :

<u>The observed streams have significantly eccentric sections by the plane of view</u>: A stream is assumed to have a circular cross section ; its section by the plane of the line of sight of the experiment is therefore an ellipse (fig. 3). The semi minor axis is $d/2$ and the semi-major axis $d/2 \sin \sigma$. The eccentricity of the elliptical section is smaller than 0.5 for the Quadrantids, Lyrids, Perseids, χ Cygnids, Giacobinids, Andromedids, Leonids and Ursids (table 1) ; those streams are actually the observed streams, except for the Giacobinids, or Leonids, which are periodic, and for the Cygnids : a stream may be young and narrow, or after diffusion, older and larger ; it is still detected by the experiment when the corresponding line of sight is long enough.

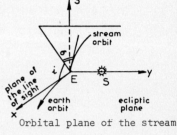
Orbital plane of the stream

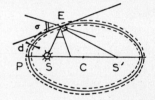
Plane of the line of sight

<u>A parallactic effect is observed during the crossing of the inhomogeneity by the Earth</u> : For instance, from the 19 to the 21st of April, period which could correspond to an encounter with the Lyrids, an increase in intensity has been detected in front of the Earth (at $\beta = 45°$, $\varepsilon = 90°$ and $\beta = 90°$). Then the signal has been decreasing at $\beta = 45°$, while it remained constant at the north ecliptic pole. At the end of April, the signal has finally been increasing behind the Earth (at $\beta = 0°$, $\varepsilon = -90°$ and $\beta = -45°$, $\varepsilon = -90°$) while it was nominal everywhere else. This observation shows an asymmetry in the inhomogeneity since the maximum of intensity occurs very soon after the beginning of the phenomenon. This asymmetry could be due to a non cylindrical symmetry of distribution of particles in the stream. A study of similar asymmetries could provide a method of observing the Poynting-Robertson effect.

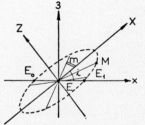

Figure 3 - INTERSECTION OF A STREAM WITH THE PLANE OF THE LINE OF SIGHT.

Shower	σ	Major ax/Minor ax	m	EE_1 10^6 km	Time of crossing(d)
Quadrantids	7.43	7.59	13.0	8.9	< 1
Lyrids	16.20	3.57	30.2		
η Aquarids	31.54	1.92	Whole ellipse	20.4	16
σ Cetids	65.13	1.09	"		
Arietids	68.78	1.07	"	24.4	19
δ Perseids	47.83	1.35	"		
β Taurids	50.31	1.29	"		
δ Aquarids S	75.60	1.03	"	23.0	18
δ Aquarids	75.38	1.03	"		
α Capricornids	37.45	1.63	"		
ι Aquarids S	60.18	1.14	"		
ι Aquarids	53.46	1.25	"		
Perséids	14.03	4.16	25.6	7.4	6
χ Cygnids	10.71	5.26	19.1		
Giacobinids	7.43	7.69	13.0		
Orionids	41.77	1.49	Whole ellipse		
Taurids S	48.87	1.33	"		
Taurids	51.80	1.26	"		
Andromédids	27.94	2.12	66.7		
Léonids	10.09	5.55	17.9		
Géminids	62.94	1.12	Whole ellipse	7.1	5
Ursids	16.22	3.57	30.2		

Table 1 - CHARACTERISTICS OF THE ELLIPTICAL SECTION BY THE PLANE OF VIEW.

III.- EXTENSION AND DENSITY OF THE INHOMOGENEITIES

Extension : During the time of observation of an inhomogeneity, the Earth moves by a few degrees on its orbit, that is to say about 10^7 km. The distance of the inhomogeneity to the Earth may be assumed to be almost equal to zero (if not, too small a fraction of the field of view would be covered). Therefore, the extension of an average inhomogeneity in the ecliptic plane is 10^7 km.

Density : For the homogeneous part of the zodiacal cloud, the intensity I at 90° elongation is given as a function of ecliptic latitude β by the usual formula :

$$I(90°,\beta) = E_\lambda \lambda^2/4\pi^2 \int_{\pi/2}^{\pi} \exp(k \sin\beta/tg\theta) \sin^\nu\theta (1-\sin^2\beta \cos^2\theta)^{-\nu/2} \sum_p N_p f_p(\theta) d\theta$$

The increase in intensity to be expected from a local inhomogeneity of length d_i is :

$$\Delta I(\varepsilon, \beta) = E_\lambda \lambda^2/4\pi^2 \int_{\pi/2}^{1/cot(-d_i)} N_i f_i(\theta) \sin^{-2}\theta \, d\theta$$

and for d_i smaller than 10^8 km, ΔI is almost proportional to d_i. For the observed fluctuations, ΔI varies from 10 to 60 S_{10}(vis). We have computed that, if particles in the cloud are small (α ∼ 20 - size distribution $\alpha^{-2.5}$) dielectric particles (GIESE, 1971) then :

$$0.1 < N_i d_i/N_p < 0.55$$

Discussion : The computed size and density seem to be in good agreement with a meteor stream hypothesis. Inversely, we can estimate the optical properties of a stream, whose physical characteristics have been well established, for instance the Quadrantids (POOLE, HUGHES and KAISER, 1972). The maximum for this stream would occur on January, 3rd at $\beta = i = 74°$. We may recall that, on January 3rd, 1972 and 1973, an increase in the zodiacal light intensity, reaching 20 S_{10}(vis) at $\beta = 80°$ has been observed in a 20° wide region of the plane of view. The radiometeoroids stream diameter d is about 1.72×10^6 km. We have computed that its section by the Earth is 1.76×10^6 km. The stream therefore could not be observed from the Earth for more than one day, as is actually observed on board D2-A. The semi major axis of the elliptical section of the stream by the plane of the line of sight is $a = 6.6 \times 10^6$ km. The angle in which the distance from the Earth to a point of the ellipse is at least $a/2$ is equal to 13°. Therefore the angular width of the stream seen from the Earth at the point of closest approach is about 25°, as we do observe. The average density is 8×10^{-24} g cm^{-3} (HUGHES, 1974). A 22 S_{10}(vis) increase in intensity would be obtained if only 10% of the total mass of the stream was due to small particles similar to the ones described previously.

This confirms our previous conclusion : the zodiacal cloud consists of an homogeneous material and of a collection of meteor streams.

REFERENCES

G.B. BURNETT, J.G. SPARROW, E.P. NEY - Nature, 249, p. 639 (1974)
R.H. GIESE - Max Planck Institut PAE München (1971)
D.W. HUGHES - Space Res. XIV p. 709 (1974)
A.C. LEVASSEUR - Journées "Les poussières en Astrophysique" IAP Paris p.66 (1974)
A.C. LEVASSEUR - Service d'Aéronomie du CNRS, Verrières le Buisson (1975)
A.C. LEVASSEUR, J.E. BLAMONT - Nature 246, p. 26 (1973)
A.C. LEVASSEUR, J.E. BLAMONT - Space Res. XV, p. 295 (1974-1975)
L.M.G. POOLE, D.W. HUGHES, T.R. KAISER - Mon. Not. R. Astr. Soc. 156, p.223 (1972)
J.G. SPARROW, E.P. NEY - Astrophys. J. 174, p. 705 (1972)

1.1.11

THE ULTRAVIOLET SCATTERING EFFICIENCY OF INTERPLANETARY DUST GRAINS

Charles F. Lillie
University of Colorado
Boulder, Colorado 80302

Surface brightness photometry of the night sky with experiments aboard several rockets and spacecraft indicates a significant enhancement of the brightness of the zodiacal light relative to the sun in the 1700 to 3000 A spectral region. This enhancement is most likely due to Mie scattering by non-absorbing (dielectric) particles with a mean radius of $\sim 0.05\mu$ and a real index of refraction which increases rapidly from ~ 1.4 to ~ 2.0 at ~ 2000 A, where most optical materials have an absorption edge (and exhibit a similar phenomenon). Assuming the visible zodiacal light is produced by 10 to 30μ particles, the number density of 0.5μ particles must be $\sim 3 \times 10^5$ times greater, in good agreement with size distributions from crater counts and space probe particle detectors. Ultraviolet observations of comets indicate the enhancement is not due to a bulk property of unmodified cometary dust grains. The most probable source of submicron particles is the breakup of large agglomerates of small (~ 1000 A) spherical particles like those found by particle collection experiments.

Full paper to be published in "Astronomy and Astrophysics".

1.1.12
SUMMARY OF OBSERVATIONS OF THE SOLAR CORONA/INNER ZODIACAL LIGHT
FROM APOLLO 15, 16, AND 17

C.L. Ross
High Altitude Observatory,
National Center for Atmospheric Research,
Boulder, Col./USA

The orbital track of the Apollo Command Service Module (CSM) around the moon during the flights of Apollo 15, 16, and 17 provided unique opportunities to observe the solar corona and zodiacal light. Photographs of the lunar sunrise and sunset were made with a 70 mm electric Hasselblad camera, 80 mm lens at f/2.8, using Eastman Kodak 2485 High Speed Recording film.

The K + F coronal radiance in the 3.0 R_o to 55 R_o area for the three Apollo missions is summarized and compared with the results from previous ground observations. Evidence that the symmetry axis of the radiance of the solar corona/zodiacal light does indeed have an annual variation in displacement from the ecliptic will be presented from the time spacing of the three missions, July 1971 (Apollo 15), April 1972 (Apollo 16), and December 1972 (Apollo 17).

Finally, the photographs from Apollo 15 reveal solar coronal electron forms to 50 R_o from sun center, while the Apollo 16 data contained no coronal forms. Results of image enhancement of the Apollo 17 data and their correlation with ground observations of the inner K-corona will be presented.

1.1.13

A TEMPORAL STUDY OF THE RADIANCE OF THE F-CORONA CLOSE TO THE SUN

R.H. Munro
High Altitude Observatory,
National Center for Atmospheric Research,
Boulder, Col./USA

During the Skylab mission - May 1973 through February 1974 - the High Altitude Observatory's white light coronagraph observed the sum of the F-corona, electron scattered K-corona, and instrumental stray light between 0.4 and 1.6 degrees from the sun. In searching for temporal variations in the F-corona, measurements were confined to the solar polar regions to minimize the effects of the K-coronal component. Changes in instrumental stray light were eliminated by restricting measurements to a single region within the instruments' field of view. The largest source of error is the photometric calibration of the individual rolls of film. Frames were specifically selected to encompass periods of time ranging from a few days to eight months. Generally no variation in the total radiance greater than three percent was detected for intervals on the order of a few weeks. This level of stability holds for most of the eight-month period, excepting a few instances when deviations of up to eight percent were observed where the calibration is most uncertain. A preliminary study of the asymmetry in the F-corona close to the sun and the possible effect of solar eruptions (e.g., flares and prominences) upon the F-corona will be discussed.

1.1.14   MEASUREMENTS OF THE F-CORONA
         FROM DAILY OSO-7 OBSERVATIONS
         R. A. Howard and M. J. Koomen
         Naval Research Laboratory
         Washington, D. C. 20375

The white light corona (K + F) between 3 and 10 R_s was observed from October, 1971 through June, 1974 by the NRL coronagraph orbiting on OSO-7. Daily images from October, 1971 through December, 1972 have been calibrated to yield the daily variation of the coronal brightness as measured through a segmented polarizing plate. This plate had segments of tangential and radial polarization which enabled the separation of the observed brightness into its polarized and unpolarized components at about 5 and 7 R_s. Each image was scanned at 35 heights above the solar limb between 3 and 10 R_s and at every 1° in heliographic position angle. The brightness of selected points in the corona was plotted to obtain the time history of the variations. Curvilinear regression of the radial brightness distributions for each month shows fluctuations in the gradient, magnitude and standard deviation from regression associated with solar activity. Taking means over the entire set of observations yields curves representing the average, the minimum and the maximum coronal brightnesses. The minimum curve derived for each position angle is interpreted as an upper limit of the F coronal brightness. The equatorial minimum curve agrees extremely well with the F model of Saito (1970) in both shape and absolute value after a corrector for stray light has been subtracted. However the minimum curve for the polar regions lies approximately 3 standard deviations (30%) below that for the equator. This implies that the F-corona is not independent of azimuth or alternatively, the minimum equatorial curve may contain a contribution from the K-corona.

Reference:

Saito, K., 1970, Ann. Tokyo Astron.Obs. 12, 53.

1.1.15 THE THERMAL EMISSION OF THE DUST CORONA
DURING THE ECLIPSE OF JUNE 30, 1973

P. Léna and Y. Viala
Université Paris VII et Observatoire de Meudon

D. Hall
Kitt Peak National Observatory

A. Soufflot
Laboratoire de Physique Stellaire et Planétaire

Abstract. Observations of the F-corona have been made in the 10 µ region of the infrared during the eclipse of June 30, 1973. Use of the supersonic aircraft "Concorde 001" permitted 74 min of observation during totality and greatly reduced problems due to sky noise. The plane of the ecliptic was scanned over heliocentric distances of from 3 to 19 solar radii off the east limb. Bright features previously observed at shorter wavelengths, notably emission at 4 solar radii are evident on the 10 µ scans, strongly indicating that the radiation is due to thermal emission by dust. The specific intensity in the 4 $R_\odot$ feature is 5 µ W cm^{-2} sterad^{-1} µ^{-1} higher than the intensity 22 arcmin above the ecliptic. Spectra were taken at one region in the ecliptic and tentatively attributed to silicate-type material. Complete details may be found in Astron. & Astrophys. 37, 75-79 and 81-86, 1974.

1.1.16 THE COLOR CHARACTERISTICS OF THE
EARTH-MOON LIBRATION CLOUDS

J. R. ROACH
Ball Brothers Research Corporation
Boulder, Colorado

ABSTRACT

The OSO-6 Zodiacal Light experiment has provided evidence for the backscatter of light from the Earth-Moon libration regions at 5000 A. Additional measured data at 4000 A and 6100 A show similar characteristics. The combined three-color data lead to libration region color indices slightly bluer than the average surrounding sky background. The differential b-v for the L4 region is approximately -0.05.

INTRODUCTION

The photometric results obtained from the OSO-6 Zodiacal Light experiment have been shown to be a sensitive measure of the sky brightness in the anti-sun region by Roach, et.al., (1973). Using these data, a detailed analysis of variations in the counterglow direction has shown a photometric enhancement in the region of the Earth-Moon libration point by Roach (1975). This analysis has been extended to include the three wavelength bands of data available at: 1) 4000 A $\pm$ 200 A; 2) 5000 A $\pm$ 200 A; and 3) 6100 A $\pm$ 200 A.

OBSERVATIONS

The observations were obtained over a period of sixteen successive lunations (7200 orbits from September 1969 to February 1971). Data were taken every orbit in each of the three wavelength bands for a series of solar elongation angles. This analysis is from data taken in the anti-sun direction, a portion of which is shown in Figure 1. These linear data counts are uncorrected for the instrument sensitivity at each wavelength.

The perturbations observed in these data have been identified as due to the South Atlantic Anomaly effects on the instrument; discrete stars in the field of view, which have a characteristic signature 30 orbits wide; and discrete perturbations attributed to the counterglow

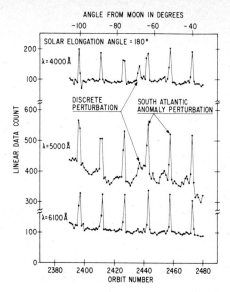

Figure 1 - A Portion of the Linear Count Data from the Instrument for Three Wavelength Bands; Lunation Number 5, L_4.

from Earth-Moon libration clouds. One of these latter perturbations is shown in Figure 1 at 68 degrees from the Moon. It is quite evident in the 4000 A and 5000 A data and is just a suggestion in the 6100 A data.

If the environmental effects and discrete stars are removed from the total sky observation and the background sky is removed by underdrawing, then a residual count remains. Previously I showed that accumulation of this residual count at 5000 A for the sixteen lunations gave an enhancement at the L4 and L5 60 degree positions from the Moon. Inspection of the 4000 A and 6100 A data show similar enhanced characteristics for these libration regions. Rather than analyzing each wavelength band independently I have used a sky temperature parameter based on the color indices for this instrument.

ANALYSIS

In the previous work with these data, discrete stars and the counterglow data have been used to calibrate the instrument photometric characteristics; Roach, et. al., (1973) and Roach (1975). In this analysis, I have used a calibration developed by F. E. Roach (1975) from discrete stars observed, to provide the color calibration of the instrument. This calibration is given in Table 1.

Table 1

STARS USED FOR COLOR INDEX CALIBRATION

Star	b-v (OSO-6)	B-V (literature)
α Librae	-0.021	-0.195
β Sco	-0.031	-0.08
α Sco	2.575	1.80
α Leo	0.177	-0.12
φ Sgr	0.146	-0.10
σ Sgr	0.019	-0.20
γ Sgr	0.983	1.18
σ Sco	0.019	0.14
δ Vir	1.185	1.57
ξ Vir	0.145	0.11

In Table 1, the color index b-v for the OSO-6 instrument is compared to the general astronomical color index B-V from the literature.

F. E. Roach has derived the following relationships for the instrument color calibration using these data and setting the indices equal to zero at a black body temperature of 11000 degrees:

$$b-v = 2.5 \log \frac{R(5000)}{R(4000)} - 0.781$$

$$b-r = 2.5 \log \frac{R(6100)}{R(4000)} + 0.894$$

$$v-r = 2.5 \log \frac{R(6100)}{R(5000)} + 1.675$$

These color indices have been used to develop an effective sky temperature calibration based on black body calculations. Respective sky temperatures were calculated for each index as well as an average sky temperature for the three indices. The average background sky temperature is approximately that of the Sun in the 5500 to 6000 degree range. The analysis to date has been based primarily on the sky temperature derived from the b-v index, but all four temperature parameters show similar characteristics.

Using the underdrawing technique to eliminate discrete stars, the environmental effects and the background, I have accumulated the residual sky temperature for the sixteen successive lunations in the L_4 region. This is shown in Figure 2 and the average increase in sky

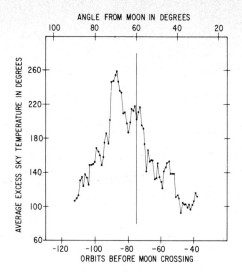

Figure 2 - The Residual Sky Temperature Accumulated and Averaged for Sixteen Lunations in the L_4 Region. Data Have Been Smoothed by Running 7's. The Vertical Line is the Location of the Libration Point L_4.

temperature for the counterglow from the libration cloud region is approximately 120 degrees. The width of this enhanced sky temperature region is approximately 30 degrees at the half maximum position, similar to that of the 5000 A enhanced region. The secondary peak at about 70 degrees from the Moon is the effect of the two major perturbations at orbit 2440 (shown in Figure 1) and orbit 6917 (Roach, 1975).

This increase in the residual sky temperature parameter around the L_4 60 degree point from the Moon, is twice that of the residual away from this region and is, therefore, significant. An increase in effective sky temperature indicates that the excess counterglow from the libration cloud region is slightly bluer than the background counterglow. The equivalent color excess in the b-v index is approximately -0.05. This slightly blue signature for the libration clouds suggests that relatively small particles are the major constituents of the clouds. However, these measurements were made looking through the Earth's penumbra and Schmidt and Kovar (1967) have shown that the Earth's atmosphere causes a focusing of blue sunlight at about the lunar orbit distance. Therefore, the blue libration cloud counterglow may be due to this blue sunlight reflecting from larger and longer-lived particles.

REFERENCES

ROACH, F. E., CARROLL, B., ALLER, L. H., and ROACH, J. R. (1973). "A Photometric Study of the Counterglow from Space." Planetary and Space Science, 21, 1179-1184.

ROACH, F. E. (1975). Private Communication.

ROACH, J. R. (1975). "Counterglow from the Earth-Moon Libration Points." Planetary and Space Science, 23, 173-181.

SCHMIDT, H. and KOVAR, N. S. (1967). "A Possible Model for the Gegenschein." AJ, 72, 827.

1.1.17 A SEARCH FOR FORWARD SCATTERING OF SUNLIGHT FROM THE LUNAR LIBRATION CLOUDS

C.L. Ross
High Altitude Observatory,
National Center for Atmospheric Research,
Boulder, Col./USA

Observations to determine the radiance of forward scattered sunlight from particles in lunar libration regions have been attempted with the white light coronagraph on Skylab. The libration regions could not be distinguished against the solar K + F coronal background; the upper limit to the libration cloud radiance is determined to be $2.5 \times 10^{-11} B_o$, where B_o is the radiance of the mean solar disk. Employing models of the particle type and size distribution in the libration clouds, density enhancements have been calculated on the basis of the upper limit of the forward scattered radiance presented herein, and on the basis of earlier observations of the libration region backscattered radiance. The cases where the power law particle size distribution exponent K and complex index of refraction m are 2.5, 1.33-0.05i and 2.5, 1.50-0.05i, respectively, are inconsistent with the forward and backscatter observations. Finally, the brightness contrast of remaining possible models of the libration clouds with respect to the K- and F-coronal background is calculated, and is shown to be a maximum in the vicinity of elongation angle $\sim 30°$.

1.1.18   PRESENTATION OF ZODIACAL LIGHT INSTRUMENT ABOARD

THE D2B ASTRONOMICAL SATELLITE

M. Maucherat and P. Cruvellier
Laboratoire d'Astronomie Spatiale, Marseille

Abstract. The french astronomical satellite D2B is to be launched on September 1975. During the dial of the orbit, one axis of the satellite points to the Sun, with an accuracy of ± 30 arcmin; during the night of the orbit, the satellite keeps this stabilization by gyroscopic effect. The perpendicular axes -one of them is the optical axis of the Z.L. (Zodiacal Light) instrument- roll around the Earth-Sun direction, once every 4 minutes. Because of the Earth's motion around the Sun, the celestrial sphere will be scanned along ecliptic meridians, every orbits; the complete exploration of the sky will be achieved after six months. The Z.L. instrument is a photometer analyzing galactic and zodiacal fluxes, in the range $\lambda\lambda$ 3100-750, in five spectral intervals 200-500 wide, with a field of 2.8°□. For a single scan, the limit of detection by the instrument will be the 6th visual magnitude for early-type stars (O-B).

1. Introduction

The french astronomical satellite D2B is to be launched from French Guyana, on September 1975. The orbit of the spacecraft is inclined 37° to the ecliptic plane, with a perigée of 500 km, an apogée about 700 km height, and a period of revolution of 96.5 min. The Zodiacal Light instrument is used only during the night eclipse behind the Earth. The satellite is stabilized in one axis: during the dial of its orbit, one axis points to the Sun, with an accuracy of ± 30 arcmin; during the night, it keeps its stabilization by gyroscopic effect. The perpendicular axes (one of them is the optical axis of the Zodiacal Light instrument) roll around the Earth-Sun direction, once every 4 minutes. Because of the Sun's motion on the ecliptic, each displaced by 4 arcmin, every orbit, the complete exploration of the sky will be achieved after six months, the presumed lifetime of the spacecraft.

Figure 1 presents the position of the Zodiacal Light instrument inside the spacecraft, and its scanning of the sky, at 90° elongation. The region of the north ecliptic pole will be analyzed every rotation of the vehicle.

2. Instrumental design

A scheme of the optical design of the instrument is presented in Figure 2. An objective grating, with an aperture ratio of F/3 analyses stellar fluxes in a field

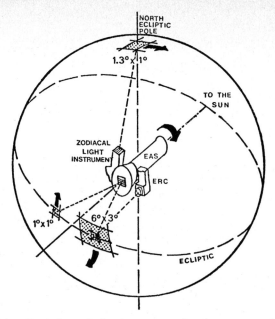

Fig. 1. Position of the instrument in the spacecraft

of 1.3° limited by a baffle; in order to observe faint emission while exposed to earthlight or moonlight, a diffraction baffle allows only components of second or higher order diffraction to enter the photometer. The spectrum is formed in the focal surface of the photometer with a dispersion of 68 Å mm^{-1}. Behind the exit slits, about 4 mm wide, set in a plane close to the focal surface is a pulse-counting photomultiplier detector. A change of the orientation of the grating makes possible an exploration of the spectrum in five passbands. The motion of the vehicle causes the spectrum image to move over the exit spectral slit, realizing a scanning without any mechanical need. The spectrum is analysed in the range $\lambda 750$, $Ly\alpha$, $\lambda 1650$, $\lambda 2200$, $\lambda 3100$, with passbands 200-500 Å wide. The field of view is 2.8°□ due to the motion of the vehicle during an integration counting. Figure 3 presents the wavelength calibration curves obtained at the Laboratory in the range $Ly\alpha$, $\lambda 3100$. These are preliminary values of absolute instrumental efficiency.

3. Sensitivity of the instrument

The background consists of the diffuse UV sky background and of the dark background of the detector. The sky background on the far ultra-violet is 3 to $6 \cdot 10^{-12}$ erg s^{-1} cm^{-2} $Å^{-1}$ $°□^{-1}$ (Lillie, 1972; Henry, 1973). The dark background of the detector is equivalent to a diffuse emission of intensity less than $1 \cdot 10^{-13}$ erg s^{-1} cm^{-2} $Å^{-1}$ $°□^{-1}$, and may be negligible. The limit of the flux from the source is

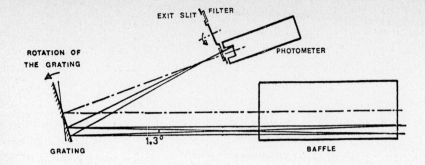

Fig. 2. Optical diagram of the instrument

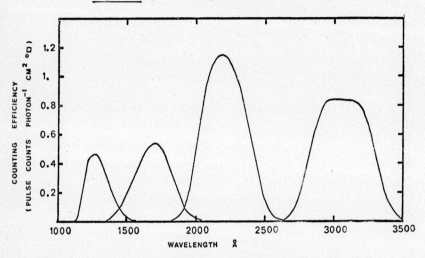

Fig. 3. Preliminary values of the instrumental efficiency

$1 \cdot 10^{-12}$ erg s^{-1} cm^{-2} Å^{-1} °☐$^{-1}$. So this instrument is capable of an essentially complete survey of the intensity of Zodiacal Light and of galactic light. As a matter of fact, knowing that the intensity of the Zodiacal Light is of the same order at λ4250 and at λ1680 (Lillie quoted by Davidsen et al, 1974) from the measurements of its intensity in the blue color (Dumont et al, 1966; 1973; Frey et al, 1974) its presumed flux at λ1650 at the north ecliptic pole is $8 \cdot 10^{-12}$ erg s^{-1} cm^{-2} Å^{-1} °☐$^{-1}$, on the ecliptic at 90° elongation $3 \cdot 10^{-11}$ erg s^{-1} cm^{-2} Å^{-1} °☐$^{-1}$.

Taking into account that the airglow intensity is of the same order than the limiting flux, the intensity of zodiacal light will be at least 8 times brighter than the limiting flux. Thus the sensitivity of the zodiacal light instrument is adequate to meet the objectives of this experiment. Moreover, the mean intensity of direct starlight in the solar neighbourhood at $\lambda 1400$ is $3 \cdot 10^{-11}$ erg s^{-1} cm^{-2} Å^{-1} °$\square^{-1}$ from the calculations of Habing (1968) making the galactic flux perfectly detectable. A complete survey of the diffuse galactic light intensity would perhaps be possible with this instrument, with an optimized field of view, small enough to eliminate a substantial contribution from hot stars, but large enough for a good sensitivity to diffuse radiation.

References

Davidsen,A., Bowyer,S., Lampton,M. 1974, Nature 247, 513.

Dumont,R., Sanchez-Martinez,F. 1966, Ann. Astrophys. 29, 113.

Dumont,R., Sanchez-Martinez,F. 1973, Astron. Astrophys. 22, 321.

Frey,A., Hofmann,W., Lemke,D., Thum,C. 1974, Astron. Astrophys. 36, 447.

Habing,H.J. 1968, Bull. Astr. Inst. Netherlands 19, 421.

Henry,R.C. 1973, Astrophys. J. 179, 97.

Lillie,C.F. 1972, Scientific Results from OAO-2, 95 (NASA, 1972).

VISIBLE AND UV PHOTOMETRY OF THE GEGENSCHEIN AND THE MILKY WAY

Llebaria, A.
Laboratoire d'Astronomie Spatiale
Marseille/France

The French astronomical satellite D2B is to be launched on September 1975. The spin axis is constantly oriented towards the sun with an accuracy of ± 30 arc min. The ERC experiment (field recognition) will provide images in three spectral intervals of a region centered at the anti-sun point and having a diameter of 10°. The scanning of this region is realized by a combination of the spin motion of the satellite and internal electronic scanning. This experiment includes a catadioptric objective (optimized Kern telescope) with a field of 3° x 6° and an image dissector which analyses the image.

This experiment will achieve a mapping of the Gegenschein and of local regions of the Milky Way with a resolution of 40' x 40' in three spectral bands centered at 2800, 3600 and 4200 Å.

The "Laboratoire d'Astronomie Spatiale (L.A.S.)" is participating in the research programme of the scientific satellite D2B to be launched next september. L.A.S. is in charge of three experiments placed on board the satellite. Their purpose is the extended sources study and the photometry of hot stars.

The E.A.S. and E.L.Z. (antisolar and zodiacal light experiments) are spectrophotometers whose objective operates for wavelengths between 70 nm and 340 nm. The E.R.C. (field recognition experiment) is a photometer with very large bands, sensitive in three wavelengths from 220 nm to 650 nm. The E.R.C. should give back a photometric image with weak resolution of the antisolar zone. The identification of stars in the stard field will provide, once processed, a precise indication of the satellite's altitude.

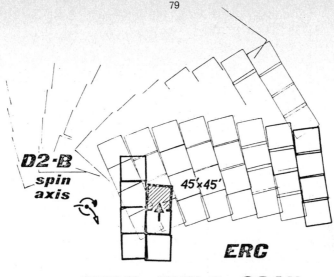

fig. 1

The E.R.C. whose optical axis is shifted 3,5° with respect to the satellite's spin axis, will constantly explore an area of the sky, 20 degrees in diameter, around the antisolar point, while the satellite is constantly pointed in this direction. Thus the actual rotation of the satellite will give us is a E.R.C. photometric measurements of the antisolar point during the satellite's duration, as well as the variations of the sky background with relation to its position on the ecliptic. Cruvellier and Maucherat (1), (2) give a detailed description of the satellite and of the other experiments. Here we will describe the E.R.C.

The E.R.C. is a bi-dimensional photometer with a space resolution of 45' x 45' and an overall field of view of 6° x 3° (rectangular). The rectangle's principal axis is in the same plane as the spin axis. As already stated, as the satellite turns, the E.R.C. explores an area of the sky that is crown-shaped around the spin axis. This axis has a residual movement of precession caused by parasite couples and a slight excess of transverse cinetic energy that makes this exploration practically uniform.

The E.R.C. is made up of an objective and a bidimensional photometric detector, the objective calculated by Detaille (3) is 90° bent, because of crowding. Its optical diagram is that of a perfected Kern telescope. It has three mirrors and three diopters. A parabolic mirror with a Schmidt blade in the entrance pupil forms the image in the centre of the field. A spherical mirror transports the image to the detector's photocathode, with the help of a 45° plane mirror and a field lens that takes up the image at the photocathode's curb. In a ± 3,5° field the image spot is less than 80 μm (for 90% of the energy), in the spectral region 220 μm to 660 μm.

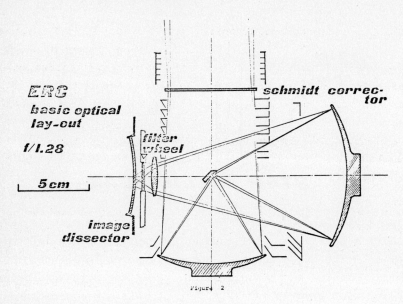

Figure 2

Two low pass filters reduce this band pass to 290 nm and 390 nm respectively, while the third leaves it uncharged. Thus 3 band passes are obtained, two of which are obtained by subtraction of the characteristics given in the adjoining figure.

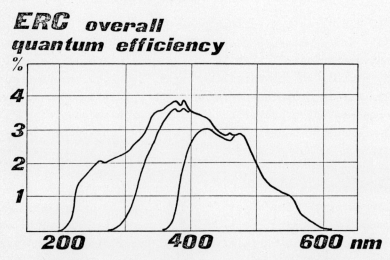

Figure 3

The detector used is an E.M.R. 459 image dissector with a square slit, 1 mm x 1 mm, for exploration of an 8 x 4 mm^2 field. The principle characteristics of this detector are high quantum efficiency (> 20%), a 5% photocathode uniformity and a negligible dark current (1 count/second).

The field is thus decomposed into 32 elementary adjacent zones making up 4 lines and 8 columns. The analog electronics commands gradual exploration, column by column, in the opposite direction to that of the satellite's rolling and at 0,5 sec per elementary zone (since the satellite's rotation speed is 1 revolution/4 minutes, the apparatus scans 16 fields per revolution). Thus the apparatus scanning, the satellite's rotation and the displacement of its spin axis according to the antisolar point and along the year allow us to explore a zone of $\pm$ 10° on either side of the ecliptic.

The E.R.C.'s flight models were calibrate at L.A.S. and at the Foux d'Allos Observatory (2600m). The stability of their geometric and photometric characteristics are regularly checked. The calibrations done at the laboratory included:
. Electronic centering and measurement of geometric characteristics,
. Measurements of band passes,
. Uniformity of the field,
. Rate of diffusion of parasitic light,
. Checks for long-range stability.

Measurements of the sky at Foux d'Allos Observatory (2600 m) included:
. Photometric sensitivity,
. Dynamic simulations.

The expected flux, according to the results obtained, is indicated in the adjoining table.

E. R. C. COUNTS PER $S_{10}(V)$

Channel		I	II	III
Spectral Type	A0	79	106	125
	F0	65	86	98
	G5	45	57	61

Counting period : 0.5 sec.

For a sky background with a G5V spectral type and $S_{10}(V) = 100$ intensity, the average number of impulses counted per 0.5 sec will be 4700, 5700 and 6100 for channels I, II and III respectively.

At the present time, the programmes for altitude restitution are complete, and we are preparing the programmes for photometric restitution. The latter must include:
. Qualification of light measurements,
. Centralization of measurements in a data base,
. Summary processing allowing the obtention of results within a brief time period (1 day),
. More evolved processing working on 15 days measurements periods.

For the processing, the following are needed:
. Detection of possible variations of the Gegenschein UV flux,
. One must provide a Gegenschein and an ecliptic mapping in three band passes (I-II, II-III and III).

References:

(1) Maucherat, M., Cruvellier, P. 1976, Presentation of Zodiacal Light Instrument aboard the D2B french astronomical satellite, this volume.

(2) Cruvellier, P., Maucherat, M., Maucherat, J. Le satellite D2B. Colloque d'Aussois du 24-28 avril 1972. Editions de l'IAP. Paris.

(3) Detaille, M., Saisse, M. Les télescopes de Schmidt anastigmates. 5ème Journées d'Optique Spatiale. Marseille 1975. Edition du CNES. Paris.

1 ZODIACAL LIGHT

1.2 GROUNDBASED OBSERVATIONS

GROUND-BASED OBSERVATIONS OF THE ZODIACAL LIGHT

René DUMONT

Observatoire de Bordeaux
33270 FLOIRAC (France)

I. OBSERVATIONS AND THEIR DEGREE OF RELIABILITY

At the present time, when space experiments bring us more and more information of increasing quality, it might appear questionable whether ground-based zodiacal light observations are still of interest. Nevertheless, any detailed examination of the available data (see, for example, the review paper of Leinert 1975) shows that a considerable part of our present optical knowledge of interplanetary dust has been contributed by ground-based programmes. Moreover, new available parameters arising from space data, mainly in the field of heliocentric dependence of brightness, open new important abilities for a better interpretation of ground-based observations.

Shortcomings and advantages of ground-based compared to space zodiacal light observations

Airglow is the main drawback of ground-based data, and it has been in the past responsible for the lack of consistency of many ground-based results with each other and with space results, especially off the ecliptic. Most of the figures proposed before 1967 for off-ecliptic brightness and polarization degree are nowadays reckoned to be largely erroneous, viz. overestimated.

The difficulties of ground-based observations are even greater when we go from the blue-green range of the spectrum, either towards the red, where OH bands of the nightglow are increasingly disturbing, or towards the $0.3 - 0.4$ μm range, where tropospheric corrections become large. A consequence is that colorimetric measurements from ground are scarce and not fully reliable. Extensions to IR and UV domains are practically impossible from ground.

As emphasized below, telluric disturbances to polarimetry from ground are more or less worrying according to the direction, and may become totally unacceptable near the horizon or the antisun.

Several doubtful results concerning z.l. intrinsic variations, especially short-timed ones, may be ascribed at least partially to these various sources of errors, particularly to airglow inhomogeneities and variations.

On the other hand, ground-based observations have in some cases no serious disadvantage compared to space experiments; on a few particular points they may even be credited for a true superiority. In the case of Fabry-Perot interferometry, disturbances coming from atmosphere are moderate, and in a space experiment the only major improvement expected would be a better coverage of small elongations. In the case of photopolarimetry, the advantages of space are much more obvious; still, ground-based data keep the best (up to now) with respect to the following rather important items:

1) better abilities to reduce the integrated starlight correction. The well-known uncertainty of this term can be lowered by a considerable ratio if the light-collector has a sufficient diameter, so that stars down to an advanced limit-magnitude can be excluded from the field by a visual and manual operation - typically irrelevant to spatial constraints.

2) the possibility of a very long time span of observation. Only a few space programmes have lasted a year or more, and many ones have lasted some days or even minutes, while ground-based programmes like those of Haleakala, of Pic-du-Midi and of Tenerife have been carried out for periods of 10 years or more. Now, problems about stability or evolution of the zodiacal cloud throughout a solar cycle obviously demand long programmes, with stable instrumentation.

3) ground-based observations cover the whole sky except small elongations, not only more or less restricted regions. Perhaps the most striking feature of our presently available collection of space results is its " mosaic " character: each concerns limited areas or lines, and the resulting density of the information remains rather patchy over the sky. This shortcoming will certainly be removed in the future, but to-day it contrasts highly with the bulk of homogeneous data obtained from ground over 90 percent of the celestial sphere.

Some remarks about the reliability
of ground-based z.l. photometry and polarimetry

The various limitations to the accuracy of ground-based data arising from airglow and from tropospheric scattering largely depend upon wavelength, celestial direction and zenith distance, so that the situation cannot be summarized briefly.

The fact that this complexity has not always been borne in mind has sometimes resulted in excessive opinions, in both senses, with respect to the accuracy to be hoped - either a total suspicion of all ground-based results, or too much confidence in them. In order to clarify that problem, a rather extensive study of the accuracy has been attempted (Sánchez 1969; Dumont and Sánchez 1973, 1975a) and has led to the following conclusions:

a) in ground-based z.l. photometry, airglow is the main

disturbance, except in the bright cones. All kinds of ways to minimize its intervention must be sought (lowest possible latitude and zenith distance; careful choice of the spectral range...) One such which we have adopted at Tenerife (Dumont 1965, 1967; Dumont and Sánchez 1975a) is to make use of the green oxygen line [OI] 5577 Å as a photometric indicator for the nightglow continuum variations, since a fair correlation appears most of the time between them. The existence of that correlation above all observatories remains controversial, but its usefulness in those ones where it is seldom failing, is beyond doubt. The accuracy provided by this refinement is of the order of ±10 percent in the cones down to ±25 percent in the dim off-ecliptic regions, but it remains possible to perform a reliable photometry up to the ecliptical pole.

b) ground-based z.l. polarimetry has, in addition, to take account of the false polarization originating in tropospheric transfers. The three celestial sources (z.l., airglow and stars) scattered along the line of sight introduce three parasitic totally polarized components j, which disturb the true

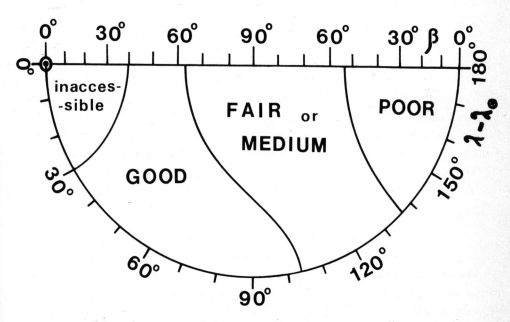

Fig. 1. Distribution over the sky in helioecliptic coordinates (with conservation of areas) of the expected reliability in ground-based polarimetry.

J component ($J = PZ$, where Z is the brightness and P the polarization degree of the z.l.). The resulting disturbance is of the order of 1 S_{10} near 30° zenith distance, up to 5 S_{10} near 70°; its effect is weak in the regions where J is strong, but it completely distorts the results when J is only a few S_{10}.

The sky may be divided into a few zones of different reliability for z.l. polarimetry (fig. 1). We see that an antisolar cap of about 45° radius is an area of very poor reliability from ground. This is the reason why, in my opinion, the problem of neutral points and slight negative polarization on the wings of the gegenschein is a typical aim for space experiments. Another consequence is the questionable credibility of some works (Bandermann and Wolstencroft 1974) which purport to show variations in gegenschein polarization, on the basis of observed J variations at levels of 1 S_{10} (see also Sparrow and Weinberg 1975). Despite these various difficulties in ground-based polarimetry, satisfying figures for the polarization degree P can be obtained over the major part of the sky.

The difficulties to compute the polarization degree P led several authors to restrict the problem to the determination of J = PZ. Obviously this quantity, although being disturbed by low-atmosphere scattering, is considerably safer to be computed from ground - and even from space - than P, since the bothers arising from the diluting sources (airglow and integrated starlight, whose brightnesses are uncertain but polarization is roughly negligible) are avoided as far as J alone is concerned. Unfortunately, J is a hybrid quantity, and its determination without any independent determination of Z and P - in other terms, the 2nd and 3d Stokes' parameters being known and the first unknown - remains a much less fundamental step towards the optical knowledge of the dust. Contrary to the opinion of Wolstencroft and Brandt (1972), I think that moderately accurate measurements of brightness Z and of polarization P are more useful than accurate data on the totally polarized component J = PZ, if interpretations in terms of dust distribution, size, optical properties, and nature, are the final purpose.

II. RESULTS AND INTERPRETATION:
OPTICAL PROPERTIES AND DISTRIBUTION OF THE GRAINS

The attempt in this section is to summarize the results and to extract which kind of information they contain about interplanetary dust, with emphasis on photopolarimetry over the whole sky.

Doppler-shifts measurements and the motion of the grains

The most evident conclusion arising from the recent works in this field (Reay and Ring 1968; James and Smeethe 1970; Hicks et al. 1974) is that almost all the grains move along prograde orbits. Obviously, much more information may be expected from the diagrams wavelength shift vs. elongation, especially, as pointed out by James (1969), about the size distribution of the grains.

Nevertheless, serious difficulties of interpretation seem to arise (in addition to the observational problems due to the weakness of the expected and observed Doppler-Shifts), and important discrepancies remain between the results, and if they are compared to the theoretical curves. The variety of the interpretations suggested to explain these discrepancies - a circumterrestrial component (Vanysek and Harwit 1970), eccentric orbits (Bandermann and Wolstencroft 1969) or interstellar dust streaming through the solar system (Hicks et al. 1974) - means that, up to now, unambiguous information can hardly be extracted from the available data in this field.

Photopolarimetric surveys of the ecliptic and the phase function of an elementary volume

It is obvious before any calculation that a lot of concealed information about the scattering functions of the grains and their distribution within the zodiacal cloud is contained in the surveys providing the brightness Z and the polarization degree P with a very wide coverage of the sky.

Measurements along the ecliptic are of special interest because of the rotational symmetry that we may ascribe to the zodiacal cloud, at least as an outline. The eccentricity of the earth's orbit, and the fact that the invariable plane of the solar system is more and more generally accepted to be the true symmetry plane, are able to introduce small seasonal effects, some of which are detected in Tenerife data (Dumont and Sánchez 1968), in Pic-du-Midi data (Robley 1973) and by the D2A satellite experiment (Levasseur and Blamont 1974). However, these seasonal changes are slight enough (a few percent) to ensure that observations in the ecliptic from the earth are practically equivalent with observations in the true symmetry plane and from 1 A.U. heliocentric distance exactly.

Considering that partial surveys of the ecliptic - some points, or a restricted range of a few tens degrees - are not so readily useful for interpretations as wide surveys on a considerable range of elongations are, only have been selected here the available data extending to at least 2/3 of the ecliptic (i.e. an elongation range $\Delta\epsilon \geqslant 120°$, since no significant east-west dissymmetries have been reported). It is noteworthy that, for brightness (fig. 2), only four published surveys satisfy this condition - three of them being ground-based and one from a balloon - and for polarization (fig. 3) only two surveys - both ground-based - presently offer this wide coverage.

Brightness along the ecliptic.

Fig. 2 shows two groups of curves; the upper pair gives from 35 to 110° elongation stronger values of Z than the lower pair by a nearly constant factor of 1.2. In the far zodiacal band, the four curves are more separated, the maximal discrepancy being almost a factor of 1.7. The elongation of the minimum ranges from 125 to 140°, and the ratio Z(antisun)/Z(minimum) ranges from 1.25 to 1.60. There is a fair agreement on the value of the derivative $dZ/d\epsilon$ at $\epsilon = 90°$, which is directly related to the scattering efficiency of a unit-volume of interplanetary space (Dumont 1972, 1973).

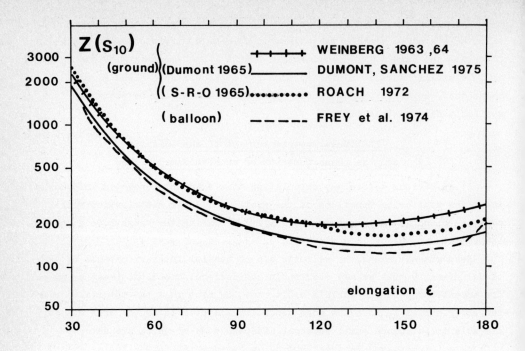

Fig. 2. Zodiacal light brightness along the ecliptic, according to the available surveys covering an elongation range $\Delta\epsilon \geq 120°$.

Polarization along the ecliptic.

The only two available surveys of the whole ecliptic give the same general shape of curve (fig. 3). The maximum is 0.229 at $\epsilon = 70°$ according to Weinberg 1964, and 0.177 at $\epsilon = 62°$ according to Dumont and Sánchez 1975b (similar values already given in Dumont 1965). The greatest discrepancy is 6 percent polarization near $\epsilon = 80°$. Both curves have a moderate slope at $\epsilon = 30°$, suggesting that P remains rather high when the line of sight approaches the sun. This is in agreement with the photometric results obtained in the inner z.l. by the rocket experiment of Leinert et al. (1974).

Both in Z and in P these differences between Haleakala and Tenerife are rather important, and they exceed the minimal errors inherent to ground-based observations. However, they are not dramatic, and they probably do not prevent the extraction of conclusions in general agreement on dust properties and distribution.

The scattering phase function.

The possibility of deriving the phase function in arbitrary units, over the same range of scattering angle θ as the available observed range of elongation ϵ, has been emphasized by Dumont 1973 ; see also Leinert 1975, and Dumont (1976). The expression of the phase function for a unit-volume situated at 1 AU from the sun is

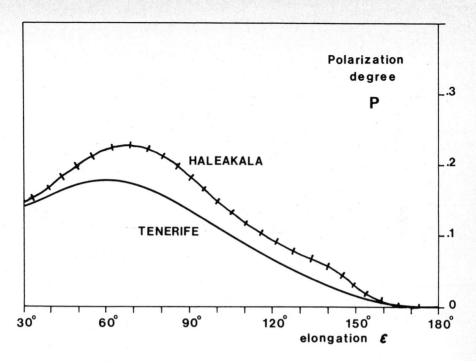

Fig. 3. Polarization degree of the zodiacal light along the ecliptic, according to the available surveys covering an elongation range $\Delta\varepsilon \geqslant 120°$.

after inversion of the brightness integral along the line of sight:

$$\sigma(\theta = \varepsilon) = -(1+n)\cos\varepsilon\, Z(\varepsilon) - \sin\varepsilon\, \frac{dZ}{d\varepsilon} \qquad (1)$$

if the space density in the symmetry plane of the zodiacal cloud (practically, in the ecliptic) is assumed to be proportional to r^{-n}, at least in the range of r the most efficient for producing the z.l. observable from ground ($0.5 < r < 2$ A.U.). The fact that the distribution law seems to be broken down to zero in the asteroidal belt cannot be seriously argued against the validity of this formula (Dumont, 1976).

Very different values of n have been suggested in various models, but the plausible range is presently much more restricted. Most of the preliminary results of Pioneer 10 (optical data) are in favour of $n \cong 1$ (Hanner and Weinberg 1973a 1973b; Soberman et al. 1974). Still, according to the weak residual brightness due to the z.l. at r= 2.4 A.U. reported by Hanner et al. (1974), n could be >1, perhaps of the order of 1.5.

Introducing high values of n (of the order of 2 or even 1.6) in eq. (1) leads, on the basis of our knowledge of Z along the ecliptic, to negative and therefore meaningless values of σ . Values >1.5 being eliminated by these con-

siderations, and values < 1 conflicting with Pioneer 10's results, I suggest that
n = 1.2 ± 0.3 is presently the best evaluation of this parameter (see footnote).

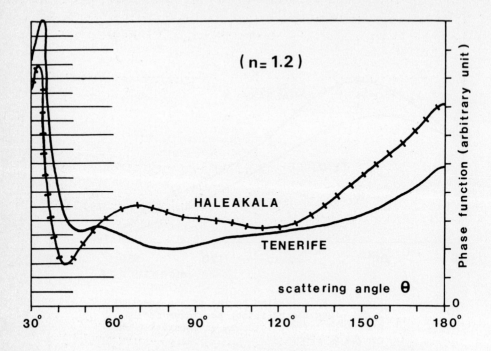

Fig. 4. The phase function of interplanetary dust, as resulting from eq. (1) and from Haleakala and Tenerife photometric surveys. A space density $\sim r^{-1.2}$ in the ecliptic is assumed. The left side is poorly reliable.

Fig. 4 shows the phase functions obtained when supplying eq. (1) with the $Z(\varepsilon)$ data of Haleakala (Weinberg 1964) and of Tenerife (Dumont 1965; Dumont and Sánchez 1975b), assuming n = 1.2. We must notice that the accuracy of the method is decreasing with decreasing θ, due to the fact that the two terms of eq. (1)'s right hand side increase rapidly, with absolute values of the same order and opposite signs. The peaks near θ = 33°, although conspicuous in both curves, must therefore be considered with caution. Rightward of $\theta \cong 50°$ the reliability becomes fair, and both curves show a rough isotropy from 50 to 130° and they climb in the backscattering domain.

These curves are rather sensitive to the value of n adopted, but not enough

---

- The evolution with heliocentric distance of the brightnesses observed by Helios between 1 and 0.3 A.U., as reported by Link et al. (1976) is in favour of n = 1.3, therefore in very good agreement with the above conclusion.

to lose their conspicuous trend to isotropy for medium values of θ, as far as values of n in the range 1.0 - 1.5 are assumed (see Dumont and Sánchez 1975b, fig. 3). The variations of σ from $\theta = 50°$ to $\theta = 130°$ are probably within a factor of 2, and this conflicts with many theoretical scattering functions which show variations by a factor of 5 to 10 in the quoted range of scattering angle.

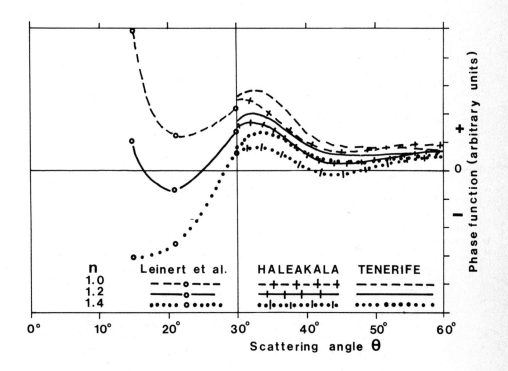

Fig. 5. Forward scattering range of the phase function of interplanetary dust, as resulting from eq.(1). A density law $\sim r^{-n}$ (n = 1.0; 1.2; 1.4) is assumed in the ecliptic. Classical z.l. ground-based surveys rightwards of $\theta = 30°$; inner z.l. rocket results leftwards of that scattering angle. The level of accuracy given by eq. (1) decreases rapidly when $\theta \rightarrow$ zero.

Fig. 5 concentrates on the forward-scattering range ($\theta < 60°$). Even if we keep in mind the lower level of accuracy achieved when θ decreases, we may notice the good junction of the ground-based curves with those derived from the rocket data of Leinert et al. (1974) at the elongations 15, 21 and 30°. We see that:

- high values of n also conflict with the latter observations;

- the general trend to isotropy extends to the inner z.l. if values of n in the range 1.0 - 1.2 are assumed (waves of moderate amplitude, such as the slight negative range of σ around $\theta = 20°$ for n = 1.2, must of course be disregarded, since in this domain only the general appearance of the curves remains significant).

Indeed, the variations of σ seem to be within a factor of 5 in the $20° < \theta < 180°$ range. Such a flatness can agree with few of the scattering functions reported by Wickramasinghe (1973) for Mie particles, or by Giese (1970, 1971, 1974) for elaborate mixtures of homogeneous or mantle-core particles: most of these curves exhibit a much stronger forward scattering. Some agreement may perhaps be sought with absorbing particles, for which the ratio $\sigma_{max}/\sigma_{min}$ is generally of the order of 10 in the $20° < \theta < 180°$ range, but can be as low as 2 or 3. On the other hand, the curves of figs 4 and 5 seem very hard to reconcile with dielectric particles, for which the same ratio is between 20 and 200 or more.

Polarized components of the phase function.

On the assumption that the vibration plane of the scattered light does not deviate from the plane perpendicular to the scattering plane (or, if some negative polarization occurs, from the scattering plane itself), eq. (1) can be duplicated

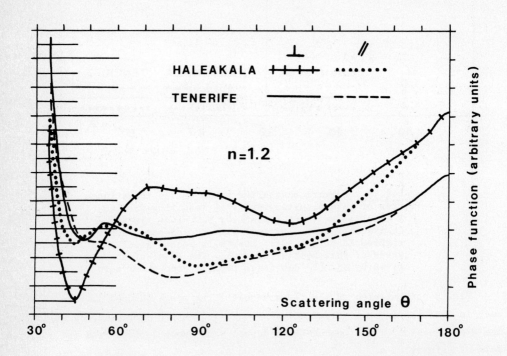

Fig. 6. Polarized components (electric vector perpendicular to: $\perp$, and lying in: $\parallel$, the scattering plane) of the phase function of interplanetary dust, given by a duplication of eq. (1). A density law $\sim r^{-1.2}$ is assumed in the ecliptic. The left side is poorly reliable.

for the two components of the scattered light, corresponding to the Fresnel or electric vector perpendicular to (1), or lying in (2), the scattering plane.

Fig. 6 shows, for Haleakala and for Tenerife data, the polarized components of the phase function. The trend to isotropic scattering is especially conspicuous upon the component 1 for Tenerife results, since no variations greater than 10 percent of the mean value occur between $\theta = 44°$ and $\theta = 142°$. The component 2 has a minimum at a level of about half the mean level of the component 1. The backscattering efficiency is higher than the mean scattering efficiency for the first component by a factor of 1.7 (Tenerife) or 1.9 (Haleakala). In the left (poorly reliable) side, some negative polarization appears, but it is probably not genuine.

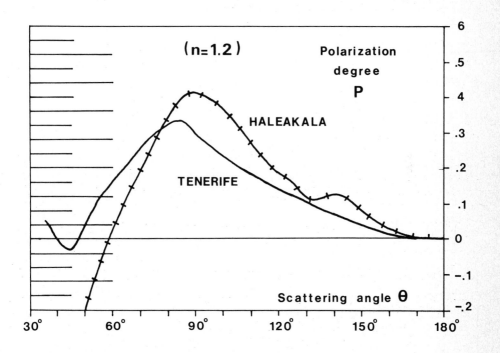

Fig. 7. The polarization curve of interplanetary dust (polarization degree vs. scattering angle θ) obtained by a duplication of eq. (1) and assumption of a density law $\sim r^{-1.2}$ in the ecliptic. The left side is poorly reliable.

The polarization degree of the scattered sunlight, $\mathcal{P} = (\sigma_1 - \sigma_2)/(\sigma_1 + \sigma_2)$ is given by fig. 7, still under the assumption that n = 1.2. Contrary to the photometric curves, the polarization curve is weakly sensitive to the value of n adopted. This is due to the fact (Dumont 1972, 1973) that two points are rigorously independent of n ; they are $\theta = 90°$ and $\theta = \varepsilon_M$, i.e. the elongation of the maximum of observed polarization P (60 to 70°); in the latter case, we have $\mathcal{P}$ = P. As far as the polarization curve is expected to be symmetrical with respect to θ = 90°, these two points are more or less sufficient to determine the whole curve, so that the curves really obtained for very different values of n (see Dumont and Sánchez 1975b, fig. 5) will not differ a great deal from the curve corresponding to n = 1.2. The maximum of the true local polarization $\mathcal{P}$ (0.41 at θ = 88° from Haleakala data; 0.33 at θ = 82° from Tenerife data) is stronger than the corresponding maximum of P observed in the z.l. by almost a factor of 2.

Off-ecliptic photopolarimetry and the oblateness of the zodiacal cloud

Rather few data are available in the field of photometry and/or polarimetry over large off-ecliptic sky areas. Photometric data are given by Roach (1972) on the whole sky; by Dumont (1965) on 90 percent of the sky; by Frey et al. (1974) (balloon) on 80 percent of the sky. All other data, from ground or space, are of fragmentary nature with respect to the coverage of the sky. Concerning off-ecliptic z.l. polarimetry, Tenerife results (Dumont 1965; Dumont and Sánchez 1966; Sánchez 1967) are the only extended ones.

For the simplicity of the presentation and discussion, we shall concentrate here on the circle sun-ecliptic pole-antisun, which cuts the zodiacal cloud practically along a meridian plane. This circle provides the largest differences in the observations, compared to the ecliptic.

Off-ecliptic brightness.

Fig. 8 gives the brightness from 15° to 180° elongation (= angular distance to the sun), according to Roach's compilation (1972), and from 40° to 180°, according to Tenerife results (Dumont 1965; Dumont and Sánchez 1973; the provisional values for a paper in preparation about off-ecliptic results are also taken into account). Also plotted are the rocket data at 15, 21 and 30° elongation of Leinert et al. (1974).

The minimum, which is at the ecliptical pole according to Roach 1972, is significantly lower in Tenerife results, and it is shifted by 20° towards the antisun ($\lambda - \lambda_\odot$ = 180°, β = 70°). Let us recall that several space determinations of Z at the ecliptic pole agree on 50 - 60 S_{10}, i.e. slightly less than the brightness found at Tenerife (65 S_{10}). The latter figure leads to 0.32 for the ratio Z (90, 90)/Z (90, 0), frequently used in zodiacal cloud models and theoretical works.

Comparisons with an ellipsoidal model of the zodiacal cloud.

The most direct and simple assumption for the zodiacal cloud is an ellipsoidal (oblate) shape. The model we wish to propose (Dumont and Sánchez, to be published) has the following parameters:

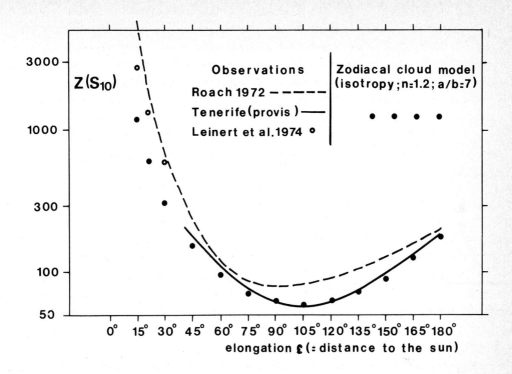

Fig. 8. Off-ecliptic z.l. brightness (in the plane sun-ecliptic pole-antisun) and comparison with the ellipsoidal model of the zodiacal cloud proposed in the text.

a) in the symmetry plane, the run of the space density is $r^{-1.2}$.

b) the dust has the same optical properties outside as inside the ecliptic, i.e. the phase function $\sigma(\theta)$ found in the ecliptic (fig. 4), its polarized components (fig. 6), and its polarization curve (fig. 7) are valid in all directions. Within $\theta = 45°$, viz. where σ is poorly known, we have assumed an isotropy.

c) the isodense surfaces are ellipsoids, with a ratio of oblateness a/b, to be determined by the observations.

Brightness Z and polarization P in each direction are easily computed for such a model by integrating along the line of sight the local values $\mathcal{J}(\theta)$ (= intensity scattered under the scattering angle θ by a unit volume of space) and $\mathcal{P}(\theta)$ (= polarization degree of the scattered light).

The value a/b = 7 has been chosen for the oblateness ratio, since it fits the observed ratio Z(90, 90)/Z(90, 0).

On fig. 8, also appear the theoretical values of Z corresponding to our model. Although the rocket data for the inner z.l. fit better Roach's values than our model, it can be seen that this model is in very satisfying agreement with all Z values obtained at Tenerife; especially, it reproduces the shift of the minimum towards the antisun by a score of degrees.

Off-ecliptic polarization.

Similar agreement arises between our observational data and the above model, with respect to the polarization degree P. Fig. 9 shows the theoretical and obser-

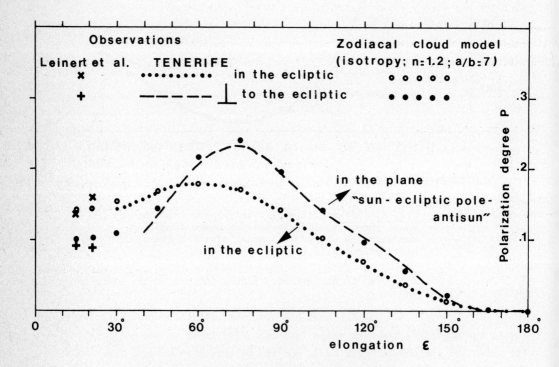

Fig. 9. Off-ecliptic polarization of the z.l. (in the plane sun-ecliptic pole-antisun) compared to the results along the ecliptic. Also plotted are the polarization degrees given for both planes by the ellipsoidal model of the zodiacal cloud proposed in the text (note the good agreement of the model with Tenerife data and with inner z.l. rocket data).

ved P values along the plane sun-ecliptic pole-antisun, compared to the same values along the ecliptic. Our results along the former plane are still provisional, awaiting a thorough reduction of all observations made since 1964. Near the elongation $\varepsilon = 50°$ the polarization degree is the same (0.17) at all inclinations; P is stronger off the ecliptic than along it for greater elongations, weaker for smaller elongations. Part of this result was already implied in the oblate isopolarimetric curves given in the antisolar hemisphere by the former Tenerife results (Dumont 1965, fig. VII-5). A similar trend is reported in the same region by the preliminary results of Skylab in its study of low light level phenomena (Weinberg and Hahn 1975).

An excellent agreement arises between the polarization degrees given by our model in the inner z.l. and those found along circles of 15 and 21° radii around the sun by the rocket experiment of Leinert et al. (1974). Our model predicts that the polarization degree P remains rather high when the line of sight approaches the sun ($\varepsilon \rightarrow$ zero), with a limit nearly equal to P (ε = 90°), viz. about 0.15, and a plain geometrical explanation of this fact can be found (Dumont and Sánchez 1975b). Perhaps the most convincing test of validity for the model is that it reproduces very well the change of sign for the difference of polarization degree between the two planes, near 50° elongation.

CONCLUSION

A considerable part of the observed photopolarimetric features of the zodiacal light appear to be simultaneously fitted by a rather simple model, the outlines of which are:

- an ellipsoidal dust cloud, flattened in a ratio of about 7;

- a space density decreasing with heliocentric distance slightly steeper than $1/r$;

- a scattering phase function and polarization curve exhibiting a quasi-isotropy for the normal component (except some backscattering excess, and a possible but not proven enhancement near θ = 30°), with a loss of 50 percent at right angle (or somewhat before) for the parallel-component.

On the basis of the phase functions obtained (figs. 4 to 7), the field remains open to determine which mixtures of grains (size spectrum, refractive and absorptive indices, nature) are candidates to be the interplanetary " dust " producing the zodiacal light.

REFERENCES

Bandermann L.W., Wolstencroft R.D. 1969, Nature, 221, 251.
Bandermann L.W., Wolstencroft R.D. 1974, Nature 252, 215.
Dumont R. 1965 (thesis), Ann. Astrophys. 28, 265.
Dumont R. 1967, in " The zodiacal light and the interplanetary medium ", Honolulu Symposium, ed. J.L. Weinberg, NASA Sp-150, 63.
Dumont R. 1972, C.R. Acad. Sci. Paris 275 B-765.
Dumont R. 1973, Planet. Space Sci. 21, 2149.
Dumont, R. 1976, this Volume.
Dumont, R., Sánchez, F. 1966, Ann. Astrophys. 29, 113.

Dumont R., Sánchez F. 1968, Ann. Astrophys. 31, 293.

Dumont R., Sánchez F. 1973, Astron. & Astrophys. 22, 321.

Dumont R., Sánchez F. 1975a, Astron. & Astrophys. 38, 397.

Dumont R., Sánchez F. 1975b, Astron. & Astrophys. 38, 405.

Frey A., Hofmann W., Lemke D., Thum C. 1974, Astron. & Astrophys. 36, 447.

Giese R.H. 1970, " Tabellen von Mie-Streufunktionen I " (Max Planck Institut, München, MPI-PAE extraterr. 40).

Giese R.H. 1971, " Tabellen von Mie-Streufunktionen II " (Max Planck Institut, München, MPI-PAE extraterr. 58).

Giese R.H., Schwehm G., Zerull R. 1974, " Grundlagenuntersuchungen zur Interpretation extraterrestrischer Zodiakallichtmessungen und Lichtstreuung von Staubpartikeln verschiedener Formen ", Bochum, BMFT-FB W 74-10.

Hanner M.S., Weinberg J.L. 1973a, Sky & Telesc. 45, 217.

Hanner M.S., Weinberg J.L. 1973b, paper C 3.8 presented at XVI. COSPAR Meeting, Konstanz, Germany.

Hanner M.S., Weinberg J.L., DeShields L.M., Green B.A., Toller G.N. 1974, J. Geophys. Res. 79, 3671.

Hicks T.R., May B.H., Reay N.K. 1974, Mon. Not. Roy. Astron. Soc. 166, 439.

James M.F. 1969, Mon. Not. Roy. Astron. Soc. 142, 45.

James M.F., Smeethe M.J. 1970, Nature 227, 588.

Leinert C. 1975, " Zodiacal Light - a measure of the interplanetary environment " Space Sci. Rev. (in press).

Leinert C., Link H., Pitz E. 1974, Astron. & Astrophys. 30, 411.

Link, H., Leinert, C., Pitz, E. and Salm, N. 1976, this Volume.

Levasseur A-C., Blamont J.E. 1974, paper IIIC-1.5 presented at XVII. COSPAR Meeting, Sao Paulo, Brazil.

Reay N.K., Ring J. 1968, Nature 219, 710.

Roach F.E. 1972, Astronom. J. 77, 887.

Robley R. 1973, Ann. Geophys. 29, 321.

Sánchez F. 1967, in " The zodiacal light and the interplanetary medium ", Honolulu Symposium, ed. J.L. Weinberg, NASA Sp-150, 71.

Sánchez F. 1969 (thesis) Urania 269-270.

Soberman R.K., Neste L.S., Lichtenfeld K. 1974, J. Geophys. Res. 79, 3685.

Sparrow J.G., Weinberg J.L. 1975, Astron. & Astrophys. (in press).

Vanysek V., Harwit M. 1970, Nature 225, 1231.

Weinberg J.L. 1963 (thesis), Univ. of Colorado.

Weinberg J.L. 1964, Ann. Astrophys. 27, 718.

Weinberg J.L., Hahn R.C. 1975, " Multicolor photometry of low light level phenomena from Skylab " (proc. AIAA/AGU Conference on scientific experiments of Skylab, Huntsville, Alabama, in press).

Wickramasinghe N.C. 1973, " Light scattering functions for small particles ", ed. Hilger, London.

Wolstencroft R.D., Brandt J.C. 1972, " Planets, stars and Nebulae studied with photopolarimetry " (Tucson Symposium, ed. T. Gehrels), 768.

1.2.2   POLARIMETRY OF THE ZODIACAL LIGHT AND MILKY WAY FROM HAWAII

R. D. Wolstencroft*, Royal Observatory, Edinburgh

L. W. Bandermann, Institute for Astronomy, University of Hawaii

I. INTRODUCTION

This paper describes the results of a program of observations of the polarized intensity vector of the night sky in the anti-solar hemisphere obtained on 72 nights between May 1973 and November 1974. The observations were made with the night sky polarimeter at Mt. Haleakala, Maui, Hawaii, using a 5300A interference filter, 62A wide, to define the passband. They were made at fixed altitudes from $35°$ to $90°$ along the north-south meridian and repeated in sequence throughout the night. The diameter of the field of view was $6.0°$ and the integration time per point was either 30 or 120 seconds. Polarization was detected using a rotating polaroid (11 1/4 revs $\sec^{-1}$); the signal from the polaroid drive generator provided a phase trigger allowing the photomultiplier signal to accumulate and be digitized for each of the four $90°$ phase intervals of the polarization modulation cycle. This process continued for the entire integration period. The digitization and data handling and recording (paper tape and teletype) was carried out by a PDP-8I computer, and calibrated Stokes parameters (I, Q, U) and the polarized intensity pI were calculated later. Intensity calibration in S_{10}(V,G2V) units was obtained using a C_{14} low brightness source, and the amplitude and phase of the polarization modulation were checked each night.

II. DATA ANALYSIS

The polarized intensity vectors of the observations for a given night were plotted in a solar ecliptic coordinate diagram. Averages of these vectors for a given new moon period were calculated at standard points in (β, $\lambda-\lambda_\odot$) by averaging Q and U over small equal areas (appr. 64 sq. deg.). Next, data for all months at galactic latitudes $|b^{II}|>45°$ (where we suppose the influence of Milky Way polarization is negligible) were averaged similarly to obtain the average zodiacal light (AZL), i.e., the zodiacal light uncontaminated by galactic polarization. Finally we calculated the polarized intensity of the Milky Way by subtraction of the Stokes parameters (Q, U) of the AZL from those of the nightly observations. The resulting Stokes parameters of the Milky Way were area-averaged.

The potential sources of error in our observations are photon noise, red leak of the filter, tropospheric scattering and instrumental polarization. (a) For a typical sky brightness at 5300A of I=500 S_{10}(V) the photon count--including photons from the red leak discussed in (b)--for a 30 second integration time was 2×10^7 corresponding to a standard error in pI of 0.14 S_{10}(V) and the polarization orientation angle χ of $2.8°$ for a typical low polarized intensity of 2 S_{10}(V). (b) Concerning the red leak, our 5300Å filter is only partially blocked at wavelengths longer than 7000Å. OH airglow dominates the sky brightness at these wavelengths and although the effective

---

*On sabbatical leave from the Institute for Astronomy, University of Hawaii

red response is down by a factor of 12 in the red (mainly due to the decreasing sensitivity of our EMI 9558 QAM photomultiplier) the total sky brightness is 5 times higher in the red so that we calculate that 30% of the measured intensity signal is contributed by this red leak. Analysis of night sky observations with this filter and a narrower blocked 5300Å filter yield 20% contribution. The corresponding contribution of the red leak to the measured polarized intensity signal is fortunately much less, namely 8%. This is because the ratio of the polarized intensities rather than the ratio of the straight intensities (at 4300Å and in the red) is involved, these two ratios being 1:1 and 5:1 respectively. (c) We have studied the importance of tropospheric scattering at 5300Å by following regions of fixed solar ecliptic coordinates through a wide range of zenith distances. In over 90% of the cases studied the orientation angle, χ, deviated from the value close to the zenith by less than $\theta=\pm 10°$. The trospospheric component associated with this deviation, namely (pI) trop $\tilde{\sim}$ $(pI)_{ZL}$ $\tan\theta$, can amount to as much as $3\, S_{10}(V)$ at $35°$ altitude. During the course of the year the tropospheric component at a fixed point in solar ecliptic coordinates changes in magnitude and direction (relative to the direction of the north ecliptic pole) because of the change in altitude and in the relative orientation of the ecliptic, galactic and alt-azimuth coordinate systems. By averaging our data over the year we expect the influence of the tropospheric component to be reduced to a much smaller level than $3\, S_{10}(V)$. We take the standard errors of the pI values of the AZL to be a measure of this reduction: for $pI < 20\, S_{10}(V)$ the error is $0.5\, S_{10}(V)$, and we take $0.5\, S_{10}(V)$ to be a typical value of the residual tropospheric component in the AZL data. This in turn corresponds to a tropospheric influence on χ of $7°$ $(0.7°)$ for $pI=2\, (20)\, S_{10}(V)$. (d) The instrumental polarization of 0.16% introduces a polarized component typically of $1.0\, S_{10}(V)$. Again, because of changes in the direction of the instrumental polarization vector (fixed in alt-azimuth coordinates) relative to the direction of the north ecliptic pole during the course of a year, our data averaging over the year reduces this to a small value.

III. RESULTS

The average zodiacal light is shown in Fig. 1. Below we list its main features. 1. The angle, $\Delta\chi$, between the observed direction of the polarization vector and the direction for positive polarization is large relative to its standard error at many points within about $40°$ of the anti-solar point, and in these regions the polarization is generally neither positive nor negative. Smaller but statistically significant deviations occur elsewhere. 2. $\Delta\chi$ averaged over the entire anti-solar hemisphere is essentially zero. 3. Along the anti-solar meridian the pI distribution with respect to the ecliptic is asymmetric, being negative between $\beta= -30°$ and $+10°$ with a minimum of $pI= -3\, S_{10}(V)$ at $\beta= -15°$. 4. South of the ecliptic there is symmetry in the orientation of the pI vectors relative to the anti-solar meridian. 5. Along the ecliptic pI never becomes negative although significant rotations away from the positive polarization direction occur at $180° < \lambda - \lambda_\odot < 220°$. 6. Some representative values of pI and $\Delta\chi$ are shown in the table (n is the number of observations).

The Average Zodiacal Light at 5300A

pI	$S_{10}(V)$	19	6.2	4.9	18	7.7	2.6	0.7	3.3	0.3	5.4
σpI	"	0.4	0.3	0.2	0.4	0.3	0.5	0.4	0.3	0.4	0.6
Δχ	(°)	2.6	0.1	-4.5	-1.6	2.5	7.3	53	85	85	1.8
σΔχ	"	1.0	3.6	2.4	0.8	2.1	14	14	2.7	29	2.4
β	"	0.0	0.0	0.0	0.0	47	33	13	-12.9	-33	-47
λ-λ$_o$	"	121	147	213	239	180	180	180	180	180	180
n		13	28	31	17	25	24	20	36	24	12

7. At the anti-solar point pI=1.5±0.5 $S_{10}(V)$ (n=30) and the polarization plane is oriented along a direction 21°±10° counterclockwise relative to the direction to the north ecliptic pole, as seen from outside the celestial sphere.

We do not propose to discuss the interpretation of these results here except to comment on point 1. If the interplanetary dust particles were spherical or non-spherical and randomly oriented, then the polarization of the zodiacal light would be either positive (Δχ=0°) or negative (Δχ= ±90°). The existence of intermediate values of Δχ, as pointed out in 1, requires that a significant fraction of the particles are non-spherical and partially oriented.

The pI-vectors of the Milky Way are shown in Fig. 2. They are predominently oriented normal to the galactic plane as expected for the diffuse galactic light. By using tabular values of the integrated starlight (Roach & Gordon 1974), assuming the polarized intensity of the integrated starlight is much less than that of the diffuse galactic light and adopting an albedo of 0.5 for the interstellar grains, we deduce the degree of polarization of the diffuse galactic light: at ℓ^{II}=17° we obtain values ranging from 2.1% to 6.9% at -33°<b^{II}<+33°. A region of enhanced pI is present centered at b^{II}=20°, ℓ^{II}=105°. Although the polarization orientation is consistent with that of nearby vectors the large values of pI are difficult to understand. The region is not far from the north ecliptic pole (b^{II}=28°, ℓ^{II}=123°) and consequently observations of the zodiacal light at high ecliptic latitudes must be corrected for Milky Way polarization. This has not been realized formerly and may account for some of the differences in published values of pI at the north ecliptic pole.

For a full description of this work see (Bandermann & Wolstencroft 1975).

References

Bandermann, L. W. and Wolstencroft, R. D. 1975. Mem. Royal Astr. Soc. (in press).
Roach, F. E. and Gordon, J. L. 1974, "The Light of the Night Sky", D. Reidel Publishing Company, Dortrecht-Holland.

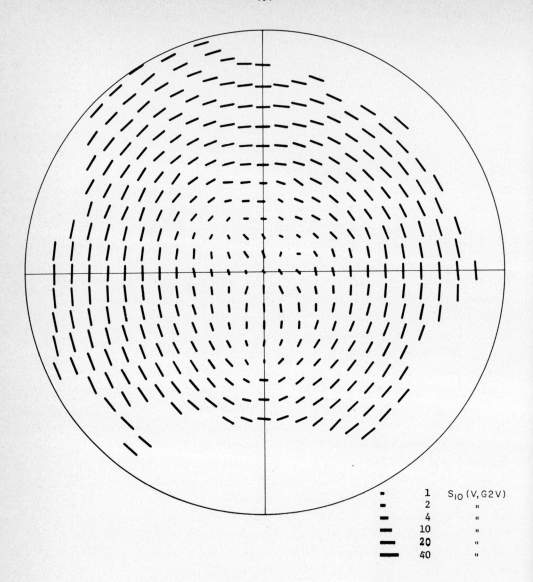

Fig. 1 The average zodiacal light at 5300A in the anti-solar hemisphere: May 1973 to November 1974. The length of each vector is proportional to $\log(1 + pI)$ where pI is the polarized intensity in S_{10}(V, G2V) units. The solid circle represents $90°$ solar elongation; the horizontal line is the ecliptic and the vertical line the anti-solar meridian. The north-ecliptic pole is at the top; $\lambda - \lambda_o$ increases from left to right.

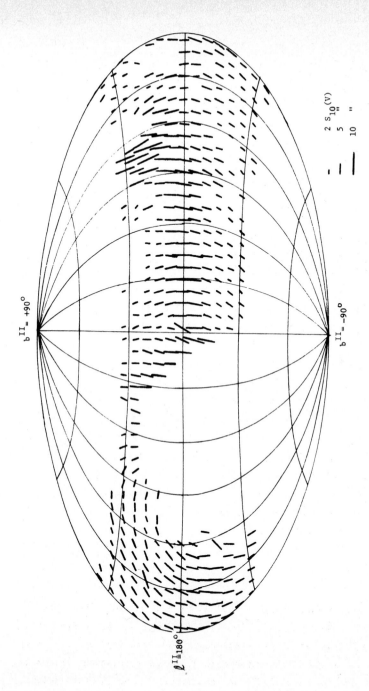

Fig. 2 The polarized intensity of the Milky Way at 5300A. The length of each vector is proportional to pI where pI is the polarized intensity in $S_{10}(V, G2V)$ units. Galactic longitude increases from left to right, the galactic center being at the center of the diagram.

1.2.3   SCATTERING IN THE EARTH'S ATMOSPHERE: CALCULATIONS FOR
        MILKY WAY AND ZODIACAL LIGHT AS EXTENDED SOURCES

H.J. Staude
Max-Planck-Institut für Astronomie

D-6900 Heidelberg-Königstuhl

Abstract:

The results of detailed calculations on first order Rayleigh- and Mie-scattering in the Earth's atmosphere illuminated by the Milky Way and the Zodiacal Light are presented. The influence of various independent parameters, as optical depth of the Rayleigh- and Mie-component and position of Milky Way and Zodiacal Light, is discussed. Linear and circular polarization of the scattered light are considered. The results are compared with the procedures commonly applied in reducing photometric observations of extended sources of the night sky. It is shown that a substantial part of the discrepancies between the brightness distributions of Zodiacal Light resp. Milky Way given by different authors is due to an inaccurate treatment of the atmospheric scattered light.

A comprehensive account of the assumptions made in the present calculations and a qualitative discussion of the results are given in Astron. & Astrophys. 39, 325 - 333 (1975).

Since these computations may be useful for future reductions of observational work, they are available on request in tabulated form.

1.2.4 SCATTERING LAYER OF COSMIC DUST IN THE UPPER ATMOSPHERE

François Link

Institut d'Astrophysique

75014 Paris, France

The upper atmosphere acts as a giant collector of cosmic dust swept by the Earth on its travel through space. The transient presence of cosmic dust in the upper atmosphere was detected by the rise of the luminance of the twilight sky during the activity of several meteoric showers either by balloon photometers launched by C.N.E.S. at Aire sur l'Adour (Fehrenbach et al., 1972 a) or by the similar photometer at the Pic du Midi Observatory (Link, 1973).

The permanent presence of cosmic dust in the upper atmosphere is more difficult to prove. We introduced therefore a new technique of measurements by balloon photometers which enables us to obtain during the same ascent the twilight curves at small (30°) and at large (85°) zenithal distances (Fehrenbach and Link, 1974). We can write for the luminances

$$b = R+r+M \quad \text{at zenithal distance} \quad z=30° \quad \text{and}$$
$$b = R'+r'+M' \quad \text{at zenithal distance} \quad z'=85° , \quad (1)$$

where R, R' stand for the computed primary components, r,r' for the multiple scattering components and M,M' for the dust components measured at solar depressions U and U'.
Both twilight curves log b(U) and log b'(U') were obtained for the ascent on June 29th 1973 at 35 km i.e. outside any important shower activity (Fig.1). In the same graph are drawn the computed curves log R(U) and log R'(U') according to Link and Weill (1975).

On both curves of R and R' we pick up the points A and A' where R=R' at solar depressions U and U' and we compare there the

observed luminances b and b'. We find b' > b. As according to (1)
$$b' - b = M' - M + r' - r \qquad (2)$$
this excess of light near the horizon might be due either to the meteoric component M' or to the multiple scattering component r'. As the latter is concerned, our method (Fehrenbach et al., 1972 b) based on azimuthal profile gives r'/R' < 0,1 and the meteoric origin of the light excess can be accepted.

Now two effects may be the origin of it :

1) The shape of the scattering indicatrix of dust particles and

2) their stratification in the upper atmosphere.

In the first eventuality if the sedimentation of cosmic dust in the upper atmosphere were governed only by the gravitation and the resistance of the quiet atmosphere, the concentration of the dust particles should be proportional to the number density of the air (Link, 1950) which enters in both integrals of R and R'. In consequence
$$M = c(\gamma) R \quad \text{and} \quad M' = c(\gamma') R' = c(\gamma') R \qquad (3)$$
where $c(\gamma') > c(\gamma)$ are the function of both scatterings angles γ and γ'. Therefore the ratio obtained at different solar depressions
$$\frac{b' - b}{b} = \frac{c(\gamma') - c(\gamma')}{1 + c(\gamma)} = \beta \qquad (4)$$
should be nearly constant.

Actually we find for β a curve (Fig.2) with a maximum near U' = 14°. At this moment the edge of the Earth's shadow on the line of sight passes near the 110 km level (Fig.2). We can interpret it as a consequence of the stratification of the scattering medium at this level.

We reach a similar conclusion for the azimuthal profiles of the sky obtained at z = 85° during the same ascent i.e.
$$B(a) = b(a)/b(0°) \quad \text{compared with the theoretical profile} \qquad (5)$$
$$F(a) = R(a)/R(0°) \qquad (6)$$
where a is the azimuth computed from solar vertical plane where a=0°. For the ratio of both we obtain with the aid of (1)
$$k = \frac{B(a)}{F(a)} = \frac{1 + M(a)/R(a)}{1 + M(0°)/R(0°)} \qquad (7)$$

In this manner we obtained the curves of k represented on Fig.3. The ratio k rises or falls with the azimuth a if
$$dk/da \gtrless 0 \qquad (8)$$

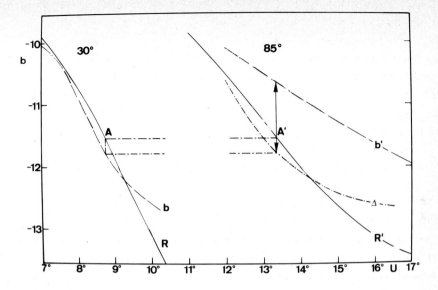

Fig. 1. Twilight curves at small (30°) and large (85°) zenithal distances in solar vertical (a=0°) for balloon ascent in June 1973 at Aire sur l'Adour.

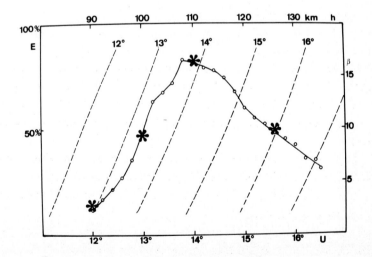

Fig.2. Curve of β (—) as function of solar depression U. Curves (---) of solar illumination E for solar depressions 12°-15° as function of the height h on the line of sight z =85° and a =0°. Asterisks are the values of β computed for a model of a homogeneous scattering layer between 110 and 130 km.

or if

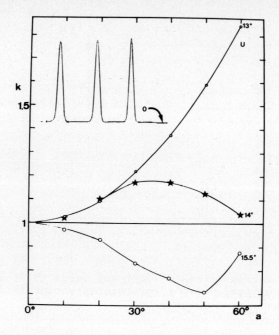

Fig.3. Curves of the variation of k with the azimuth a for different solar depressions U. Inset : Specimen of the azimuthal profile b(a) at z = 85° ; for a> 60° the intensity falls practically to zero (obscurity) marked on the left by an arrow. The smallness of r' is therefore obvious.

$$d \log M/da \gtrless d \log R/da \qquad (9)$$

Returning provisorily to the assumption (3) we get for (9)

$$d \log c(\gamma)/da \gtrless 0 \qquad (10)$$

as the condition of rising or falling k with a.

As a matter of fact in the interval $10° < \gamma < 70°$ concerned by our measurements the above ratio remains constantly (Giese,1970)

$$d \log c(\gamma)/da < 0$$

and there cannot exist the rise of k. On the curve of k at U =14° the fall starts at about a = 30-40° and at this moment the shadow passes near the 120 km level. Below this level where k rises the condition (3) cannot be fulfilled and the stratification of the scattering medium takes place.

Summing up we reach the following conclusions :

a) The excess of light of the twilight sky near the horizon reveals the permanent presence of cosmic dust (outside the activity of major meteoric showers) in the upper atmosphere.

       b) Its distribution is different from the gravitational sedimentation.

       c) The level 110-120 km is critical for this distribution.

It was suggested (Burnett, Lillie) at the presentation of this paper at the Heidelberg IAU Colloquium that the airglow continuum can intervene in our observations. This possibility demands a detailed discussion regarding the amount of airglow continuum and the much larger twilight luminance. Also the expected results of recent launching during (5.V.1975)and after (9.V.1975) the activity of eta-Aquarids may throw some light onto our problem.

R E F E R E N C E S .

Fehrenbach,M.,Frimout,D.,Link,F. and Lippens,C. 1972 a, Ann. Géophys. $\underline{28}$,363

Fehrenbach,M.,Frimout,D.,Link,F.,Lippens,C. and Weill,G. 1972 b, Compt. Rendus. Acad. Sci. Paris $\underline{B-275}$,223.

Fehrenbach,M. and Link,F. 1974, Compt. Rend. Acad. Sci. Paris $\underline{B-279}$, 687

Giese,R.H. 1970, Max Planck Inst. PAE/Extraterr. 70, München

Link,F. 1950, Bull. Astron. Inst. Czechoslov.$\underline{2}$,1

Link,F. 1973, Space Research $\underline{XIV}$,703

Link,F. and Weill,G. submitted for publication to Geophys. Ann.

1 ZODIACAL LIGHT

1.3 MODELS AND INTERPRETATION

1.3.1

SOME FORMULAE TO INTERPRET ZODIACAL LIGHT PHOTOPOLARIMETRIC DATA IN THE ECLIPTIC FROM GROUND OR SPACE

René DUMONT

Observatoire de Bordeaux
33270 FLOIRAC (France)

A fundamental step towards the knowledge of interplanetary matter from zodiacal light photometry is to eliminate the integral along the line of sight — an intrinsic, cumbersome feature of all z.l. observations — so as to reach, by inversion, the local optical properties of elementary volumes of space.

Let $\mathcal{J}(r, \theta)$ be the intensity (per steradian) of sunlight scattered, under the scattering angle θ, by a unit-volume of space situated at r A.U. from the sun. Let r_o be the heliocentric distance of a space probe containing a photometer, which aims in some direction whatever; the z.l. observed will be the integral

$$Z = \int \mathcal{J} \, d\ell \qquad (1)$$

for the whole line of sight, where ℓ is the distance of the probe to an elementary current slice of the exploring narrow cone. Eq. (1), of course, is also valid near $r_o = 1$ A.U., viz. in the case of ground-based, balloon-borne or satellite-borne experiments.

SPACE DENSITY RUN IN THE ECLIPTIC AND PHASE FUNCTION (GENERAL FORMULAE)

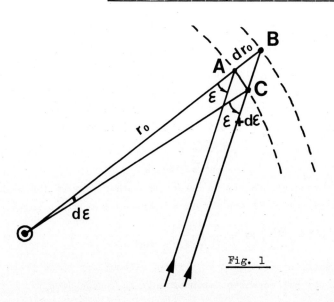

Fig. 1

Suppose the photometer to be and to aim in the symmetry plane of the zodiacal cloud. Assuming the local properties of dust in that plane to depend upon r only (rings or gaps of dust possible, but no deviation to circular symmetry around the sun), the photometer will observe the z.l. brightness Z, function of the two variables r_o and ε (elongation). Fig. 1 allows us to introduce elementary differences between the three neighbouring locations A,B,C assumed for the probe.

When going from A to B without change of elongation, we lose in zodiacal brightness:

$$Z(A) - Z(B) = -\left(\frac{\partial Z}{\partial r_o}\right)_{\varepsilon, r_o} \cdot AB \qquad (2)$$

(the partial derivative is taken at ε = cst and r_o variable). When going from B to C with the same line of sight, the loss will be, according to eq. (1) :

$$Z(B) - Z(C) = \mathcal{J}(r = r_o, \theta = \varepsilon) \cdot BC \qquad (3)$$

When returning in A, the increase will be:

$$Z(A) - Z(C) = -\left(\frac{\partial Z}{\partial \varepsilon}\right)_{r_o, \varepsilon} \cdot d\varepsilon \qquad (4)$$

If we replace in (2) AB by dr_o ; in (3) BC by $dr_o \sec \varepsilon$; and in (4) $d\varepsilon$ by $dr_o \, tg\varepsilon/r_o$, the loop ABCA leads us to write:

$$\mathcal{J}(r = r_o, \theta = \varepsilon) = \cos \varepsilon \left(\frac{\partial Z}{\partial r_o}\right)_{\varepsilon, r_o} - \frac{1}{r_o} \sin \varepsilon \left(\frac{\partial Z}{\partial \varepsilon}\right)_{r_o, \varepsilon} \qquad (5)$$

Let $\Phi(r)$ be the total energy scattered by a unit-volume in all directions; we may write the intensity:

$$\mathcal{J}(r = r_o, \theta = \varepsilon) = \Phi(r) \cdot \sigma(\theta) \qquad (6)$$

where $\sigma(\theta)$ is the phase function, normalized to unity for the whole sphere. If the properties of interplanetary dust are assumed to be the same everywhere, then $\sigma(\theta)$ does not depend on r, so that $\Phi(r)$ is proportional to the space density $\rho(r)$ and to the solar flux. Therefore, the space density near the probe may be written, in arbitrary units:

$$\rho(r_o) = r_o^2 \cos \varepsilon \left(\frac{\partial Z}{\partial r_o}\right)_{\varepsilon, r_o} - r_o \sin \varepsilon \left(\frac{\partial Z}{\partial \varepsilon}\right)_{r_o, \varepsilon} \qquad (7)$$

It would be optimistic to expect from the available as well as from the forth-coming space probe data a complete coverage of the field $Z(r_o, \varepsilon)$; however, we see that a photometer continuously aiming at the antisun can provide the gradient $\partial Z/\partial r_o$, from which the space density (in arbitrary units) is directly derived:

$$\rho(r_o) = - r_o^2 \left(\frac{\partial Z}{\partial r_o}\right)_{180°, r_o} \qquad (8)$$

On the other hand, if in eq.(5) we try to cancel the other term of the right-hand side, we obtain a determination of the intensity scattered at right angle:

$$\mathcal{J}(r_o, 90°) = -\frac{1}{r_o} \left(\frac{\partial Z}{\partial \varepsilon}\right)_{r_o, 90°} \qquad (9)$$

so that, for $r_o \simeq 1$ A.U., this intensity can be derived without any space probe data, nor assumption about the heliocentric dependence of the space density.

Since we have assumed $\sigma(\theta)$ to be the same everywhere, and especially at 1 A.U., we have, in arbitrary units:

$$\sigma(\theta=\varepsilon) = \cos\varepsilon \left(\frac{\partial Z}{\partial r_o}\right)_{\varepsilon,1} - \sin\varepsilon \left(\frac{\partial Z}{\partial \varepsilon}\right)_{1,\varepsilon} \tag{10}$$

Up to now, in order to obtain the phase function over a wide range of θ, not only earthbound observations ($r_o = 1$) of Z over the same wide range of ε seem to be required, but moreover the gradient in r_o - a quantity presently known only for a few elongations.

GRADIENT OF Z.L. WITH HELIOCENTRIC DISTANCE
AND PRACTICAL FORMULA FOR THE PHASE FUNCTION

Fortunately, a simplification arises if we assume the space density ρ to follow a regular law and to be proportional to r_o^{-n}.

Consider (fig. 2) two locations of the photometer, aligned with the sun, and two parallel lines of sight. Consider a secant pivoting around the sun, and let it carve the two beams in corresponding elements denoted M' (length: dl') and M" (length: dl"). We have dl"/dl' = 1 + (dr$_o$/r$_o$). Since the scattering angle is the same, eq. (6) shows that the ratio of the intensities scattered towards B and A by unit-volumes situated at M" and at M' is $\mathcal{J}''/\mathcal{J}'$ = $\Phi(\odot M'')/\Phi(\odot M')$ = $[1 + (dr_o/r_o)]^{-(2+n)}$. When integrating along the two beams we obtain, according to eq.
(1), Z(B)/Z(A) = ($\mathcal{J}''/\mathcal{J}'$).(dl"/dl') = $[1 + (dr_o/r_o)]^{-(1+n)}$ = $1 - (1+n)(dr_o/r_o)$.
Therefore,

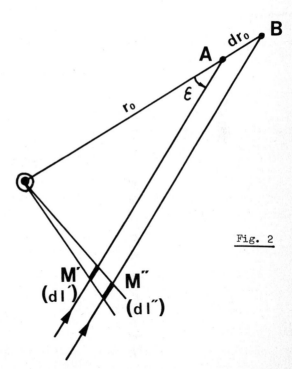

Fig. 2

$$\left(\frac{\partial Z}{\partial r_o}\right)_{\varepsilon,r_o} = -\frac{1+n}{r_o} Z(r_o,\varepsilon) \tag{11}$$

so that eq. (10) becomes (at 1 A.U., and still in arbitrary units):

$$\sigma(\theta=\varepsilon) = -(1+n)\cos\varepsilon\, Z(\varepsilon) - \sin\varepsilon\, \frac{dZ}{d\varepsilon} \tag{12}$$

In the above assumption, and in so far as the parameter n can be determined (essentially with space probe data), the phase function can be derived from a photometric survey along the ecliptic at 1 A.U. In another paper (Dumont 1976) we derive the phase function from $\theta = 15°$ to the antisun, according to the $Z(\varepsilon)$ data of Leinert et al. 1974 (rocket), Frey et al. 1974 (balloon), and to the ground-based data of Haleakala (Weinberg 1964) and of Tenerife (Dumont and Sánchez 1975). We assume the most probable value of n to be 1.2.

It might be argued against the validity of eqs. (11) and (12) that a perfectly smooth law such as r^{-n}, even if fitting acceptably the true run of the space density in the inner solar system, is more and more unlikely very far from the sun when $r \to \infty$. The fact that a complete fall of zodiacal brightness, therefore of density, is reported by Pioneer 10 as crossing the asteroidal belt, reinforces such a criticism. However, the weakness of the residual brightness when entering the belt (Hanner et al. 1974) allows to think that the lack of dust beyond 3.3 A.U. can only bring minor disturbances to the $\sigma(\theta)$ function provided by eq. (12). Moreover, if we concentrate upon the rotating secant of fig. 2, we notice that a zero-level of dust beyond a given heliocentric distance would leave the above geometrical derivation of eqs. (11) and (12) still valid (since an integration along a segment of the double-beam, instead of an infinite double-beam, would provide $Z(B)/Z(A)$ without any change to the preceding formulae).

POLARIMETRIC FORMULAE

Eqs. (1) to (6) and (9) to (12) may be written in the polarimetric fashion, i.e. separately for each Fresnel vector (1 = perpendicular to, and 2 = lying in, the scattering plane). The corresponding components of the z.l. are $Z_1(r_o, \varepsilon)$, $Z_2(r_o, \varepsilon)$, those of the intensity are $\mathcal{J}_1(r, \theta)$, $\mathcal{J}_2(r, \theta)$. The observed degree of polarization is $P = (Z_1 - Z_2)/Z$; the true local degree of polarization will be $\mathcal{P} = (\mathcal{J}_1 - \mathcal{J}_2)/\mathcal{J}$.

From a double formulation of eq. (12), and omitting here the intermediate steps (see Dumont 1973; Leinert 1975), we are led to the following expression of $\mathcal{P}$, which generally differs from P :

$$\mathcal{P}(r=r_o, \theta=\varepsilon) = P(r_o,\varepsilon) - \frac{1}{\sigma(\theta=\varepsilon)} \sin\varepsilon \, Z(r_o,\varepsilon) \left(\frac{\partial P}{\partial \varepsilon}\right)_{r_o,\varepsilon} \qquad (13)$$

where n, and its uncertainty, only appear through σ. Therefore, at $\theta = \varepsilon = 90°$, n vanishes from eq. (13) since it vanishes from eq. (12). A second value of θ ruling n out will be $\theta = \varepsilon_M$, i.e. the elongation of the maximum of observed polarization P. The existence of those two particular values of θ allowing to compute $\mathcal{P}$ independently of n involves that the whole function $\mathcal{P}(\theta)$ is rather weakly sensitive to the value of n adopted (Dumont 1976).

Our assumption that dust properties, except its space density, do not depend on r, implies that $\mathcal{P}$ also is independent of r (at a given constant scattering angle θ). Besides, a double formulation of eq. (11) shows that P also has to be independent of r_o as far as the density law r^{-n} is valid, because calling J the quantity $Z_1-Z_2 = PZ$, and omitting the indices ε and r_o in the derivatives, we have

$$\frac{\partial J}{\partial r_o} = -\frac{1+n}{r_o}J$$

$$\frac{\partial P}{\partial r_o} = \frac{1}{Z^2}\left[Z\frac{\partial J}{\partial r_o} - J\frac{\partial Z}{\partial r_o}\right] = \frac{1}{Z^2}\left[-\frac{1+n}{r_o}ZJ + \frac{1+n}{r_o}ZJ\right]$$

so that, at a constant elongation,

$$\frac{\partial P}{\partial r_o} = 0 \qquad (14)$$

This result, compared to the forthcoming data of deep space probes, could be a test of validity for the assumptions that have been made.

In practice, eqs. (7), (8), (10) and (11) could be of interest when interpreting these space probe data; eqs. (9) and (12) are able to extract valuable informations from the observational data near 1 A.U.; eq. (13) provides the polarization curve of the scatterers, which according to eq. (14) is expected to be independent of heliocentric distance.

References:

Dumont, R. 1973, Planet. Space Sci. **21**, 2149.

Dumont, R. 1976, Ground based Observations of the Zodiacal Light, this volume.

Dumont, R., Sánchez, F. 1975, Astron. and Astrophys. **38**, 405.

Frey, A., Hofmann, W., Lemke, D., Thum, C. 1974, Astron. and Astrophys. **36**, 447.

Hanner, M.S., Weinberg, J.L., DeShields II L.M., Green, B.A., Toller, G.N. 1974, J. Geophys. Res. **79**, 3671.

Leinert, C. 1975, "Zodiacal Light - a measure of the interplanetary environment" Space Sci. Rev. (in press).

Leinert, C., Link, H., Pitz, E. 1974, Astron. and Astrophys. **30**, 411.

Weinberg, J.L. 1964, Ann.Astrophys. **27**, 718.

1.3.2

DISCUSSION OF THE ROCKET PHOTOMETRY
OF THE ZODIACAL LIGHT

Leinert,C., Link,H. and Pitz,E.
Max-Planck-Institut für Astronomie

D 6900 Heidelberg - Königstuhl

Abstract:

The measurements of the zodiacal light at elongations $\varepsilon = 15°$ to $30°$ performed by the ESRO rocket experiment R - 214 were discussed on the basis of model calculations.

The total amount and the shape of the brightness decrease outside the ecliptic plane were used to test assumptions on the three dimensional spatial distribution of interplanetary dust. Models which give the decrease of dust density as a function of the height above the ecliptic only are unacceptable as well as models for which the lines of equal spatial density are ellipsoids. To represent the inner zodiacal light a "fan - like" spatial distribution is required, where the decrease of dust density is a function of the angular distance from the ecliptic as seen from the sun. A fit to the observation yields
$n(r,z) \sim r^{-1} \exp[-2.6 \, (z/r)^{1.3}]$.

The observed deviation of the brightness maximum from the ecliptic plane can be quantitatively explained by symmetry of the interplanetary dust inside 1 A.U. with respect to the plane $\Omega = 66 \pm 12°$, $i = 3.7 \pm 0.6°$. This is a significant deviation from the invariant plane of the solar system, generally believed to be the plane of symmetry, towards the planes of the inner planets and the solar equator.

The reddening observed in the inner zodiacal light is only compatible with a flat size distribution of interplanetary dust, e.g. $n(a) \, da \sim a^{-2.5} \, da$.

The comparatively high degree of polarization at $15°$ is compatible with scattering by micronsized slightly absorbing silicates ($m = 1.5 - 0.03 \, i$)

A full paper is to be published in Astronomy and Astrophysics.

1.3.3 Consequences of the Inclination of the Zodiacal Cloud on the Ecliptic

R. Robley
U.E.R. Observatoires du Pic du Midi et de Toulouse/France

Assuming that the decrease in the density of the interplanetary dust follows an exponential distribution both in the transverse and radial direction, we can write $n = n_o \, \mathrm{Exp}(-(h/H)-(r-1/R))$, where h is the distance from the ecliptic plane and r the heliocentric distance both expressed in astronomical units (a.u.); then we show that the modulation of the radiance B(90, 0) of the zodiacal light observed at the ecliptic pole defines the parameter H as a function of the inclination angle ß between the zodiacal cloud and the ecliptic plane; moreover, the experimental value of the ratio B(90, 0)/B(90, 90) defines the parameter R. It can be deduced that the flatness of the zodiacal cloud, expressed by R/H, is < 5 and that the plane of symmetry of the zodiacal cloud is very close to that of the invariant plane of the solar system (ß < 2°).

Assuming that the composition of the interplanetary medium is mainly Fe and SiO_2 grains (not including ice for which the estimated life time is too short), we show that only the dust grains with a size distribution favorising the submicronic grains give a ratio B(90, 0)/B(90, 90) close to the observed value.

The degrees of polarization computed for some mixtures of Fe and SiO_2 do not agree very well with the observations; using an inverse power law for the size distribution of the radii, say $n(a) \sim a^{-p}$, it seems that the law would be better with p = 3 or 3.5 than with p = 4.

The full paper will be published in Astron. and Astrophys.

1.3.4 METHOD FOR THE DETERMINATION OF THE INTENSITY OF SCATTERED SUNLIGHT PER UNIT-VOLUME OF THE INTERPLANETARY MEDIUM.

A. Mujica and F. Sánchez
Instituto Universitario de Astrofísica
Universidad de La Laguna-Islas Canarias

The knowledge of the functions which give the intensity of the scattered sunlight is important in order to establish the models of the zodiacal cloud.

This paper pretends following the idea exposed by Dumont(1973), to develop a method for the calculation of the functions which yield the intensity of the ligth scattered by each unit-volume of the interplanetary medium. The functions obtained will account for the averaged local properties of the zodiacal cloud (Scattered phase function and spatial density).

The quantities used are the luminance Z and the degree of polarization, P.

Two assumptions are made:

a) On the symmetrical plane of the zodiacal dust cloud (practically the ecliptic plane) the properties of the interplanetary medium depend only on the distance to the sun.

b) The luminance, $Z(r,\varepsilon)$, is a differentiable function.

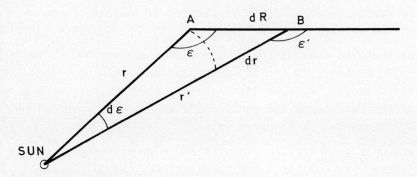

For a photometer situated on A and angular elongation ε measurement of the luminance $Z(r,\varepsilon)$, is obtained. If the point of the observation is now taken to B, a close point to A, on the same line of observation, another measurement will be obtained $Z(r',\varepsilon')$. The difference between these two measurements is caused by the scattered light in the medium between

A and B, for a scattering angle which is identical to the elongation angle. Therefore

$$dZ = -J\, dR \qquad (1)$$

where J is the intensity scattered by a unit-volume of interplanetary matter.

On the other hand, $Z(r,\varepsilon)$ being differentiable

$$dZ = \frac{\partial Z(r,\varepsilon)}{\partial r} dr + \frac{\partial Z(r,\varepsilon)}{\partial \varepsilon} d\varepsilon$$

and the relations

$$dr = -\cos\varepsilon\, dR$$
$$d\varepsilon = \frac{\sin\varepsilon}{r} dR$$

give

$$dZ = \left(-\cos\varepsilon \frac{\partial Z(r,\varepsilon)}{\partial r} + \frac{\sin\varepsilon}{r} \frac{\partial Z(r,\varepsilon)}{\partial \varepsilon}\right) dR \qquad (2)$$

Comparing (1) and (2):

$$J(r,\varepsilon) = \cos\varepsilon \frac{\partial Z(r,\varepsilon)}{\partial r} - \frac{\sin\varepsilon}{r} \frac{\partial Z(r,\varepsilon)}{\partial \varepsilon}$$

If r and ε are looked upon as the polar coordinates in a plane, then $Z(r,\varepsilon)$ can be considered as a scalar field, and one can write:

$$J = \operatorname{grad} Z \cdot \vec{u} \qquad \vec{u}(\cos\varepsilon, -\sin\varepsilon)$$

Therefore the calculation of J depends on the knowledge of Z. We can measure $Z(1,\varepsilon)$ and we know also, by experiments of Pionner X, how the function $Z(r,\varepsilon)$ varies when ε takes two fixed values. It seems reasonable to put.

$$Z(r,\varepsilon) = Z(1,\varepsilon) f(r) \qquad f(r) = Z(r,C^{te})$$

With this Z we obtain the function J. To check this Z we can calculate theoretically the luminance with the expression

$$Z = \int_0^\infty \operatorname{grad} Z \cdot d\vec{R}$$

The values obtained with this integral are in good agreement with the experimental ones. Therefore the assumption on Z seems well founded. In a similar way we can obtain the local degree of polarization

References:
Dumont,R.(1973) Planet Space Sci. 21. 2149
Hanner, M.S.and Weinberg,J.L.(1973)Sky and Telescope 45, 217
Hanner, M.S.and Weinberg,J.L.(1973)Cospar Meeting.May 1973,Konstanz.

ON THE VISIBILITY OF THE LIBRATION CLOUDS

S. Röser

Max-Planck-Institut für Astronomie

Heidelberg-Königstuhl, FRG

Introduction

In recent years satellite observations of the region of the Earth-Moon libration points L_4 and L_5 became available, which showed different results. Roach (1975) has measured perturbations of the Zodiacal Light intensity near the libration points L_4 and L_5 with a photometer on board the OSO-6 satellite. He interpreted the results as counterglow of a cloud of particles, having an angular diameter of 6 degrees and an average brightness of 20 $S_{10\ V}$, whereas observations by Burnett et al. (1974) set an upper limit of 10 $S_{10\ blue}$. Earth bound observations made by Bruman (1969) with the 48-inch Palomar Schmidt telescope gave no indication for discrete objects nor for clouds.

Looking at the dynamical aspects of the matter, the circumstances are not favourable for a straightforward solution. In celestial mechanics there are great difficulties in treating a restricted ($m_4 = 0$) four-body problem analytically, and though several papers are now present confirming the existence of periodic orbits around the libration points, even if eccentricity of the orbits of the primaries and the effect of radiation pressure are admitted (e.g. Matas, 1974), this does not confirm the existence of a cloud. Therefore we undertook a numerical approach to the problem, which enabled us to compare computed values for the mass supply of the cloud with measurements.

Assumptions and calculations

Besides the gravitational attraction of the Sun, the Earth and the Moon, we included in our calculations the effect of the Sun's radiation pressure using formulas given by Robertson (1937). For the equations of motion we used cartesian relative coordinates originating in the Earth's centre. In order to get a good approximation to the real situation we made the following assumptions for the motion of the Sun and the Moon. The Sun moves around the Earth on an elliptic orbit with eccentricity 0,0167.

The Moon's orbit has an inclination to the ecliptic of 5.°1, an eccentricity of 0.0549, and the rotation of perigee and node are included. The initial positions of Sun, Earth and Moon were collinear, but this did not influence the results as has been shown by test calculations with different initial conditions.

The position of the libration point L_4 was that of the elliptic restricted three-body problem, that means 60 degrees in advance of the Moon in the Moon's plane of orbit, and Earth, Moon and L_4 form an equilateral triangle at any time. The probe began its motion at various positions at and near L_4, the initial velocities being varied. Starting with these initial conditions at time t = 0 the integration was performed in positive direction of time and was stopped, whenever the particle left the Earth-Moon system or hit one of the main bodies. Beginning at the same initial position and integrating backwards in time we found the origin of particles that could possibly contribute to a libration cloud.

For the numerical treatment of the problem a fourth order Runge-Kutta scheme with automatical step-size control has been applied. To insure the correctness of the integration procedure we reversed the direction of time at the final point of the integration and pursued the path back again until we reached t = 0. We only used the results if no coordinate of the particle deviated from its original value more than 10 per cent.

The lifetime of the particles

The influence of radiation pressure on a particle depends on its size, mass-density and chemistry. For the calculations we adopted the parameter $1 - \mu = \frac{\text{radiation pressure}}{\text{gravity}}$, which can be transformed into particle sizes via the rule of thumb

$$\rho d = \frac{C}{1-\mu}$$

where d = particle diameter, ρ = mass-density and $C = 1.19 \times 10^{-4}$ g cm^{-2}. We made calculations for 8 values of $1-\mu$, the largest being 0.1. Test computations showed that for larger values of this parameter the perturbations were too strong and the particles left the region around L_4 immediately. From this we conclude that particles with a diameter smaller than about 3 μm ($\rho = 3.5$ g cm^{-3}) cannot contribute to a libration cloud. For each value of $1-\mu$ 59 different initial conditions in position and velocity of the probe have been considered.

Table 1 Life - times of the particles near the Earth - Moon libration point. T_{tot} is the total life-time of the particles, T_1 the time of residence in an area of 50 x 26 degrees^2 around L_4 at distances between .8 and 1.2 (unit of length = mean distance Moon - Earth) and T_2 the time of residence in a 1 degree^2 central part around L_4. The case $1 - \mu = 0$ is shown for comparison.

$1 - \mu$	.1	.01	5×10^{-3}	10^{-3}	5×10^{-4}	10^{-4}	10^{-5}	0
r [mm] (ρ=3.5)	1.5×10^{-3}	1.1×10^{-2}	1.9×10^{-2}	7.2×10^{-2}	.13	.49	3.3	-
m [g] (ρ=3.5)	6×10^{-11}	2×10^{-8}	10^{-7}	6×10^{-6}	3.4×10^{-5}	1.8×10^{-3}	5.6×10^{-1}	-
T_{tot} [d]	65	1091	2985	2066	5663	6303	6642	6448
T_1 [d]	15	113	246	749	3696	5362	5747	5767
T_2 [d]	0.4	1.4	1.9	7.0	24.7	48.8	52.5	54.6

In Table 1 we list so-called lifetimes of the particles in the libration region. For the different values of $1-\mu$ the corresponding particle radii and masses are given for $\rho = 3.5$ g cm^{-3}. T_{tot} is the total lifetime of the particle until it leaves the system or hits the Earth or the Moon. T_1 gives the time, the particle spends within an area of 50 x 26 degrees^2 centered at L_4, having distances from the Earth between 0.8 and 1.2 (unit of length = mean distance Earth-Moon). The times of residence in a 1 degree^2 central region around L_4 are called T_2. They are used to compute the surface brightness.

Origin of the particles

We computed the origin of the particles using only 227 exactly known trajectories. The major part (62 per cent) of the particles had interplanetary origin, 30 per cent came from the Moon, and 8 per cent of the trajectories started at the Earth. Of the particles with interplanetary origin we computed the heliocentric orbital elements outside the Earth's sphere of influence. The resulting distribution in the (a,e) - plane is shown in Figure 1.

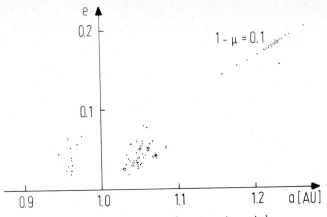

Figure 1. Original values of a (semi-major axis) and e (eccentricity) for the libration particles

It is remarkable, that the majority of the particles have semi-major axes greater than unity; this can be explained by the action of the Poynting-Robertson drag, which tends to diminish angular momentum and total energy. Of the particles having a < 1 all exept one had 1-μ values less or equal 10^{-4}; these are not very much influenced by radiation pressure, as is confirmed by their lifetimes. The remaining particle suffered a near passage by the Moon, which led to a drastic change in its orbital elements.

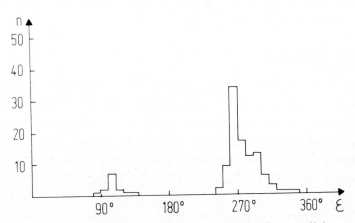

Figure 2. Directions of origin of the libration particles (ε = Elongation from the Sun)

The histogram in Figure 2 gives the distribution of the directions at the sphere, from which the particles must originate, if they are to reach the libration point L_4. The abscissa shows elongation from the Sun and n is the number of particles coming from a 10 degrees intervall of elongation. The two clearly distinct peaks at the apex and antapex directions of the Earth's orbit are in exact correlation with the results on the orbital elements. In the neighbourhood of the Earth's orbit particles having semi-major axes greater than unity must have velocities which are greater than that of the Earth. They must come from the antapex direction, whereas the contrary is true for the lower energetic particles. The velocities relative to the Earth were very low in all cases. For the smallest particles we found mean relative velocities of about 1.8 km sec^{-1}, for all others the velocities ranged between 0.8 and 0.9 km sec^{-1}.

The surface brightness of the libration cloud

For the supply of particles to the libration point region we discuss two possible sources, the interplanetary dust particles and the lunar ejecta. We have used recent flux measurements to compute the surface brightness near L_4, which is proportional to
  i) the lifetime of the particles near L_4 as given in Table 1.
  ii) the brightness of a single particle. Instead of a complicated scattering function we assumed totally reflecting particles, which gives an upper limit for the brightness of the cloud.
  iii) the amount of mass supplied by the different sources.

For the interplanetary particles we adopted the flux measurements given by Fechtig et al. (1974). The directions at the sphere, from which the particles must originate, are according to Figure 2. To avoid incompleteness we allowed an area of origin of 500$\square^°$ around the antapex and of 200 $\square^°$ around the apex direction. We made no restriction with respect to the velocity vector, so that the brightness is overestimated considerably. Under these assumptions a surface brightness of 2 $S_{10\ V}$ at L_4 has been found, a value which can only be understood as a very high upper bound.

The situation is even worse, if we want to discuss the mass supply by the Moon. At present time it is not known, if the Moon gains or looses mass. Nevertheless a rough estimation can be made. From the statements i) and ii) given above, we can compute the mass supply necessary to maintain a surface brightness of 1 S_{10} within the central area of 1$\square^°$

around L_4. A simple calculation yields

$$M \ [\text{kg d}^{-1}] = 7.5 \times [T_2 \times (1-\mu)]^{-1}$$

The masses needed to compensate the loss amount to 0.8 tons per day for the smallest particles ($\sim 10^{-8}$ g) that can come from the Moon, up to 12.5 tons per day for the largest ($\sim 10^{-1}$ g). According to Hughes (1973) the present influx rate of interplanetary dust to the Earth is 5.7×10^9 g yr^{-1}. Applied to the Moon, this means an influx rate of 4 tons per day. Assuming that the Moon looses mass at about the same amount, which is not established by measurements, only between 10 and 20 per cent of it will remain within the Earth-Moon system (Shapiro et al., 1966). Again we did not make any rectriction with respect to velocities, which would diminish the mass supply. So it seems evident that the Moon is out of question as a source.

Conclusion

Calculations as presented here will always be at a state of incompleteness. The great variety of parameters in the phase space prevents a comprehensive treatment of the problem. Besides these theoretical difficulties the experimental data are not yet at a satisfactory level. Nevertheless even the rough estimations in the last chapter indicate, that the conditions are not favourable for libration clouds, so that we are in doubt if they really exist.

References

Burnett, G.B.; Sparrow, J.G. and Ney, E.P.; Is the Zodiacal Light intensity steady? Nature 249, 639 (1974).

Bruman, J.R.; A Lunar Libration Point Experiment. Icarus 10, 197 (1969).

Fechtig, H.; Gentner, W.; Hartung, J.B.; Nagel, K.; Neukum, G.; Schneider E.; and Storzer, D.; Microcraters on lunar samples. Soviet-American Conference on Cosmochemistry of the moon and the planets, Moscow, 1974.

Hughes, D.W.; Interplanetary Dust and its influx to the Earth's Surface, XVI Plenary Meeting of Cospar, Konstanz, 1973.

Matas, V.; Periodic Solutions of a Disturbed Elliptic Restricted Three - Body - Problem. BACz, 25, 129 (1974).

Roach, J.R.; Counterglow from the Earth - Moon Libration Points, Planetary and Space Science, 23, 173 (1975).

Robertson, H.P.; Dynamical effects of radiation in the solar system MNRAS, 97, 423 (1937).

Shapiro, I.I.; Lautman, D.A.; and Colombo, G.; The Earth's dust belt: Fact or fiction? JGR 71, 5695 (1966).

1.3.6   SCATTERING FUNCTIONS OF DIELECTRIC AND ABSORBING

IRREGULAR PARTICLES

R. Zerull

Bereich Extraterrestrische Physik, Ruhr-Universität Bochum, FRG

Introduction

Most model computations for the analysis of interplanetary dust are based on scattering functions of homogeneous spheres (Mie-theory). To investigate the influence of this restriction, scattering properties of nonspherical particles were studied by microwave analog experiments. The results of earlier investigations on dielectric rough spheres, cubes and octahedrons (Zerull, 1973; Zerull and Giese, 1974) may be shortly summarized as follows: Roughness ($\sim \frac{1}{10}$ wavelength) does not significantly change the scattering behaviour of spheres, whereas cubes and octahedrons scatter evidently different than equivalent spheres. The dominant differences are increased intensity at medium scattering angles but no further increase towards backscattering, and nearly neutral polarization. These results encouraged to continue the investigations with really irregular particles of dielectric as well as absorbing material.

Definitions

The scattering properties of irregular particles are illustrated by plotting the averaged scattering functions I_1 and I_2 or the first Stokes parameter $I = I_1 + I_2$ and the degree of linear polarization $P = (I_1 - I_2) / (I_1 + I_2)$, respectively. The size parameter $\alpha = \pi D/\lambda$ (D = diameter, λ = wavelength) is referred to spheres of equal volume, with which the results are compared. The size distribution of the particles is assumed in this paper to be a power law $dn \sim \alpha^{-2.5} d\alpha$, but other size distributions can easily be simulated using the same measuring data. The physical properties of the particles are given by the complex refractive index $m = m' - m''i$.

Dielectric Irregular Particles

The particles treated in this section are nearly purely dielectric (m=1.5 - 0.005i). There are two different types of irregular particles. Typical examples are shown in fig.1. Following van de Hulst I call them "convex" and "concave" bodies. For "convex" particles theoretical predictions concerning the averaged geometrical cross section and the reflected part of the scattered radiation can be

Fig. 1: "Convex" and "concave" particles.

made, not so for "concave" particles (van de Hulst, 1957). In fig. 2 the averaged scattering functions of polydisperse mixtures are plotted for the size interval $1.9 \leq \alpha \leq 5.9$. This size interval includes the first maximum of the extinction cross section of the equivalent spheres. This maximum does not occur in the case of irregular particles (Hodkinson, 1963). Therefore and because of the close relationship between extinction cross section and forward scattering, intensity within the diffraction lobe is lower for the irregular particles than for the equivalent spheres. For scattering

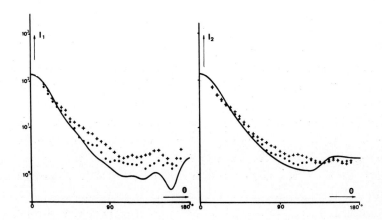

Fig. 2: Scattering functions
$1.9 \leq \alpha \leq 5.9$
$m = 1.5 - 0.005i$

—— spheres
•••• convex particles
++++ concave particles

angles $\theta \gtrsim 30°$ irregular particles show increased intensity, especially evident for the "concave" bodies. For both types of irregular particles the increase of I_1 is stronger than of I_2, so polarization is shifted to more positive values.

Fig. 3 shows the averaged scattering functions of polydisperse mixtures of the irregular particles and equivalent spheres for the size interval $5.9 \leq \alpha \leq 17.8$. There is a good correspondence in the diffraction lobe, as far as it could be measured. For scattering angles $30° \lesssim \theta \lesssim 150°$ both types of irregular particles show increased intensity, up to the factor 5. Enhanced backscattering is produced by the "convex" particles, not quite as marked as by the spheres, whereas the scattering of the "concave" particles is nearly isotropic for $\theta \gtrsim 100°$. Enhancement of intensity is much more significant for I_1 than I_2. So, by irregular particles of this size, too, polarization is shifted to more positive values. In fig. 4 the polarization properties of the "convex" and "concave" bodies are plotted separately considering sizes $1.9 \leq \alpha \leq 17.8$. No type of the irregular particles shows the strong negative polarization which is typical for the equivalent spheres. Against that, their polarization behaviour is much more neutral, small negative values dominating for the "convex" bodies, small positive values dominating for the "concave" ones.

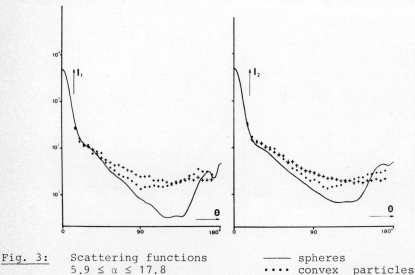

Fig. 3: Scattering functions
$5.9 \leq \alpha \leq 17.8$
$m = 1.5 - 0.005i$

——— spheres
•••• convex particles
++++ concave particles

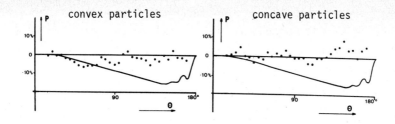

Fig. 4:   Polarization
         $1.9 \leq \alpha \leq 17.8$
         $m = 1.5 - 0.005i$

—— spheres
···· irregular particles

Absorbing Irregular Particles

There are three conditions for theoretical treatment of the scattering properties of randomly oriented irregular absorbing particles: Their shape must be convex, they must be big enough for application of geometrical optics, and absorption has to be strong enough to ensure, that the refracted part of the radiation is totally absorbed. In that case the scattering diagram is completely determined by diffraction and Fresnel reflection only.

Fig. 5: Irregular absorbing particles.

The scattering bodies discussed in this section do not satisfy the three conditions mentioned above. Typical examples are shown in fig. 5. Their shape is neither purely convex nor concave. The refractive index is $1.45 - 0.05i$, the size parameter is $9.8 \leq \alpha \leq 17.3$. In fig. 6 the scattering properties of the irregular absorbing particles are compared to equivalent absorbing spheres.

The scattered intensity of the irregular particles is higher up to factor of 3. Most interesting is the run of the polarization.

Here the irregular particles approximate the theoretical Fresnel reflection curve much better than spheres of the same size.

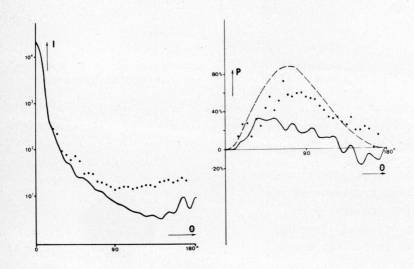

Fig. 6: Total intensity and polarization   ——— spheres
        $9.8 \leq \alpha \leq 17.3$ ···· irregular particles
        $m = 1.45 - 0.05i$ ---- Fresnel reflection

Literature

Hodkinson, J.R. : Light Scattering and Extinction by Irregular Particles Larger than the Wavelength. Electromagnetic scattering (ed. M. Kerker), Pergamon Press Oxford, London, New York, Paris, 1963.

Hulst, H.C. van de : Light scattering by small particles, John Wiley, New York, 1957.

Zerull, R. : Mikrowellenanalogieexperimente zur Lichtstreuung an Staubpartikeln, BMFT-FB W 73-18, 1973.

Zerull, R. und Giese, R.H. : Microwave Analog Studies. In: Planets, Stars and Nebulae studied with Photopolarimetry, University of Arizona Press, Tucson 1974.

1.3.7 THE COMPATIBILITY OF RECENT MICROMETEOROID FLUX CURVES WITH

OBSERVATIONS AND MODELS OF THE ZODIACAL LIGHT

R.H. Giese
Bereich Extraterrestrische Physik, Ruhr-Universität Bochum, FRG.

and

E. Grün
Max-Planck-Institut für Kernphysik, Heidelberg, FRG.

Increased sophistication in both, direct impact detectors and zodiacal light measurements encourages to discuss the compatibility of the results obtained by these quite different methods of investigating interplanetary dust. Taking recent measurements of particle fluxes and velocities obtained by the space missions of Pioneer 8/9 (Berg and Grün 1973), Heos 2 (Hoffmann et al. 1975), and comparing them with submicron-sized craters on lunar surface samples (Schneider et al. 1973, Fechtig et al. 1974) there seem to be two types of interplanetary dust populations: larger ($>10^{-12}$g) micrometeorites orbiting around the sun as the classical zodiacal dust cloud and a second component of very small ($<10^{-12}$g) particles coming radially from the direction of the sun with high velocities (>50 km/s). On the basis of the flux data referred to above and adopting for both components velocities of 10 or 50 km/s relative to the detector, respectively, a differential distribution function $n(a) \cdot da$ was found for the particle radii (a) as shown at a logarithmic scale in fig. 1. A density of 3 g/cm^3 was adopted in order to convert particle masses into radii. The regions A, B, C (see Table 1) correspond approximately to the regimes of "submicron particles", the classical zodiacal cloud particles, and the meteoritic component of the interplanetary dust complex. From this information the brightness $I(\varepsilon)$ of the zodiacal light in the ecliptic plane can be computed as a function of elongation by approximating the distribution function $n(a)$ in the different regions by simple power laws $a^{-\kappa} \cdot da$ and by adopting a resonable scattering function $\sigma(\theta)$ for the average scattering behaviour of one particle of the mixture depending on the scattering angle θ. By use of an inverse ($\nu = 1$) decrease of particle number densities $n = n_0 \cdot r^{-\nu}$ with solar distance r(AU), where n_0 is the number density at $r=1$ AU,

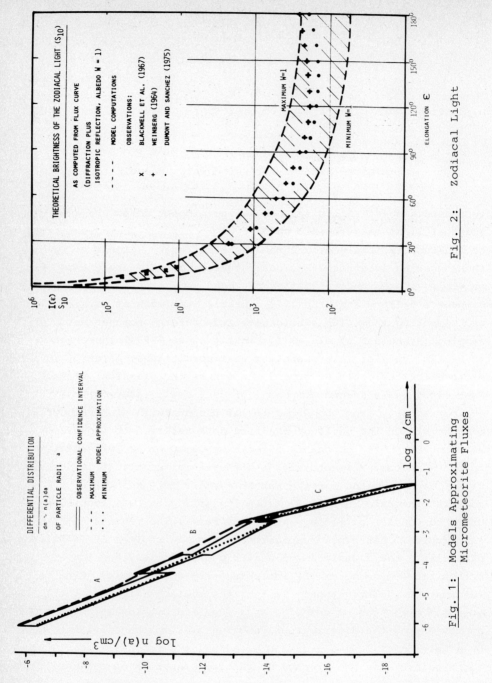

Fig. 1: Models Approximating Micrometeorite Fluxes

Fig. 2: Zodiacal Light

one obtains with a particle size distribution law $n(a)da \sim a^{-\kappa}da$
in the different intervals of sizes (Table 1) the intensity of the
zodiacal light (in stars of 10th magnitude per square degree, S_{10})
as shown in fig. 2. The two models (Maximum, Minimum) correspond to
an upper and to a lower limit of particle number densities compatible
with the in situ measurements, respectively.

Table 1

Parameters (see text) of the two Models used in Fig.2

region	Maximum - Model			Minimum - Model		
	$a/\mu m$	κ	n_o/cm^{-3}	$a/\mu m$	κ	n_o/cm^{-3}
A	0.008 to 0.16	2.7	$0.95 \cdot 10^{-12}$	0.008 to 0.16	2.7	$1.75 \cdot 10^{-13}$
B	0.16 to 28.8	2.0	$1.12 \cdot 10^{-14}$	0.40 to 21.4	2.5	$1.99 \cdot 10^{-15}$
C	28.8 to 339	4.33	$1.91 \cdot 10^{-17}$	21.4 to 339	4.33	$1.17 \cdot 10^{-17}$

The scattering function adopted for fig. 2 was a simple superposition
of diffraction plus conservative, isotropic reflection (albedo W=1).
Even with this high albedo the intensity produced by the minimum
model is too low by a factor of 2 to 4 for $30° \leq \epsilon \leq 180°$. On the
other hand, the maximum model produces with W = 1 in this range of
ϵ intensities higher than observed and approaches the observational
values with somewhat lower (W = 0.4 to 0.6) values of the albedo.
It also fits the zodiacal light fairly well in the inner regions
($\epsilon < 30°$). Therefore we conclude that the interplanetary particle
number densities derived by in situ measurements are now in good
agreement with the optical measurements.

The relative contribution of the different particle sizes to the
zodiacal light raises some important aspects in interpretation. Up
to now model computation based on Mie-theory implied a great
fraction of small particles to produce positive polarization at
the correct region of elongation. Such components were dielectric
particles of submicron-size ($\bar{a} \simeq 0.1$ µm, Weinberg 1964) or somewhat
larger ($\bar{a} \simeq 0.4$) absorbing particles or slightly absorbing particles
of ~ 10 µm-size (see Giese 1973). If, however, the number densities

of the Maximum-Model are adopted the contribution to the total
brightness at $\varepsilon = 90°$ is < 0.6 % for particles of
0.008 µm ≤ a ≤ 0.16 µm and 19 % for particles of 0.16 ≤ a ≤ 9.5 µm.
Therefore the contribution of the submicron-size particles (region
A of Table 1) is completely neglible. The largest particles (a > 29µm)
of the Maximum-Model (region C) still contribute 42 % of the total
brightness.

Due to the considerable contribution of larger particles more
information about the scattering function of irregular dust grains
in the size range above a ≃ 10 µm is needed. Here Mie-theory is of
limited use, not only because such computations are rather time
consuming. For larger dielectric particles the approximation by
spherical shapes (Mie-theory) produces extremely strong backscattering
effects and bumps in polarization (see Giese 1963), such as haze
bows, which are unrealistic for interplanetary dust particles.
Microwave analog measurements, which simulate light scattering by
practically dielectric, nonspherical particles corresponding to
micron-size (Zerull 1975, Fig. 3 and 4) suggest much more isotropic
scattering and rather neutral polarization outside the diffraction
region of the scattering diagram.

For larger (a ≥ 1.5 µm), sufficiently absorbing spheres Mie-theory
can be approximated roughly by a simple superposition of diffraction
plus Fresnel reflection (see Giese 1961). Microwave measurements
(Zerull 1976, Fig. 6) suggest that also irregulary, absorbing
particles produce strong positive polarization similar to Fresnel
reflection.

In a qualitative way we conclude, that intensity and positive
polarization observed in the zodiacal light can be produced by a
mixture of irregular particles with an appropriate mixing ratio
of dielectric and absorbing material without the need to fall back
upon a large contribution due to particles of micron or even
submicron size. There are, however, still problems to obtain
quantitative agreement, especially with the observational position
of the maximum of polarization. Analysis of the empirical scattering
function as derived directly from zodiacal light measurements
(Dumont 1975, Leinert et al. 1976) and further laboratory experiments
are needed to achieve convincing quantitative models.

Literature

Berg, O., Grün, E.: Space Res. XIII, 1047 (1973).

Blackwell, D.E., Dewhirst, D.W., Ingham, M.F.: The Zodiacal Light
in: Z. Kopal (ed.) Advances in Astronomy and Astrophysics,
Vol. V, 2-69, New York, London 1967.

Dumont, R. and Sánchez, F.: Astron. and Astrophys. 38, 405-412 (1975).

Fechtig, H., Hartung, J.B., Nagel, K., Neukum, G., Storzer, D.:
Proc. Fifth Lunar Sci. Conf., Geochim. Cosmochim. Acta
1, 2, 3, Suppl. 5, Vol. 3, p. 2463 (1974).

Giese, R.H.: Space Sci. Rev., 1, 589-611 (1963).

Giese, R.H.: Planet. Space Sci., 21, 513-521 (1973).

Hoffmann, H.J., Fechtig, H., Grün, E., Kissel, J.:
Plan. Space Sci., 23, 981 (1975).

Leinert, C., Link, H., Pitz, E., Giese, R.H. (1976), Rocket Photometry of the Inner Zodiacal Light, to be published in Astron.
and Astrophys.

Schneider, E., Storzer, D., Hartung, J.B., Fechtig, H., Gentner, W.:
Proc. Forth Lunar Sci. Conf., Geochim. Cosmochim. Acta
Suppl. 4, Vol. 3, p. 3272 (1973).

Weinberg, J..L.: Ann. d'Astrophys., 27, 718 (1964).

Zerull, R. (1976), Scattering Functions of Dielectric and Absorbing
Irregular Particles, this volume.

2 IN SITU MEASUREMENTS OF INTERPLANETARY DUST

2.1 Measurements from Satellites and Space Probes

2.1.1   In-Situ Records of Interplanetary Dust Particles -
Methods and Results

H. Fechtig

Max-Planck-Institut für Kernphysik, Heidelberg/F.R.G.

Abstract

A review is given on the techniques used to record and to quantitatively measure data of individual interplanetary dust particles. New developments in detection techniques are briefly discussed.

The main results from recent space missions at about 1 AU and in the earth-moon neighborhood are discussed and compared with the flux results from lunar microcrater studies. Spatial anisotropies and time fluctuations are found indicating that the earth is exposed to two main micrometeoroid dust populations: the "apex"-population and the ß-meteoroids. The near planet-dust enrichments measured by HEOS 2 near the earth and by the Pioneer 10/11 near Jupiter are emphasized. The experimental data strongly suggest a fragmentation process associated with the earth. The role of the moon as a dust source is discussed. The important problems in the dust field for future space missions are summarized.

I. Techniques

The active detectors used for in situ experiments to detect interplanetary dust are listed in Table 1. The indicated sensitivities have been obtained for an impact velocity of appr. 10 km/sec. Some of the techniques listed are not used any more. Some other detectors, however, which have first been used more than 10 years ago are still in use today and are gaining important new results: the pressurized cells on the Pioneer 10/11 mission (Humes et al., 1974). The use of

Table 1: Active detector techniques used for in situ experiments to detect interplanetary dust

Type	Sensitivity at 10 km/sec
Piezoelectric Microphone Systems	$10^{-11} - 10^{-9}$ g
Penetration Detectors: thin wires, photocells, pressurized cells etc.	$10^{-9} - 10^{-7}$ g
Capacitor Detectors	$10^{-12} - 10^{-10}$ g
Ionisation Detectors	$10^{-15} - 10^{-12}$ g
Semiconductor Detectors	$\sim 10^{-15}$ g

semiconductors as detectors for interplanetary dust has been investigated by Kassel (1973) and by Rauser (1974). The Langley group has employed a semiconductor detector on the MTS satellite for the first time (Alvarez et al., 1975).

The most important and most frequently used detector technique is still the ionisation detector. The plasma produced during the impact of high velocity dust particles into a solid target (preferably a heavy metal like gold or tungsten) is detected. The amplitude of the signal produced by the plasma is measured and in one case (HEOS-, Helios-experiments) its risetime. Coincident events for electrons and positive ions are detected using electrically biased collectors. Different versions of these detectors have repeatedly been described in detail earlier by Berg and Gerloff (1971), Adams and Smith (1971), Dietzel et al. (1973) and Bedford (1975). The sensitivities are ranging between 10 µ diameter particles at 1 km/sec down to 1/10 diameter particles at approximately 40 km/sec.

The technique to collect particles using thin films and metal plates as collectors has been sucessfully used by Hemenway and Hallgren (1970). The inflight shadowing technique to identify collected particles vs. contaminant particles seems to be a reliable technique. This technique, however, is limited to low relative collecting velocities and is preferably to be used in the upper atmosphere. The collection experiments in deep space are based on impact craters or penetration holes produced by high speed particles (Hemenway et al., 1974, Nagel et al., 1976).

II. Interplanetary Dust at 1 AU

The knowledge of the dust population at 1 AU from the sun has been gained mainly by the Pioneer 8/9 dust experiments and from the investigation of lunar microcraters produced by impacting interplanetary dust particles. (See also Dohnanyi, 1972, for an earlier review). <u>Pioneer 8/9 results:</u> According to Berg and Gerloff (1971) and Berg and Grün (1973) the Pioneer 8/9 dust experiments recorded dust particles intercepting the sensors from all directions in the ecliptic plane.

Table 2 shows the number of events from particles intercepting the sensors from sun-, apex-, antiapex- and antisun-direction; the apex-direction is the direction of the velocity of the orbiting satellite around the sun. The majority of these particles are hitting the sensor mainly from 2 directions: the sun- and the apex-direction. From time of flight measurements Berg and Gerloff (1971) calculated the impact velocity of the apex-particles; they obtained values in the range of 2

Table 2: Numbers of Particles recorded from Pioneer 8/9 dust experiments per year

Viewing Direction	Number of events per Year (orbit)	%
Sun	90	56
Apex	40	25
Antiapex	20	12,5
Antisun	10	6,5
Total	160	100

to 20 km/sec. Thus, a meteoroid flux for the apex-particles could be determined to be $\phi_{apex} = 2 \times 10^{-4}$ m^{-2}sec^{-1} $(2\pi \text{ster})^{-1}$ for $m \geqslant 10^{-12}$ g. For the "sun"-particles, however, Berg and Grün (1973) did not obtain any velocity measurements and, therefore, no fluxes could be calculated. The authors conclude that the "sun"-particles are extremely small ($< 1\,\mu$ in diameter) and fast moving (> 50 km/sec).

Lunar Microcraters: The lunar surface samples are exhibiting micron- and submicron-sized impact craters down to $0.2\,\mu$ in diameters (Schneider et al., 1973) and even down to 250 Å in diameters (Blanford et al., 1974). The overall micron- and submicron-sized crater statistics shows a bimodal distribution as shown in Fig. 1 for the lunar sample 15205 (Hartung and Storzer, 1974): a steep part for craters $< 5\,\mu$ in diameter and a flat part for craters $> 5\,\mu$ in diameter. Using suitable simulation results (Fechtig et al., 1974) and dated samples according to the heavy ion track dating method (Schneider et al., 1973), the crater distributions could be converted into dust particle fluxes. Fig. 2 shows the result. It was assumed that the flat part in the crater size distribution represents the "apex"-component, whereas the steep part is due to the "sun"-component. A comparison with flux data from the HEOS 2 dust experiment

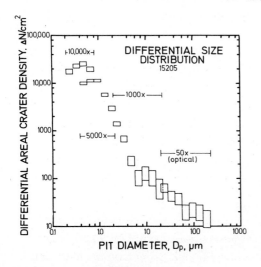

Fig. 1

Differential crater number densities on lunar sample 15205.

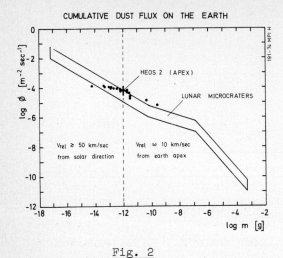

Fig. 2
Cumulative dust flux at 1 AU derived from lunar crater statistics and HEOS 2 dust experiments

(Hoffmann et al., 1975b) indeed shows an agreement with the lunar data in both absolute numbers and slope. A calculation using the lunar, Pioneer 8/9 and HEOS 2 results yield a number frequency vs. mass plot. The results are in agreement with the observed zodiacal light brightness (Giese and Grün, 1976). Detailed summaries of the results from lunar microcrater investigations are presented by Hörz et al. (1975) and by Fechtig et al. (1974).

Zook and Berg (1975), Zook (1975), Dohnanyi (1976) and Whipple (1976) have discussed a possible production mechanism of the "sun"-particles, which are called ß-meteoroids (according to a suggestion by Zook and Berg (1975)). Between 1 AU and the sun meteoroids are spiralling in towards the sun because of the Poynting-Robertson effect (Wyatt and Whipple, 1950); they frequently collide with each other and with the ß-meteoroids producing a large number of fragments. Generally the fragments are smaller than a critical minimum mass and are hence accelerated outward from the sun by radiation pressure which now overpowers the attractive force of gravity. These fragments are therefore approaching the spacecraft from the sun-direction and are in a hyperbolic orbit leaving the solar system.

A second possibility is that particles approaching the sun are vaporised until they are small enough to be accelerated outward by radiation pressure; particles consisting of different chemical components can even be fragmentated by this process because of their different melting and vaporization temperatures. Huebner (1970) has considered this mechanism when he discussed the presence of a heavy solar wind component. Kaiser (1970) has discussed this process, too, in an effort to explain observations of spectral lines in the infrared close to the sun by Peterson (1967) and McQueen (1968). Sekanina (1976) has calculated the dynamics of particles produced by this mechanism close to the sun. He found that this mechanism could account for the observed ß-meteoroids.

A third alternative, finally, is suggested by Hemenway et al. (1972): from their measurements the authors conclude that at least a substantial part of the ß-meteorites are originated directly from the surface of the sun (see Hemenway, 1976).

III. Interplanetary Dust near the Earth

An enhancement in the number density of interplanetary dust particles near the earth has for a long time been the subject of discussion. Early measurements using microphone detectors (Alexander et al., 1963) and particle collection results (Hemenway and Soberman, 1962) indicated the existence of a dust cloud around the earth. Later measurements, however, showed that part of the microphone pulses were due to noise signals (Nilsson, 1966) and part of the collected particles obviously were contaminants (Fechtig and Feuerstein, 1969). However, there still remains a considerable difference between near earth and deep space results in particle number densities (Fechtig, 1973).

The orbit of the HEOS 2 earth satellite with its high eccentric orbit above the North pole (apogee: 240.000 km, perigee: $\leq$ 3.000 km) offered a good chance to study the spatial particle distribution as a function of the distance from the earth. The main results have already been published (Hoffmann et al., 1975a, b); some of the overall results are summarized here for the whole mission period. A total of 431 particles have been recorded during the lifetime of the satellite from February 7, 1972, through August 2, 1974. Table 3 shows a summary

Table 3: HEOS 2-dust experiment: Summary of registered particles on the 100 cm^2 surface detector and measured quantities

Type of measurement	Number of particles	Measured quantities
Counted events:	174	events
Half quantitative measurements:	191	Impact produced charge $Q \sim m \cdot v^{3.5}$
Quantitative measurements:	66	Charge $Q \sim m \cdot v^{3.5}$ Rise time $t_{rise} = f(\frac{1}{v})$
Total	431	

of the detected particles: some of the 431 particles are counted, some are half-quantitatively measured (by measuring only the pulse amplitude) and others are observed from measurements of their pulse amplitudes and the pulse rise time. In Fig. 3 the number of events is plotted as a function of the time differences between subsequent events for the whole time period. These data are divided somewhat arbitrarily into 3 categories: swarms with 1 or more events per 15 minutes, groups ranging between 1 event per 15 minutes and 2 events per day, random particles with 2 or less events per day. In Table 4 the number of swarm-, group- and random particles is listed for different viewing directions of the experiment. Unfortunately it was not possible to turn the experiment into the sun-direction because of technical reasons. From the event rates we find that the apex-particles are most frequent, the ecliptic south particles are, however, almost as frequent as the apex particles. A detailed analysis, however, showed that the ecliptic south particles are much smaller and faster than the apex-particles (Hoffmann et al., 1975b). The corresponding fluxes are therefore much lower for the ecliptic south particles. It is believed that these particles are fragmentation products from larger micrometeoroids.

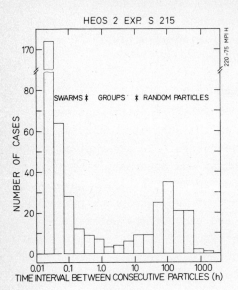

Fig. 3

HEOS 2 dust experiment: Temporal distribution of the dust particles. Each individual pair of consecutive particles is represented by a box.

The results during the activities of meteor showers are listed in Table 4 as well. No significant contributions in the dust size range could be found. This result is in agreement with the general knowledge of the size distributions in meteor showers (Millman and McIntosh, 1964, 1966, Millman,1970, Dohnanyi, 1970).

In order to further discuss these results the 3 categories of particles are plotted in Fig. 4 as a function of geocentric radius. An important overall result is that more than 80 % of all registered particles are clustered, that means they appear in swarms or groups. In the perigee region (< 67.000 km) even more than 93 % of the registered particles are clustered. In Fig. 4 one can see that the swarms are overwhelmingly

Table 4: HEOS 2-dust experiment: Number of recorded particles per 100 cm² sensor surface for different viewing directions

		Duration (days)	Swarms		Groups		Number of random particles	Particles per day
			Number	Particles	Number	Particles		
Interplanetary Region (>67,000 km)	Apex	289.0	1	32	5	11	32	0.11
	Antiapex	55.1	1	51	1	4	3	0.05
	Ecl. North	263.9	–	–	3	7	14	0.05
	Ecl. South	114.4	–	–	2	5	12	0.11
Meteor Showers	Perseids 1972	14.8	–	–	1	5	2	0.14
	Perseids 1973	19.3	–	–	1	2	4	0.27
	Draconids 1972	21.2	–	–	–	–	1	0.05
	Quadrantids 1973	29.4	–	–	–	–	4	0.14
	Quadrantids 1974	16.9	–	–	–	–	–	–
	Ursids 1973	13.8	–	–	–	–	2	0.15
Perigee Region (≤67,000 km)	all directions	69.8	11	206	9	18	16	0.23
		907.6	13	289	22	52	90	0.10
	Total:							

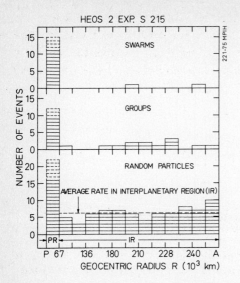

Fig. 4
Rate of swarms, groups and random particles as a function of the geocentric radius R; each box represents a group, swarm or random particle; dashed boxes are due to the bad data coverage in the PR;
P = perigee, A = apogee,
PR = perigee region.

most frequent in the perigee region, while the groups are more or less equally distributed over the whole altitude range (except for a modest enrichment in the perigee region).

The random particles clearly show an enhancement in the particle frequency of about a factor of 5 in the perigee region ($\sim$ 67.000 km) compared with the interplanetary region ($>$ 67.000 km).

In Fig. 5 the fluxes are plotted for various viewing directions in the interplanetary region. As described in detail by Hoffmann et al. (1975b) the average impact velocity of the apex particles is about 10 km/sec; this is based on a spread ranging from 1 to 20 km/sec. The corresponding flux has been calculated to be

$$\phi_{apex} = 7 \cdot 10^{-5} \text{ m}^{-2} \text{sec}^{-1} \text{ for masses} \geq 10^{-12} \text{ g}.$$

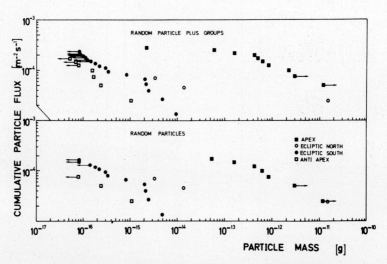

Fig. 5: HEOS 2 Experiment S 215: Uncorrected cumulative flux of random particles plus groups and random particles only as function of the particle mass.

In order to compare this flux value with the Pioneer 8/9- and the Mariner 4-fluxes (Alexander and Bohn, 1974) the HEOS 2 flux has to be expressed as a function of a viewing solid angle of 2π steradians (the given flux value ϕ_{apex} relates to the opening angle of the HEOS 2 dust experiment appr. $60°$ solid angle). The fluxes are:

HEOS 2: $\phi_{apex} = 4 \cdot 10^{-4}$ $m^{-2} sec^{-1}$ $(2\pi st)^{-1}$
Pioneer 8/9: $\phi_{apex} = 2 \cdot 10^{-4}$ $m^{-2} sec^{-1}$ $(2\pi st)^{-1}$
Mariner IV: $\phi_{apex} = 3 \cdot 10^{-4}$ $m^{-2} sec^{-1}$ $(2\pi st)^{-1}$

for masses $m \geq 10^{-12}$ g

The fluxes in the other directions are at least one order of magnitude lower and the impact speed is ≈ 20 km/sec.

Besides the enhancement of groups in the perigee region (Fig. 4), they seem to be more or less equally distributed as a function of distance from the earth in the interplanetary region. There is, however, a correlation with the position of the moon (lunar aspect angle) relative to the experiment viewing direction. As shown in Fig. 6 most of the groups in the interplanetary region appear within lunar aspect angles smaller $60°$, which is in agreement with a lunar origin as shown by detailed trajectory calculations by Hoffmann et al. (1975b). (The groups within the perigee region and during meteor showers are not to be considered to originate from the moon since the corresponding angle distribution is no criterion.) In contrary random particle events show an overall distribution independent from the lunar aspect angle and are therefore not correlated with the moon. Almost all swarms appear in the perigee region. Laboratory investigations by Eichhorn (1976) using the impact light flash technique have shown that even at low impact velocities ($\sim$ 5 km/sec) a relatively small amount of submicronsized ejecta are produced at low angles relative to

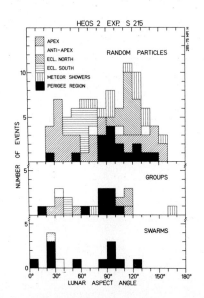

Fig. 6
Distribution of the directions under which the random particles and groups would encounter the detector assuming a lunar origin.

the target and at velocities as high as 30 km/sec. Schneider (1975) has performed light gas guns experiments and concluded that under lunar conditions a small fraction of the impact produced ejecta even leave the lunar gravitational field (also see Gault et al., 1963). A theoretical analysis by Dohnanyi (1975) shows that this lunar origin is in agreement with an ejecta origin caused by impacting meteorites in the kg-size range on the lunar surface.

The swarms do not show any correlation with the position of the moon. They seem, however, to be associated with the perigee region. Kaiser (Hughes, 1974) has suggested that meteors at grazing incident angles are producing similar dust. A corresponding observation has been published by Rawcliff et al. (1974) and by Bigg and Thomson (1969).

The Prospero dust satellite experiment supports the HEOS 2 results of the clustered dust particles near the earth (Bedford et al., 1975). As a result of a detailed analysis these authors give a comparison of fluxes between HEOS 2 and Prospero as a function of the geocentric distance. The diagram is given in Fig. 7. The fluxes are plotted for masses $\geq 10^{-14}$ g and $\geq 10^{-15}$ g for Prospero, HEOS-near earth, HEOS-interplanetary and some lunar data (Fechtig et al., 1974; Smith et al., 1974). This interpretation also allows one to understand other discrepancies: the results by Nazarova and Rybakov (1974) from the soviet spacecrafts Cosmos and Intercosmos reasonably well correspond to a near earth distance

$$\phi_{cosmos} = 2 \cdot 10^{-3} \text{ m}^{-2}\text{sec}^{-1} (2\pi st)^{-1} \text{ for}$$
$$\text{masses m} \geq 10^{-11} \text{ g and}$$
$$\phi_{intercosmos} = 10^{-4} \text{ m}^{-2}\text{sec}^{-1} (2\pi st)^{-1} \text{ for}$$
$$\text{masses m} \geq 10^{-9} \text{ g}$$

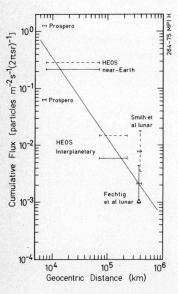

Fig. 7

Cumulative micrometeoroid flux for masses $\geq 10^{-14}$ g (solid lines) and $\geq 10^{-15}$ g (dotted lines) as a function of geocentric distance according to Bedford et al. (1975).

(Nazarova and Rybakov, 1974). New measurements of these authors (Nazarova and Rybakov, 1975) report much lower fluxes measured by Lunar Orbiter Luna 22 ($\phi_{sporadic} = 5 \cdot 10^{-5} \text{ m}^{-2}\text{sec}^{-1} (2\pi st)^{-1}$ for masses m $\geq 10^{-11}$ g) in good agreement with the results on the Lunar Explorer 35

(Alexander et al., 1972) as well as with results on the Mars 7 spacecraft ($\phi = 2\cdot 10^{-5}$ m^{-2}sec^{-1} (2πst)$^{-1}$) (Nazarova and Rybakov, 1975). Because of their lunar or even larger distances from the earth, the low fluxes of those experiments agree well with the interplanetary fluxex found by Pioneer 8/9 and HEOS 2. Humes et al. (1974) have reported a particle enhancement of 2 orders of magnitude from the Pioneer 10-11 dust experiments when approaching the Planet Jupiter.

The moon is obviously a source for small dust. The upper atmosphere also might produce dust from grazing meteors. Two or even three orders of magnitudes, however, can not easily be explained by these two sources alone. A comparison of these results with collection experiment results (Hemenway, 1973) lead to the conclusion that an effective, but unknown fragmentation process causes this enhancement. In another paper (Fechtig and Hemenway, 1976) the evidences for this fragmentation process are summarized and possible fragmentation mechanisms are discussed. It looks very likely that extremly porous meteors are producing dust by their disintegration somewhere between the moon and the earth's upper atmosphere. This problem is still open, however, and should be further explored in future experiments.

IV. Future Techniques and Scientific Problems

Future Techniques:

The techniques to detect interplanetary dust particles in deep space are low level type measurements. The event rates range from 1 event per month to 1 event per hour. Only in special cases, for example close to a planet, the event rate may be higher. To differentiate real events from electrical and oscillating noise it is therefore essential to use coincidence techniques. The HEOS 2 detector takes advantage of the positive and negative charge pulses which were measured in coincidence. The Helios detector uses a third signal: the mass spectrum of positive ions.

Eichhorn (1976) has investigated the mass- and velocity-dependences of the impact light flash during high velocity impacts. The maximum of the differential light intensity reads:

$$I_{diff.} \sim m\, v^{4,1}$$

while the integral light flash depends differently:

$$I_{int.} \sim m\, v^{3,2}$$

From these relations it is possible to determine mass m and impact velocity v of a projectile. As shown in detail by Eichhorn (1976) the sensitivity of this technique is comparable to the sensitivity of the impact plasma detector. Therefore, this technique is suitable to be used as a further coincidence signal.

Auer (1975) has published the principles of a detector which is based on the fact that charged particles are influencing charges on a conductor. The proposed technique works, if the interplanetary dust grains are sufficiently charged up. This method is of especially great interest for orbit determinations.

Not only for technical reasons should future dust experiments be of coincidental type, however. Because of scientific reasons, one should more and more combine different types of measurements. Optical and direct measurements as well as joint detection and collection experiments should be performed to explore the various parameters for the same dust particles.

Scientific Problems:
More precise and quantitative measurements are needed to understand the dynamics and the evolution of the interplanetary dust cloud of the solar system. The following topics should be further explored:
ß-meteoroids: besides the existence of the ß-meteoroids and their mass range we do not know much more. The important problems like origin (collissions, vaporization, direct sun origin?), dynamics and composition are still unsolved. Any interplanetary or near earth mission is suitable for these measurements.
Near planets dust dynamics: the HEOS 2- and the Prospero-results strongly suggest an enhancement of dust near the earth. The Pioneer 10/11 results (Humes et al., 1974) show a strong enhancement of the dust population near Jupiter. The mechanisms of gravitational enhancements, fragmentation processes, impacting induced ejecta production or other still unknown enhancement mechanisms have to be investigated. The future missions Lunar Orbiter, Jupiter Orbiter and Space shuttle/Space Lab are suitable missions.
Sources: one still does not know from direct measurements which sources contribute to the solar dust cloud and to what extent. Any possible mission should be used to directly study the sources. The most interesting mission presently in discussion is a cometary mission for example, to Comet Encke (Yeates et al., 1976).

A lot of interesting scientific work ought to be done in the future.

References:

Adams, N.G., and Smith, D. (1971), "Studies of Microparticle Impact Phenomena leading to the Development of a highly sensitive Micrometeoroid Detector", Planet. Space Sci. 19, 195.

Alexander, W.M., Arthur, C.W., Bohn, J.L., Johnson, J.H., and Farmer, B.J. (1972), "Lunar Explorer 35: 1970 Dust Particle Data and Shower related picogram Ejecta Orbits", Space Research X, 287.

Alexander, W.M., and Bohn, J.L. (1974), "Mariner 4: A Study of Cumulative Flux of Dust Particles over a Heliocentric Range of 1 - 1,56 AU 1964-1967", Space Research XIV, 749.

Alexander, W.M., McCracken, C.W., Secretan, L., and Berg, O.E. (1963), "Review of direct Measurements of Interplanetary Dust from Satellites and Probes", Space Research III, 891.

Alvarez, J.M. (1976), "The Cosmic Dust Environment at Earth, Jupiter and Interplanetary Space: Results from Langley Experiments on MTS, Pioneer 10, and Pioneer 11", This Volume.

Auer, S. (1975), "Two high Resolution Velocity Vector Analyzers for Cosmic Dust Particles", Rev. Sci. Instrum. 46, 127.

Bedford, D.K. (1975), "Observations of the Micrometeoroid Flux from Prospero", Proc. Roy. Soc. A 343, 277.

Bedford, D.K., Adams, N.G., and Smith, D. (1975), "The Flux and Spatial Distribution of Micrometeoroids in the Near-Earth Environment", Planet. Space Sci. 23, 1451.

Berg, O.E., and Gerloff, U. (1971), "More than two Years of Micrometeorite Data from two Pioneer Satellites", Space Research XI, 225.

Berg, O.E., and Grün, E. (1973), "Evidence of Hyperbolic Cosmic Dust Particles", Space Research XIII, 1047.

Bigg, E.K., and Thomson, W.J. (1969), "Daytime Photograph of a Group of Meteor Trails", Nature 222, 156.

Blanford, G.E., Fruland, R.M., McKay, D.S., and Morrison, D.A. (1974), "Lunar Surface Phenomena: Solar Flare Track Gradients, Microcraters, and Acretionary Particles", Proc. Fifth Lunar Sci. Conf., Geochim. Cosmochim. Acta, Suppl. 5, Vol. 3, p. 2501.

Dietzel, H., Eichhorn, G., Fechtig, H., Grün, E., Hoffmann, H.-J., and Kissel, J. (1973), "The HEOS 2 and Helios Micrometeoroid Experiments", J. Phys. E: Sci. Instrum. 6, 209.

Dohnanyi, J.S. (1970), "On the Origin and Distribution of Meteoroids", J. Geophys. Res. 75, 3468.

Dohnanyi, J.S. (1972), "Interplanetary Objects in Review: Statistics of their Masses and Dynamics", Icarus 17, 1.

Dohnanyi, J.S. (1975), "Gruppen von Mikrometeoriten im Erde-Mond System", Jahrestagung der Astronomischen Gesellschaft, Berlin.

Dohnanyi, J.S. (1976), "Sources of Interplanetary Dust: Asteroids", This Volume.

Eichhorn, G. (1976), "Analysis of the Hypervelocity Impact Process from Impact Flash Measurements", Planet. Space Sci., in press.

Fechtig, H. (1973), "Cosmic Dust in the Atmosphere and in the Interplanetary Space at 1 AU Today and in the Early Solar System" in "Evolutionary and Physical Properties of Meteoroids", ed. C.L. Hemenway, P.M. Millman, A.F. Cook, p. 209, NASA SP-319.

Fechtig, H., and Feuerstein, M. (1970), "Particle Collection Results from a Rocket Flight on August 1, 1968", J. Geophys. Res. 75, 6736.

Fechtig, H., Gentner, W., Hartung, J.B., Nagel, K., Neukum, G., Schneider, E., and Storzer, D. (1974), "Microcraters on Lunar Samples", Soviet-American Conference on Cosmochemistry of the Moon and the Planets, Moscow.

Fechtig, H., Hartung, J.B., Nagel, K., Neukum, G., and Storzer, D. (1974), "Lunar Microcrater Studies, derived Meteoroid Fluxes, and Comparison with Satellite-borne Experiments", Proc. Fifth Lunar Sci. Conf., Geochim. Cosmochim. Acta, Suppl. 5, Vol. 3, p. 2463.

Fechtig, H., and Hemenway, C.L. (1976), "Near Earth Fragmentation of Cosmic Dust", This Volume.

Gault, D.E., Shoemaker, E.M., and Moore, H.J. (1963), "Spray Ejected from the Lunar Surface by Meteoroid Impact", NASA Report NASA TND-1767.

Giese, R.H., and Grün, E. (1976), "The Compatibility of recent Micrometeorite Flux Curves with Observations and Models of the Zodiacal Light", This Volume.

Hartung, J.B., and Storzer, D. (1974), "Meteoroid Mass Distributions and Fluxes from Microcraters on Lunar Sample 15205", Space Research XIV, 719.

Hemenway, C.L. (1973), "Collections of Cosmic Dust", Whipple-Symposium, in press.

Hemenway, C.L. (1976), "Submicron Particles from the Sun", This Volume.

Hemenway, C.L., and Hallgren, D.S. (1970), "Time Variation of the Altitude Distribution of the Cosmic Dust Layer in the Upper Atmosphere", Space Research X, 272.

Hemenway, C.L., Hallgren, D.S., and Schmalberger, D.C. (1972), "Stardust", Nature 238, 256.

Hemenway, C.L., Hallgren, D.S., and Tackett, C.D. (1974), "Near Earth Cosmic Dust Results from S-149", AIAA/AGU Conference on Scientific Experiments of Skylab, Huntsville, Alabama.

Hemenway, C.L., and Soberman, R.K. (1962), "Studies of Micrometeorites obtained from a recoverable Sounding Rocket", Astron. J. 67, 256.

Hoffmann, H.-J., Fechtig, H., Grün, E., and Kissel, J. (1975a), "First Results of the Micrometeoroid Experiment S 215 on the HEOS 2 Satellite", Planet. Space Sci. 23, 215.

Hoffmann, H.-J., Fechtig, H., Grün, E., and Kissel, J. (1975b), "Temporal Fluctuations and Anisotropy of the Micrometeoroid Flux in the Earth-Moon System", Planet. Space Sci. 23, 985.

Hörz, F., Brownlee, D.E., Fechtig, H., Hartung, J.B., Morrison, D.A., Neukum, G., Schneider, E., Vedder, J.F., and Gault, D.E. (1975), "Lunar Microcraters: Implications for the Micrometeoroid Complex", Planet. Space Sci. 23, 151.

Huebner, W.F. (1970), "Dust from Cometary Nuclei", Astron. Astrophys. 5, 286.

Hughes, D.W. (1974), "Meteorites which "bounce" off the Earth", Nature 247, 423.

Humes, D.H., Alvarez, J.M., O'Neal, R.L., and Kinard, W.H. (1974), "The Interplanetary and Near Jupiter Meteoroid Environments", J. Geophys. Res. 25, 3677.

Kaiser, C.B. (1970), "The Thermal Emission of the F-Corona", Astrophys. J. 159, 77.

Kassel, Jr., P.C. (1973), "Characteristics of Capacitor-Type Micrometeoroid Flux Detectors when Impacted with Simulated Micrometeoroids", NASA Technical Report TN D-7359.

Mac Queen, R.M. (1968), "Infrared Observations of the Outer Solar Corona", Astrophys. J. 154, 1059.

Millman, P.M. (1970), "Meteor Showers and Interplanetary Dust", Space Research X, 260.

Millman, P.M., and McIntosh, B.A. (1964), "Meteor Radar Statistics I", Canadian J. Phys. 42, 1730.

Millman, P.M., and McIntosh, B.A. (1966), "Meteor Radar Statistics II", Canadian J. Phys. 44, 1593.

Nagel, K., Fechtig, H., Schneider, E., and Neukum, G. (1976), "Micrometeorite Impact Craters on Skylab Experiment S 149", This Volume.

Nazarova, T., and Rybakov, A. (1974), "The Meteoric Particle Space Density Near the Earth and the Moon, according to Data obtained by Simultaneous Observations of Space Vehicles", Space Research XIV, 773.

Nazarova, T., and Rybakov, A. (1975), "The Meteoric Matter Investigations on Mars-7 and Luna-22 Space Probes", COSPAR-Meeting 1975, Varna/Bulgaria.

Nilsson, C.S. (1966), "Some Doubts about the Earth's Dust Cloud", Science 153, 1242.

Peterson, A.W. (1967), "Multicolor Photometry of the Zodiacal Light", in "The Zodiacal Light and the Interplanetary Medium", ed. J.L. Weinberg, p. 23, NASA SP-150.

Rauser, P. (1974), "Microparticle Detector based on the Energy Gap Disappearance of Semiconductors (Se, I, Te, Bi, Ge, Sn, Si and InSb) at high Pressure", J. Appl. Phys. 45, 4869.

Rawcliffe, R.D., Bartky, C.D., Li, F., Gordon, E., and Carta, D. (1974), "Meteor of August 10, 1972", Nature 247, 449.

Schneider, E. (1975), "Impact Ejecta Exceeding Lunar Escape Velocity", The Moon 13, 173.

Schneider, E., Storzer, D., Hartung, J.B., Fechtig, H., and Gentner, W. (1973), "Microcraters on Apollo 15 and 16 Samples and corresponding Cosmic Dust Fluxes", Proc. Fourth Lunar Sci. Conf., Geochim. Cosmochim. Acta, Suppl. 4, Vol. 3, p. 3277.

Sekanina, Z. (1976), "Modeling of the Orbital Evolution of Vaporizing Dust Particles Near the Sun", This Volume.

Smith, D., Adams, N.G., and Khan, H.A. (1974), "Flux and Composition of Micrometeoroids in the Diameter Range 1-10 μm", Nature 252, 101.

Whipple, F.L. (1976), "Sources of Interplanetary Dust", This Volume.

Wyatt, S.P., and Whipple, F.L. (1950), "The Poynting Robertson Effect on Meteor Orbits", Astrophys. J. __111__, 134.

Yeates, C.M., Nock, K.T., and Newburn, R.L. (1976), "Mariner Mission to Encke 1980", This Volume.

Zook, H.A. (1975), "Hyperbolic Cosmic Dust: Its Origin and its Astrophysical Significance", Planet. Space Sci. __23__, 1391.

Zook, H.A., and Berg, O.E. (1975), "A Source for Hyperbolic Cosmic Dust Particles", Planet. Space Sci. __23__, 183.

2.1.2
Preliminary Results of the Micrometeoroid Experiment on Board Helios A

E. Grün, J. Kissel, H. Fechtig and P. Gammelin
Max-Planck-Institut für Kernphysik, Heidelberg, F.R.G.
and H.-J. Hoffmann
Physikalisches Institut, Rheinische Friedrich-Wilhelms-Universität
Bonn/F.R.G.

Abstract. For the first time in situ measurements of interplanetary dust have been performed between 0.3 AU and 1 AU from the sun by the micrometeoroid experiment on board Helios A. The measured particle masses are between 10^{-15} g and 10^{-8} g and their measured speeds are between 2 km/sec and 20 km/sec. Particle impacts are identified by the time-of-flight spectra of the ions released upon impact. 15 large particles (m $\geq 10^{-12}$ g) were detected from Dec. 15, 1974 to Sept. 5, 1975. They show a strong increase of the impact rate (appr. a factor of 10) between 1 AU and 0.3 AU. The directions from which they impacted the sensor are concentrated between the solar direction and the apex direction of the Helios spacecraft.

The Helios A spacecraft was launched on December 10, 1974 into an eccentric orbit around the sun. Helios A explores the interplanetary space between 0.98 AU and 0.31 AU from the sun. The micrometeoroid experiment (E-10) on board registers individual dust particles crossing the orbit of Helios if they hit the sensitive parts of the sensors. Fig. 1 shows a cross-section of the spacecraft with the mounting of the two sensors and their field of view (one sensor rotated by 90° in azimuth). The spin axis of Helios is perpendicular to its orbit plane which is identical with the ecliptic plane. While the spacecraft spins once a second around its axis the two sensors scan a full circle in azimuth. Two sensors are installed in order to allow a rough determination of the inclination of the particles' orbits within two channels. One sensor (south-sensor) is facing the southern ecliptic hemisphere detecting particles which have elevations from -90° to -10°. The other sensor (ecliptic-sensor) detects particles with elevations from -45° to +55° with respect to the ecliptic plane.

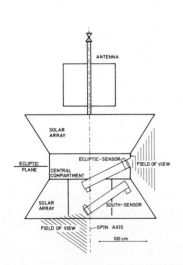

Fig. 1: Helios spacecraft with mounting of the micrometeoroid experiment.

Fig. 2 shows a cross-section of the south-sensor. The sensor consists of the solar wind protection system, the impact ionization detector and the time-of-flight spectrometer. Two small electronic boxes containing preamplifiers and high-voltage power supplies are directly

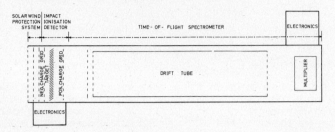

Fig. 2: Helios micrometeoroid experiment, south-sensor.

attached to the sensor. A detailed description has been given by Dietzel et al. (1973). Five quantities are generally measured if a micrometeoroid hits the venetian blind type target: (1) the total negative charge (electrons) and (2) the total positive charge (ions) released upon the impact, (3) the rise-time of the negative charge pulse, (4) the rise-time of the positive charge pulse, and (5) the time-of-flight spectrum of the ions. The instrument is triggered when a signal exceeds the threshold of either the positive or the negative charge channel. With the south-sensor additionally the charge (6) of the dust particles is measured by the charge induced on a grid in front of the target. The ecliptic-sensor is covered by a thin film (3000 Å parylene coated with 750 Å aluminium) as protection against solar radiation. Here the time (7) is measured between the penetration of the film and the impact on the target.

Besides these parameters measured directly from the impact, additional information is gathered and transmitted to earth: (8) various coincidences between the measured signals, enabling one to discriminate between noise and "probable" impacts, (9) the time at which the event occurred, and (10) the pointing direction (azimuth) of the sensor. If a "probable" impact is indicated by proper coincidences, the count on one out of four registers is increased by one, this register is selected according to the amplitude of the positive charge signal (IA). By this method one obtains from the four counters the number of "probable" impacts within 4 positive charge intervals roughly corresponding to 4 different mass intervals of micrometeoroids. All the information on one event is contained in an experiment-data-frame of 256 bits which are transmitted to earth once every 20 sec to 20 min (depending on the Helios-earth distance). To date (September 1975) the experiment worked satisfactorily, except for a total of 26 days during the first three months of the mission when the experiment was blocked in a non-measuring mode.

As only 10 % of the data have arrived at the experimentors' facility at the time of writing, the results have to be regarded as preliminary. The most complete set of data is available for larger particles (m $\geq 10^{-12}$ g at v = 10 km/sec). By evaluating the event-counters for the 3 upper charge intervals it was possible to locate all "probable" impacts. Using special data processing at the German Space Operations Center (GSOC) the complete data of 19 out of 21 "probable" impacts could be picked out of real-time-data. Evaluating these data, it was found that 15 events thereof are micrometeoroid impacts. The real impacts were identified by the presence of a time-of-flight spectrum.

The data cover the time between December 15, 1974 (experiment switch-on) and September 5, 1975. This corresponds to one complete orbit of Helios and the second inward part to 0.4 AU.

The data do not show a significant difference between the impact rate at the ecliptic sensor (6 impacts) and at the south sensor (9 impacts).

Fig. 3 shows the impact rate as detected by the Helios micrometeoroid experiment per 0.1 AU interval as a function of the heliocentric distance R. The error bars indicate the statistical one-σ error and the 0.1 AU interval, respectively. The data are not yet corrected for the instrumental deadtime, which is estimated to be, at maximum, 30 %. The impact rate increases steeply with decreasing heliocentric distance. It is more than a factor 10 greater at 0.3-0.4 AU than at 0.9-1.0 AU. There is qualitative agreement with the result of the zodiacal light experiment on Helios which found an increase of the dust density towards the sun with $R^{-1.3}$ (Link et al., 1976). Since the average impact speed increases with decreasing heliocentric distance,

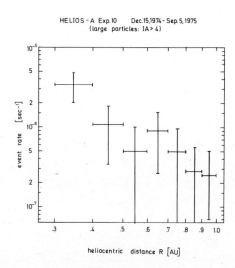

Fig. 3: Radial distribution of the impact rate.

the rate increases faster than the particle number density.

In Fig. 4 the pointing of the sensor at the time of the impact is shown as a function of the heliocentric distance. Each impact is plotted at its heliocentric distance R and the sensor pointing φ. The error bars represent the mean width of the sensor's field of view. Only two impacts were recorded during the outward path of the Helios orbit compared with 13 impacts inward. Since the inbound path of the orbit was almost completed twice and the outbound path only once this ratio is not yet regarded as significant. Most of the impacts were recorded from a direction $-120°$ (or $240°$) $\leq \varphi \leq 30°$ which was also found by the Pioneer 8 and 9 spacecrafts (Berg and Grün, 1973). Because of the high eccentricity of the Helios orbit, micrometeoroids detected at the sensor azimuth $\varphi = 0°$ must not necessarily come from the solar vicinity but also particles on circular orbits may be encountered from this direction. The conclusions drawn from this data are:

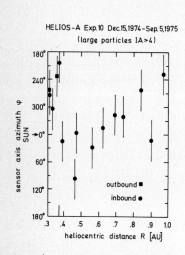

Fig. 4: Azimuth and heliocentric distance of detected impacts.

- 12 of the 15 micrometeoroids detected are compatible with particles in circular orbits. These particles would be encountered during Helios' course inward between $-90° \leq \varphi \leq 90°$ depending on the heliocentric distance R and on the influence of radiation pressure on the particles. During Helios' course outward they would be encountered from $90° \leq \varphi \leq 270°$.

- 11 of the 15 micrometeoroids can also be interpreted as particles in hyperbolic orbits moving away from the sun. They would be detected from $-90° \leq \varphi \leq 0°$ depending on the heliocentric distance, the influence of the radiation pressure and their perihelion distance.

- at least 3 of the 15 micrometeoroids had high eccentric orbits. They were recorded on Helios' course inward at $90° \leq \varphi \leq 270°$.

A more complete discussion of the data, including charge measurements and spectra, is forthcoming.

Acknowledgement

We acknowledge the excellent cooperation with RFE, GfW and GSOC. This project was supported by the Bundesministerium für Forschung und Technologie.

References:

Berg, O.E., and Grün, E. (1973), "Evidence of hyperbolic cosmic dust particles", Space Research XIII, 1047.

Dietzel, H., Eichhorn, G., Fechtig, H., Grün, E., Hoffmann, H.-J., and Kissel, J. (1973), "The HEOS and Helios Micrometeoroid Experiments", J. Phys. E: Scient. Instrum. 6, 209.

Link, H., Leinert, C., Pitz, E., and Salm, N. (1976), "Preliminary Results of the Helios A Zodiacal Light Experiment", This volume.

2.1.3 Composition of Impact-Plasma measured by a HELIOS-Micrometeoroid-Detector

B.-K. Dalmann, E. Grün and J. Kissel

Max-Planck-Institut für Kernphysik, Heidelberg/F.R.G.

The composition of the impact-plasma, produced by dust particles hitting an Au-target was measured, using a model of the HELIOS-micrometeoroid-detector. The 2 MV dust accelerators of the MPI für Kernphysik, Heidelberg, and the NASA Ames Research Center were used to accelerate particles consisting of Al, Al_2O_3, SiO_2, Soda-Lime-Glass, Polystyrene and Kaolin to velocities between 2 km/sec and 15 km/sec. Fe-projectiles could be accelerated up to 40 km/sec. The masses of the dust grains were between 10^{-15} g and 3×10^{-10} g. The experiments showed, that because of the characteristic features of the measured spectra it is possible to separate noise events from impacts even at a high noise background. The smallest particles (m 10^{-15} g) triggering the experiment produce spectra well above the noise level (more than a factor 10) because of the high sensitivity of the ion-detector (multiplier).

Two different types of mass spectra were observed, depending on material properties:

a, <u>Metals and hard dielectrics</u> Fe, Al, Al_2O_3, SiO_2, Soda-Lime-Glass. Particle constituents of low ionisation energy (7 eV, for example Na, K, Al) are dominating in the spectra of those materials at low velocities. Going to higher velocities the relative intensities change and new ions with higher ionisation energy coming from minor constituents of the projectile and target occur (e.g. C^+, H^+).

b, <u>Low hardness dielectrics</u> (Mohs' hardness 3)

The materials Polystyrene and Kaolin produced less total charge than the others did. Most of the plasma ions were molecules of the form $C_nH_m^+$ originating from adsorbed surface layers on the target or on the dust particles. The portion of elements with low ionisation potential (e.g. Li, Na, K, Al etc.) is comparatively small.

Full paper submitted to Planet. Space Sci.

2.1.4 ORBITAL ELEMENTS OF DUST PARTICLES INTERCEPTED BY
PIONEERS 8 AND 9

Henry Wolf
Analytical Mechanics Associates, Inc.
Seabrook, Maryland 20801

John Rhee
Rose-Hulman Institute of Technology
Terre Haute, Indiana 47803

Otto E. Berg
Goddard Space Flight Center
Greenbelt, Maryland 20771

INTRODUCTION

The instrument measuring particle impacts on PIONEERS 8 and 9 has been extensively described in the literature (Reference 1) and will not be repeated here. This paper concerns itself with the analysis of the data obtained.

The measurements available are front and rear film and grid ID numbers, the pulse height and the time of flight (TOF), i.e., the time delay between signals from front and rear sensors.

ANALYSIS

(1) Velocity Computation. The nominal relative velocity is calculated from the "solar aspect" of the satellite and the TOF and the impacted film and grid ID number. If front and rear film and collector ID agree the impact is termed "normal" and if one or both disagree the impact is "inclined". In both cases the nominal relative direction is parallel to the line joining the centers. The magnitude of the velocity is then obtained from the TOF and the known distance between the sensors. The nominal relative velocity vector may thus be obtained. The actual velocity vector may deviate from the nominal by about $\pm 24^\circ$. For impact angles other than nominal, the TOF will lead to different velocity magnitudes. A relative probability value is also attached to each impact value according to a method developed by Dohnanyi (Reference 2). This is a purely geometrical method including consideration of the area presented, corrected for shielding. A properly weighted average of all computed quantities can thus be calculated.

In addition to the uncertainty in impact direction, digitizing the TOF count leads to an uncertainty in the velocity magnitude. The nominal value has been adopted in the set B of trajectories and the low and high extremes in sets A and C, respectively. The relative weights of the three velocities arising were taken as equal in the probability computation.

(2) <u>Mass Computation.</u> The particle mass is calculated from the pulse height analysis (PHA) by the equation:

$$mV^{2.6} = KV_o^{1.6} 10^{c(PHA)},$$

obtained from the instrument calibration. For the nominal case K = .651, its maximum is 1., its minimum is .424. These three values of K (.424, .651 and 1.) are used in conjunction with low, nominal and high velocity values, respectively, in this analysis.

(3) <u>Element Computation.</u> The orbit of small particles is sensitive to radiation pressure and hence for a calculated mass an "effective" density (or a value of $\frac{A}{m}$) has to be assumed. The computations were carried out for six (6) densities, ∞, 8., 3., 1., .3 and .1, g/cc, respectively. The results are presented for an assumed effective density of 3., considered the most probable value. For each of these densities the maximum projected angle in and normal to the orbital plane is subdivided into nine parts, thus computing eighty one sets of elements for each of three velocity magnitudes and each density. The means and extreme values of each element are thus obtained.

Table I gives the nominal elements for the twenty (20) particles, for the nominal (and most probable) density of 3.0 g/cc.

The mean elements are close to the nominal ones given in Case B, Table I and are therefore not quoted separately.

Table I

Event Number	Date	Spacecraft		a	e	i	ω	Ω	Type of Orbit
1	March 11, 1968	8	A	0.864	0.226	0.00	0.00	0.00	
			B	0.830	0.264	0.00	0.00	0.00	Elliptic
			C	0.816	0.301	0.00	0.00	0.00	
2	March 26, 1968	8	A	0.624	0.729	0.00	0.00	0.00	
			B	0.601	0.791	0.00	0.00	0.00	Elliptic
			C	0.585	0.844	0.00	0.00	0.00	
3	April 13, 1968	8	A	0.786	0.850	0.00	0.00	0.00	
			B	0.812	0.905	0.00	0.00	0.00	Elliptic
			C	0.854	0.946	0.00	0.00	0.00	
4	April 15, 1968	8	A	0.561	0.990	180	0.00	0.00	
			B	0.616	0.866	180	0.00	0.00	Elliptic
			C	0.763	0.624	180	0.00	0.00	
5	August 25, 1968	8	A	0.675	0.626	0.00	0.00	0.00	
			B	0.652	0.685	0.00	0.00	0.00	Elliptic
			C	0.633	0.738	0.00	0.00	0.00	

(continued on next page)

Table I (continued)

Event Number	Date	Spacecraft		a	e	i	ω	Ω	Type of Orbit
6	Jan. 24, 1969	8	A	1.010	0.090	4.19	102	88.8	
			B	0.970	0.101	4.53	102	107	Elliptic
			C	0.950	0.118	4.87	102	121	
7	April 19, 1969	8	A	0.979	0.256	5.80	-177	-121	
			B	0.969	0.275	6.34	-177	-123	Elliptic
			C	0.960	0.295	6.87	-177	-125	
8	May 20, 1969	8	A	0.830	0.544	0.00	0.00	0.00	
			B	0.832	0.593	0.00	0.00	0.00	Elliptic
			C	0.830	0.639	0.00	0.00	0.00	
9	Feb. 8, 1969	9	A	-2.010	1.370	0.00	0.00	0.00	
			B	-1.330	1.540	0.00	0.00	0.00	Hyperbolic
			C	-0.984	1.710	0.00	0.00	0.00	
10	Oct. 11, 1969	9	A	0.541	0.999	0.00*	0.00	0.00	
			B	0.602	0.966	180	0.00	0.00	Elliptic
			C	0.757	0.866	180	0.00	0.00	
11	Dec. 15, 1969	9	A	-3.240	1.170	180	0.00	0.00	
			B	-0.135	6.090	180	0.00	0.00	Hyperbolic
			C	-0.051	15.10	180	0.00	0.00	
12	March 17, 1970	9	A	1.130	0.593	0.00	0.00	0.00	
			B	1.160	0.645	0.00	0.00	0.00	Elliptic
			C	1.220	0.695	0.00	0.00	0.00	
13	March 13, 1970	8	A	0.555	0.993	180	0.00	0.00	
			B	0.608	0.875	180	0.00	0.00	Elliptic
			C	0.748	0.840	180	0.00	0.00	
14	April 24, 1970	8	A	0.962	0.801	77.9	166	-147	
			B	1.620	0.827	99.6	166	-128	Elliptic
			C	5.510	0.992	113	166	-102	
15	June 11, 1970	8	A	0.545	0.993	0.00*	0.00	0.00	
			B	0.545	0.827	180	0.00	0.00	Elliptic
			C	0.553	0.993	180	0.00	0.00	
16	July 8, 1970	8	A	7.690	0.994	0.00	0.00	0.00	
			B	-2.110	1.000	0.00	0.00	0.00	Intermediate
			C	-0.825	1.000	180*	0.00	0.00	
17	Nov. 11, 1970	8	A	0.719	0.861	0.00	0.00	0.00	
			B	0.720	0.927	0.00	0.00	0.00	Elliptic
			C	0.735	0.970	0.00	0.00	0.00	
18	Nov. 12, 1970	8	A	-7.780	1.08	125	-16.1	-76.2	
			B	-0.540	2.42	133	-16.1	-51.0	Hyperbolic
			C	-0.241	4.40	138	-16.1	-42.0	
19	Feb. 23, 1970	8	A	0.744	0.519	26.7	86.1	151	
			B	0.712	0.586	31.2	86.1	155	Elliptic
			C	0.690	0.643	36.1	86.1	157	

(continued on next page)

Table I (continued)

Event Number	Date	Spacecraft	a	e	i	ω	Ω	Type of Orbit
20	Jan. 23, 1969	9	A -0.348	2.010	180	0.00	0.00	
			B -0.087	7.410	180	0.00	0.00	Hyperbolic
			C -0.039	17.10	180	0.00	0.00	

Here a is the semi-major axis (AU), e is eccentricity, i is inclination (in degrees), ω is the argument of perihelion (in degrees) and Ω is the longitude of ascending node (in degrees). The deviating inclinations denoted by "*" arise from extreme possible velocity deviations and are of extremely low probability (less than 10^{-3}).

Table II presents a classification of particles as elliptic or hyperbolic together with their relative probability for three (3) densities.

Table II

Part.	$\rho = 3.0$				$\rho = 8.0$				$\rho = 1.0$					
	Number		Probability		Number		Probability		Number			Probability		
	Ell	Hyp	Ell	Hyp	Ell	Hyp	Ell	Hyp	Ell	Hyp	HR	Ell	Hyp	HR
1	243	0	1.	0.					243	0		1.	0.	
2	243	0	1.	0.					243	0		1.	0.	
3	241	2	1.	0.					219	24		1.	0.	
4	235	8	1.	0.					180	63		.94	.06	
5	243	0	1.	0.					243	0		1.	0.	
6	243	0	1.	0.					243	0		1.	0.	
7	243	0	1.	0.					243	0		1.	0.	
8	243	0	1.	0.					243	0		1.	0.	
9	19	224	.02	.98	51	192	.11	.89	0	243		0.	1.	
10	230	13	1.	0.					175	68		.92	.08	
11	0	243	0.	1.	21	222	.21	.79	0	90	153*	0.	.45	.55*
12	239	4	1.	0.					67	176		.45	.55	
13	237	6	1.	0.					186	57		.95	.05	
14	144	99	.66	.34	171	72	.79	.21	69	174		.26	.74	
15	243	0	1.	0.					241	2		1.	0.	
16	52	191	.24	.76	77	166	.38	.62	0	243		0.	1.	
17	237	6	1.	0.					130	113		.69	.31	
18	27	213	.09	.91	40	203	.20	.80	1	242		.005	.995	

(continued on next page)

Table II (continued)

Part.	$\rho = 3.0$				$\rho = 8.0$				$\rho = 1.0$					
	Number		Probability		Number		Probability		Number			Probability		
	Ell	Hyp	Ell	Hyp	Ell	Hyp	Ell	Hyp	Ell	Hyp	HR	Ell	Hyp	HR
19	240	3	1.	0.					97	133	13*	.41	.59	0.*
20	0	243	0.	1.	0	243	0.	1.	0	90	153*	0.	45.	.55*

The last column headed "HR" stands for hyperbolic repulsive, i.e., cases where the radiation pressure exceeds the gravitational attraction.

CONCLUSIONS

(1) The statistical analysis confirms that the character of the nominal trajectory is essentially correct. The most probable elements are very close to the nominal ones (Case B, nominal velocity).

(2) The elliptic or hyperbolic character of most orbits is not usually affected by reasonable density assumptions (between 8. and 1.).

(3) Most particles are elliptic but particles 9, 11, 18 are most probably hyperbolic. Particle 20 is and remains hyperbolic under all reasonable assumptions and without resort to statistical or probabilistic arguments.

(4) The incoming asymptote of the hyperbolic orbits is consistent with the particles arriving from the apex of the solar motion.

(5) The perihelion of particle 20 is about .5 AU. This precludes evaporative and indicates interstellar origin.

REFERENCES

(1) Berg, O. E. and Richardson, F. F., "The Pioneer * Cosmic Dust Experiment", Rev. Sci. Inst. 40, October 1969.

(2) Private communication to O. E. Berg and H. Wolf.

2.1.5   FLUX OF HYPERBOLIC METEOROIDS

J.S. Dohnanyi [)]
Max-Planck-Institut für Kernphysik, Heidelberg/F.R.G.

Abstract

The production of hyperbolic meteoroids by inelastic collisions between meteoroids is estimated. It is found that, under reasonable assumptions, the calculated flux of hyperbolic meteoroids agrees with satellite data and with lunar microcrater distributions.

We have therefore obtained independent theoretical support for Zook and Berg's (1975) ß-meteoroid hypothesis and for Fechtig et al. (1974) suggestion that submicron lunar microcraters are produced by ß-meteoroids.

I. Introduction

Recent measurements (cf. Fechtig, 1976, for a review) by the Pioneer 8 and 9 satellites led Berg and Grün (1973) to conclude that a substantial flux of micrometeoroids in hyperbolic orbits originate in the region of space between the sun and earth's orbit (also cf. McDonnell, Berg and Richardson, 1975 and Grün, Berg and Dohnanyi, 1973). Zook and Berg (1975) and Zook (1975) have discussed the origin of these particles which they named ß-meteoroids. Whipple (1976) has discussed the occurrence of these particles and given further support to the hypothesis that these particles are being expelled from the solar system by radiation pressure.

We shall, in this paper, quantitatively examine the dynamics of the population of these particles as they are produced by inelastic collisions between larger "parent" particles. It will be found that with reasonable assumptions, Zook and Berg's (1975) hypothesis is in good quantitative agreement with predictions from an analysis of meteoroidal collisions.

II. ß-Meteoroids

When two meteoroids, in elliptic orbits, collide with each other, a number of fragments will be produced. Many of these fragments will be so small that the force of radiation pressure will be significant compared with the gravitational attraction of the sun. The particle velocity will, in almost every case, be comparable to the velocity of the larger parent object (cf. Gault et al., 1963 and Eichhorn, 1976,

[)] On leave from Bell Laboratories, Holmdel, N.J./USA

for a discussion of the ejecta velocity distribution). Diminishing the effective attractive force of the sun by the repulsive radiation force will also diminish the magnitude of the solar escape velocity of the effected particles. In the extreme case, for example, when the repulsive force of radiation pressure equals the attractive force of gravity, no net force at all is acting on the particle and, in that case, even the slightest speed relative the sun would cause it to escape from the solar system. It then follows that, for sufficiently small fragments, the velocity of the parent object exceeds the solar escape velocity of the fragments and the latter will be expelled from the solar system (cf. Dohnanyi, 1970, 1972). This process has originally been suggested by Harwit (1963) and the resulting fragments in hyperbolic orbits have been named ß-meteoroids by Zook and Berg (1975).

III. Distribution of Parent Objects

In order to estimate the flux of ß-meteoroids, it is necessary to estimate the space density and velocity distribution of the parent objects.

Inclinations: The distribution in inclinations of meteoroids will be taken to be similar to that of photographic meteors reduced by McCrosby and Posen (1961).

The following rough approximation to the density of inclination will be adequate for our purposes:

(1) $f_i(i)di = 2.36 \exp[-7.2\, i/\pi]di, \quad 0 \leq i \leq \pi/2$
       $= 0 \quad , \quad i \geq \pi/2$

where $f_i(i)di$ is the relative number of meteoroids having an inclination in the range of i to $i + di$ radians; $f_i(i)$ is normalized to 1.

Impact Speed: The relative speed V of a particle having a heliocentric velocity V relative to another particle with a heliocentric velocity W is

(2) $v^2 = |\vec{U} - \vec{W}|^2 = |\vec{U}|^2 + |\vec{W}|^2 - 2\,\vec{U}\cdot\vec{W}$

where the last term on the right-hand side of eq. 2 is the ordinary inner product. For sun axes we shall use the radial, transverse and z-coordinate axes where the z-direction is perpendicular to the ecliptic. Using well-known relationships (cf. e.g. Handbook of Chemistry and Physics), v^2 can be expressed in terms of the orbital elements of the colliding particles and the distance from the sun.

We then assume that most collisions between 1 AU and the sun occur near perihelion and expand v^2 in the neighbourhood of the particles' perihelion passage. The resultant formula is then weighted and averaged using the McCrosky and Posen (1961) meteors. The result for the radial dependence for the root mean square (RMS) value of relative velocity is

$$(3) \qquad \sqrt{\langle v^2 \rangle} = V_0 \, r^{-.559}$$

where V_0 is the RMS value of the relative velocity at 1 AU from the sun and r is the distance from the sun in AU.

Spacial Distribution: In order to estimate the number density of particles as a function of distance from the sun we shall assume that

$$(4) \qquad f(m, r)dm = f(m) \, dm \, r^{-b} \qquad R_0 \leq r$$
$$ = 0 \qquad\qquad\qquad r \leq R_0$$

where $f(m, r)dm$ is the number density of particles in the mass range m to m + dm, at a distance r AU from the sun, b is a parameter and R_0 is the cut-off distance from the sun; within a distance R_0 from the sun all particles are assumed vaporized or otherwise destroyed by heat so that the meteoroid population is taken to be negligible in that region.

For f(m)dm, we take (Dohnanyi, 1973)

$$(5) \qquad f(m)dm = A_0 \, m^{-11/6} \, dm \qquad m \quad 10^{-10} \, kg$$
$$ = A_1 \, m^{-3/2} \, dm \qquad m \quad 10^{-10} \, kg$$

where

$$(6) \qquad A_0 = 1.36 \times 10^{-15}$$
$$ A_1 = 10^{20/3} \, A_0.$$

IV. Collision Dynamics

In order to estimate the number density of fragments produced during collisions, we shall use a method based on experimental results by Gault et al. (1963) and discussed earlier by Dohnanyi (1969). Accordingly, we take for the ejecta spectrum

$$(7) \qquad g_M(m)dm = H(M) m^{-\eta} dm$$

where $g_M(m)dm$ is the number density of fragments produced when an object having a mass M is catastrophically disrupted. H(M) is obtained from the conservation of mass requirement:

(8) $\quad M = H(M) \int_{\mu'}^{M_b} m^{-\eta + 1} \, dm$

where M_b is the mass of the largest and μ' is that of the smallest fragment and η is a constant. Eq. 8 can then be solved for $H(M)$ in terms of M, M_b, μ' and η.

Following Dohnanyi (1969,10,12) we take

(9) $\quad M = \Lambda M$

where

(10) $\quad \Lambda = .5 \, V^2$

where the collision speed V is in km/sec.

The mass of the smallest projectile, M_p, capable of catastrophically disrupting the target mass M is taken to be (Dohnanyi, 1970)

(11) $\quad M_p = M/\gamma$

where

(12) $\quad \gamma = 250 \, V^2$, where V is in km/sec.

With this notation, the production rate (per second and per cubic meter) of fragments in the mass range of m to m + dm, due to catastrophic collisions between larger masses (M and M_2) is:

(13) $\quad g(m, r)dm = K(2-\eta) \, m^{-\eta} \Lambda^{\eta-2} \, dm \, \cdot$

$\cdot \int_{m/\Lambda}^{M_\infty/\gamma} dM \int_{M}^{\gamma M} dM_2 (M+M_2)(M^{1/3}+M_2^{1/3})^2 M^{\eta-2} f(M) f(M_2) r^{-2b}$

where $f(M)$ is given by Eq. 4, 5 and 6 and where (in standard units)

(14) $\quad K = (3 \, \sqrt{\pi}/4\rho)^{2/3} \, V$

where ρ is the average material density of the colliding objects and V is the average impact speed. For fragment mass ranges of our interest, the contribution to $g(m)dm$ of erosive collisions can be shown to be minor (cf. Dohnanyi, 1969, 1970) for distributions of parent objects given by Eq. 5.

When the integral in Eq. 13 is properly evaluated, and only the dominating terms retained, we obtain for the number of fragments produced per unit volume and unit time in the mass range m to m+dm and a distance r from the sun,

(15) $\quad g(m, r)dm \simeq K(2-\eta)\Lambda^{\eta-2} A_o^2 \mu^{\eta-4/3} m^{-\eta} dm \times (r^{-2b})$

where $\cdot \left\{ \gamma^{2.5-\eta}[\frac{-1.14}{2.5-\eta} + \frac{2}{2-\eta} + \frac{6/7}{\eta-4/3}] - \frac{(6/7)\gamma^{1/6}}{\eta-4/3}(\frac{m}{\mu\Lambda})^{\eta-4/3} \right\}$

(16) $\quad \mu = 10^{-10}$ kg

is the mass corresponding to the point where a break occurs on a double logarithmic plot of the mass-flux curve (cf. Eq. 5). The dominating contribution to this expression is the production of fragments when a projectile with a mass smaller than 10^{-10} kg catastrophically collides with a target object having a mass greater than 10^{-10} kg.

V. Flux of Fragments

When the fragment production per unit volume is integrated over a sphere of radius r with the sun at the centre, we obtain the total number of these particles, produced in this volume, every second. When these fragments are ß-meteoroids, then division of this production rate by a surface element at r would also give the flux of these ß-meteoroids through the surface element.

It was shown (Dohnanyi, 1970, 1972) that fragments having a mass of about 10^{-9} kg will become ß-meteoroids if the orbit of the parent body is highly eccentric (e > .9). For parent orbits with smaller eccentricities the limiting fragment size for hyperbolic orbits (ß-meteoroids) becomes smaller; for parent objects in circular orbits the maximum fragment mass of the ß-meteoroids is about 10^{-14} kg for a material density of 10^3 kg/m^3 and 10^{-15} kg for a material density of 3×10^3 kg/m^3. We may therefore assume that most fragments having a mass smaller than 10^{-15} kg will become ß-meteoroids. We shall, in what follows, assume that all fragments under consideration are ß-meteoroids and estimate their flux using our model.

We now proceed to calculate the flux of fragments. We use the expression g(m, r)dm, Eq. 15, for the production rate, per unit volume, of ß-meteoroids in the mass range m to m+dm at a distance r from the sun. We substitute the expression Eq. 14 for K, expression Eq. 12 for γ, expression Eq. 10 for Λ and thus we obtain the explicit functional dependence of g(m, r)dm on the average impact velocity, V. We now employ the relationship between V and r: we estimate V with the formula Eq. 3, i.e. we let

(17) $\quad V \approx \sqrt{\langle V^2 \rangle} = V_o r^{-.559}$

as given by Eq. 3 and hence obtain the explicit dependence of $g(m,r)dm$ on the distance from the sun, r.

The total number of fragments produced per second in a spherical volume with radius r and in a mass range m to $m+dm$ is

(18) $$h(m,r)dm = dm \int_{R_o}^{r} dR \int_{o}^{\pi} di \int_{o}^{2\pi} d\phi \, g(m,R) \, R^2 \, f_i(i) \sin i \, di$$

where R_o is defined by Eq. 4.

The flux of particle fragments in the mass range dm, at a distance r from the sun and per unit area in the ecliptic ($i = 0$) is then

(19) $$g(m,r)dm = \frac{h(m,r)dm}{4\pi r^2} \, f_i(o)$$

where $f_i(i)$ is defined in Eq. 1.

The cumulative flux G of these particles having a mass m or greater is then readily obtained

(20) $$G(M,r) = \int_{m}^{\infty} g(M,r)dm$$

The result for the flux at 1 AU and $= 5/3$ is:

(21) $$G(m, 1\,AU) = \frac{2 \times 10^{-18} \, V_o^2}{1.882 - 2b} \left[1 - \left(\frac{R_o}{AU}\right)^{1.882-2b} \right] m^{-2/3}, \text{ M.K.S.}$$

where we have expressed the cumulative flux of fragments per m^2 sec having a mass of m kg or greater as a function of the parameters V_o, R_o and b. This expression is also an upper limit for the production of ß-meteoroids because this flux also includes the contribution to the flux by meteoroids in bound orbits.

In Table 1 we list the numerical value of $G(m, 1AU)$ for $m = 10^{-15}$ kg and for a number of values of the parameters R_o, b and V_o. It can be seen, from Table 1, that the largest fluxes of ß-meteoroids are obtained when b is large and R_o small. This happens because for large b the concentration of particles near R_o is large compared with smaller values of b and furthermore, the closer we are to the sun, the faster are the collision speeds and hence, the more destructive are the collisions.

The expression for the flux of ß-meteoroids Eq. 21 can naturally be evaluated for a large number of other reasonable values of the various parameters. Table 1, however, provides an adequate indication of the

Table 1. Values of the flux of fragments $G(m, 1\ AU)$ for $m = 10^{-15}$ kg in units of particles per (meter^2 sec 2π sterad)

$R_o = .02$

b \ V_o	10	20	40
0	5.5×10^{-7}	2.2×10^{-6}	8.7×10^{-6}
1.0	5.1×10^{-6}	2.0×10^{-5}	8.2×10^{-5}
2.0	1.9×10^{-3}	7.7×10^{-3}	3.1×10^{-1}

$R_o = .04$

b \ V_o	10	20	40
0	5.5×10^{-7}	2.2×10^{-6}	8.7×10^{-6}
1.0	4.0×10^{-6}	1.6×10^{-5}	6.4×10^{-5}
2.0	4.5×10^{-4}	1.7×10^{-3}	7.1×10^{-3}

$R_o = .1$

b \ V_o	10	20	40
0	5.4×10^{-7}	2.1×10^{-6}	8.6×10^{-6}
1.0	3.9×10^{-6}	1.5×10^{-5}	6.2×10^{-5}
2.0	6.3×10^{-5}	2.5×10^{-4}	1.0×10^{-3}

sensitivity of the flux to the values of the parameters used. We shall employ, in what follows, a value of $\eta = 5/3$ because it is within the range of values obtained experimentally and because it will provide a slightly better fit to the lunar data than the slightly higher value of $\eta = 1.8$.

VI. Distribution of Lunar Microcraters

We shall now attempt to express lunar microcrater frequency data as a function of the mass of the projectile objects and compare it with our results for the flux of ß-meteoroids Eq. 21. This will then enable us to verify, in some detail, the suggestion of Fechtig et al. (1974) that lunar craters smaller than 1 micron are produced, mainly, by ß-meteoroids.

We use the results of Fechtig et al. (1975) for the composite relative crater cumulation frequency, Fig. 1. Using Mandeville and Vedder's (1971) calibration, (also cf. Fechtig et al., 1974) we convert crater sizes to projectile masses using the relation

$$(22) \quad M = 1.88 \times 10^4 \, D^3/V^2$$

where M is the mass of the projectile, in kg, D is the crater diameter in meters and V is the impact speed in km/sec.

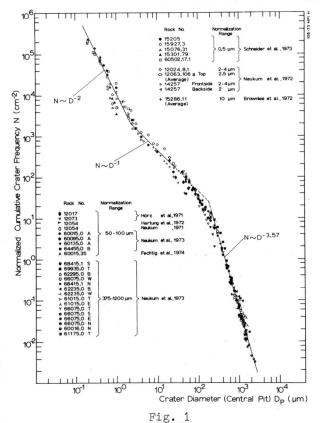

Fig. 1

Compilation of microcrater measurements from Fechtig et al. (1975). Frequencies are normalized; absolute frequency contains an arbitrary factor.

Taking an impact speed of 20 km/sec in Eq. 22 we can compute the projectile masses that correspond to given crater diameters and we identify the point, (at a crater diameter of 173 microns) in Fig. 1, with the similar break in the meteoroid flux curve at 10^{-10} kg; (cf. Eq. 5) we can then compare the lunar microcrater data with satellite flux data and with the results of our calculation. The result is Fig. 2 where we have plotted the model incoming flux, the cumulative frequency of lunar microcrater producing particles and our calculated results for the flux of ß-meteoroids.

It can be seen from Fig. 2 that reasonable values for the parameters in Eq. 21 will provide estimates of ß-meteoroids in good agreement with satellite data and with the flux curve derived from lunar microcraters. We have therefore provided independent theoretical support for Fechtig et al. (1974) suggestion.

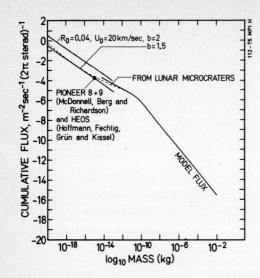

Fig. 2

Comparison of calculated and observed microparticle fluxes. Solid lines represent the calculated flux frequencies for the indicated values of the parameters R_o U_o and b (see text).

VII. Conclusion

The present study shows that using reasonable assumptions regarding the values of the physical parameters employed, disruptive collisions between meteoroids accompanied by fragmentation will produce a flux of ß-meteoroids in agreement with satellite and lunar microcrater data. We have therefore provided independent theoretical support for Zook and Berg's (1975) hypothesis of ß-meteoroids and to Fechtig et al. (1974) suggestion that lunar microcraters smaller than 1 micron are produced mostly by ß-meteoroids.

VIII. Acknowledgements

I am indebted to O.E. Berg, H. Fechtig and H.A. Zook for important suggestions and discussions.

I gratefully acknowledge that my stay at Heidelberg has been made possible by an Alexander von Humboldt Senior US Scientist Award.

References:

Berg, O.E., and Grün, E. (1973), "Evidence of hyperbolic cosmic dust particles", Space Research XIII, pp. 1046-1055.

Brownlee, D.E., Hörz, F., Hartung, J.B., and Gault, D.E. (1972), "Micrometeoroid craters smaller than 100 microns", in: The Apollo 15 Lunar Samples, pp. 407-409. The Lunar Science Institute, Houston.

Dohnanyi, J.S. (1969), "Collisional model of asteroids and their debris", J. Geophys. Res. 74, 2531-2554.

Dohnanyi, J.S. (1970), "On the origin and distribution of meteoroids", J. Geophys. Res. 75, 3468-3493.

Dohnanyi, J.S. (1972), "Interplanetary objects in review: statistics of their masses and dynamics", Icarus 17, 1-48.

Dohnanyi, J.S. (1973), "Current evolution of meteoroids", Proc. of the IAU Colloquium No. 13, NASA SP-319, pp. 363-374.

Eichhorn, G. (1976), "Impact light flash studies: Temperature, ejecta and vaporization", this Volume.

Fechtig, H. (1976), "In situ records of interplanetary dust particles - methods and results", this Volume.

Fechtig, H., Gentner, W., Hartung, J.B., Nagel, K., Neukum, G., Schneider, E., and Storzer, D. (1975), "Microcraters on lunar samples", Proc. of the Soviet-American Conference on Cosmochemistry, Pergamon Press (in press).

Fechtig, H., Hartung, J.B., Nagel, K., Neukum, G., and Storzer, D. (1974), "Microcrater Studies, Derived Meteoroid Fluxes and Comparison with Satellite-Borne Experiments", Lunar Science V, Abstract Vol. pp. 22-224 Proc. Fifth Lunar Sci. Conf. Geochim. Cosmochim. Acta Suppl. 5, Vol. 3, 4, pp. 2463-2474.

Gault, D.E., Shoemaker, E.M., and Moore, H.J. (1963), "Spray ejected from the lunar surface by meteoroid impact", NASA Rept. TND-1767.

Grün, E., Berg, O.E., and Dohnanyi, J.S. (1973), "Reliability of cosmic dust data from Pioneer 8 and 9", Space Research XIII, pp. 1057-1062.

Hartung, J.B., Hörz, F., and Gault, D.E. (1972), "Lunar microcraters and interplanetary dust", Proc. Third Lunar Sci. Conf., Geochim. Cosmochim. Acta, Suppl. 3, Vol. 3, pp. 2735-2753. MIT Press.

Harwit, M. (1963), "Origins of the zodiacal dust cloud", J. Geophys. Res. 68, 2171-2180.

Hörz, F., Hartung, J.B., and Gault, D.E. (1971), "Micrometeorite craters and lunar rock surfaces", J. Geophys. Res. 76, 5770-5798.

Hoffmann, H.J., Fechtig, H., Grün, E., and Kissel, J. (1975), "First results of the micrometeoroid experiment S-215 on the HEOS 2 satellite", Planet. Space Sci. 23, 215-224.

Mandeville, J.C., and Vedder, J.F. (1971), "Microcraters formed in glass by low density projectiles", Earth Planet. Sci. Letters 11, 297.

McCrosky, R.E., and Posen, A. (1961), "Optical elements of photographic meteors", Smithson. Contrib. Astrophys. 4, 15-84.

McDonnell, J.A.M., Berg, O.E., and Richardson, F.F. (1975), "Spatial and time variations of the interplanetary microparticle flux analysed from deep space probes Pioneers 8 and 9", Planet. Space Sci. 23, 205-214.

Neukum, G. (1971), "Untersuchungen über Einschlagskrater auf dem Mond", Ph.D. Thesis, Universität Heidelberg.

Neukum, G., Schneider, E., Mehl, A., Storzer, D., Wagner, G.A., Fechtig, H., and Bloch, M.R. (1972), "Lunar craters and exposure ages derived from crater statistics and solar flare tracks", Proc. Third Lunar Sci. Conf., Geochim. Cosmochim. Acta, Suppl. 3, Vol. 3, pp. 2793-2810. MIT Press.

Schneider, E., Storzer, D., Hartung, J.B., Fechtig, H., and Gentner, W. (1973), "Microcraters on Apollo 15 and 16 samples and corresponding Cosmic Dust Fluxes", Proc. Fourth Lunar Sci. Conf., Geochim. Cosmochim. Acta, Suppl. 4, Vol. 3, pp. 3277-3290.

Southworth, R.B., and Sekanina, Z. (1973), "Physical and dynamical studies of meteor", NASA CR-2316.

Whipple, F.L. (1976), "Sources of interplanetary dust", this Volume.

Zook, H.A. (1975), "Hyperbolic cosmic dust: its origin and its astrophysical significance", Planet. Space Sci. 23, 1391-1397.

Zook, H.A., and Berg, O.E. (1975), "A source for hyperbolic cosmic dust particles", Planet. Space Sci. 23, 183-203.

2.1.6   THE COSMIC DUST ENVIRONMENT AT EARTH,
JUPITER AND INTERPLANETARY SPACE:
RESULTS FROM LANGLEY EXPERIMENTS ON MTS, PIONEER 10, AND PIONEER 11
by
J.M. Alvarez, Langley Research Center, Hampton, Virginia 23665, USA

Langley Research Center cosmic dust experiments aboard the Earth orbiting Meteoroid Technology Satellite (MTS) and the Pioneer 10 and 11 Jupiter fly-by spacecraft are described and discussed.

Data from the cosmic dust detectors aboard the MTS indicate that:
(a) the average impact flux for -10^{-17} gram particles is $4 \times 10^{-4} m^{-2} sec^{-1}$,
(b) the average impact flux for -10^{-15} gram particles is $2 \times 10^{-4} m^{-2} sec^{-1}$,
(c) the average impact flux for -10^{-17} gram particles was much greater at the beginning of the mission. These data show that 10^{-17} gram particles do exist near Earth and that the particle number distribution function is rather flat from -10^{-17} grams to -10^{-8} grams. The much higher impact flux at the beginning of the mission suggests the spacecraft itself as a source of orbital debris in rapidly decaying orbits.

The Pioneer 10 and 11 meteoroid penetration experiments show a gap starting at 1.16 AU in the interplanetary cosmic dust environment. There was no increase in particle concentration (particles per cubic meter) detected in the region of the asteroid belt by either Pioneer 10 or 11. The large increase in penetration rate near Jupiter is due to gravitational focussing of cosmic dust in Jupiter's vicinity.

Full paper to be published in "Science".

2.1.7   DUST IN THE OUTER SOLAR SYSTEM - REVIEW OF EARLY
                   RESULTS FROM PIONEERS 10 AND 11

            R. K. Soberman[1], J. M. Alvarez[2], and J. L. Weinberg[3]

Abstract. The Pioneer 10/11 spacecraft, launched in 1972 and 1973, carried three experiments to measure cosmic dust. A comparison of these first direct measurements of dust in the outer solar system indicates that the sizes, optical properties, and spatial distribution are more complex than previously supposed.

Three interplanetary dust detectors were carried on Pioneers 10 and 11: the Imaging Photopolarimeter (IPP) in the Sky Mapping Mode, the penetration detectors of the Meteoroid Detection Experiment (MDE), and the Asteroid Meteoroid Detector (AMD). Table 1 summarizes for each instrument the measured parameters, the particle size range, and various assumptions used to derive the properties and spatial distribution of the particles. The question marks added to the size range of the zodiacal light detectors are discussed later. In the analysis of the MDE and AMD data, it was necessary to assume relative encounter velocities. From the penetration data it was concluded that the particles have circular or near-circular orbital velocities. For the AMD this was a starting assumption.

The penetration detectors indicate a constant spatial concentration with heliocentric distance, with no apparent indication of asteroid belt passage (Humes, et al., 1975). Early results from both the IPP (Hanner, et al., 1974) and the AMD zodiacal light mode (Zook and Soberman, 1974) have shown that the zodiacal light brightness decreases monotonically with increasing heliocentric distance. The IPP results indicate that the zodiacal light initially decreases faster than the inverse square of the heliocentric distance, R, then more rapidly in the asteroid

---
[1] General Electric Space Sciences Laboratory, Philadelphia, Pennsylvania, USA

[2] Langley Research Center, National Aeronautics and Space Administration, Hampton, Virginia, USA

[3] Space Astronomy Laboratory, State University of New York at Albany, Albany, New York, USA

TABLE I
COMPARISON OF PIONEER 10/II DUST EXPERIMENTS

Experiment	Measurement	Particle Diameter Range	Assumptions	Derived Results
MDE Penetration Detectors	Penetration Rate of Stainless Steel 25 μm 50 μm	~10 μm ~20 μm	Distribution of Orbital Parameters for Relative Velocity	Spatial concentration
IPP Zodiacal Light Mode	Polarization & Brightness in 2 Colors	Micron and/or Sub-micron ?	Mie Theory - Constant Size Distribution	Spatial distribution Size Shape Refractive index
AMD Zodiacal Light Mode	Brightness	Micron and/or Sub-micron-?	Mie Theory - Constant Size Distribution	Spatial distribution
AMD Individual Particle Mode	Peak Intensity Transit Time	50 μm and Larger	Circular Orbit Encounter Vel. - Average Transit Thru View Cone - Diffuse Geometrical Reflection From Spherical Particles	Size distribution Spatial concentration Zodiacal light brightness

belt, with no measurable contribution beyond 3.3 AU (Hanner et al., 1976). Based on the assumption that the scattering properties do not change significantly with heliocentric distance, these results suggest that the spatial distribution can be represented by a power law, $R^{-\gamma}$ ($\gamma \approx 1$) or by a two-component model ($\gamma \approx 1.5$) with increased dust in the asteroid belt.

The discrete particle results from the Pioneer 10 AMD (Soberman, et al., 1974) show an increase in the number of particles out to the asteroid belt. There appear to be minima in the vicinity of both the Earth's and Mars' orbits which are more pronounced for the larger particles. Beyond 3.5 AU the event rate drops below instrumental limits, the fall-off occurring first for the larger particles. The size distribution differs significantly from the 1 AU model for the larger sizes and is of the type expected for an asteroidal population (Dohnanyi, 1969). Particle sizes were obtained by assuming a value of 0.2 for the albedo in order to extrapolate to the penetration detector results.

Figure 1 shows the relative change in zodiacal light brightness with heliocentric distance as measured by the IPP and as derived from the AMD discrete particle mode. Although the relative brightnesses are in satisfactory agreement, the absolute brightness derived from the AMD results is more than an order of magnitude too large by comparison with the photometric results from the same instrument and from the IPP. For example, the AMD gives a gegenschein brightness at 1 AU of approximately 2500 S_{10} (V)[*]. This difference is believed to arise from the fact that the AMD measures peak rather than average values for particle brightness when operating in the discrete particle mode, and that the particles contain many reflecting surfaces that give off bright glints of light as the particles rotate (such as observed from sunlit particles in the vicinity of Earth-orbiting vehicles). Because of this glint effect, the planned orbital measurements could not be made with the AMD, and it was necessary to assume particle velocities relative to the instrument to derive sizes and heliocentric variations.

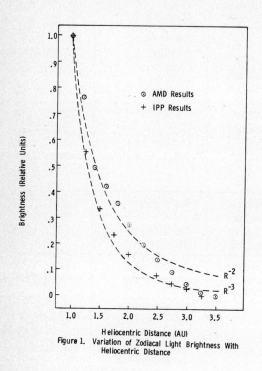

Figure 1. Variation of Zodiacal Light Brightness With Heliocentric Distance

The results from the three dust experiments on Pioneers 10 and 11 seem to be completely discordant. The zodiacal light results indicate that the concentration of dust decreases initially at least as fast as the inverse heliocentric distance and then more rapidly while passing through the asteroid belt. The penetration detectors indicate a uniform spatial concentration with the exception of the gap regions (Humes, et al., 1975). The discrete particle results of the AMD indicate a varying concentration going outward, peaking in the asteroid belt and then dropping off to a negligible value at approximately 3.5 AU. The simplest explanation for this

---

[*]Equivalent number of tenth magnitude (V) stars of solar spectral type, per square degree.

divergence would be that the three sensors were measuring in three different size domains as was indicated in Table 1. This simple explanation cannot be ruled out, although it is not likely that the two extreme sizes are similar in concentration but different from the concentration of the intermediate sizes.

A further question is whether micron or submicron particles contribute appreciably to the zodiacal light. If the concentration of 10 and 20 micron particles measured by the penetration detectors does not change with heliocentric distance, then these particles probably do not contribute significantly to the zodiacal light. Comparing the cross-sections (assuming that the albedo is not a strong function of size), the concentration of one micron particles would have to be two orders of magnitude higher than the concentration derived from the penetration detector results to yield even an equal brightness contribution. Such a concentration of one micron particles is not consistent with the results from Pioneers 8 and 9 (Berg and Grün, 1973), MTS (Alvarez, 1976), HEOS 2 (Hoffmann et al., 1975), and the Lunar Cratering Results (Neukum, 1974). The situation becomes worse if one relies on submicron particles. If we are to believe that the zodiacal light is produced primarily by large particles (radius > 50 microns) additional theoretical calculations and laboratory measurements are required to demonstrate that these particles can produce the observed distribution of polarization with elongation, including polarization reversal.

The zodiacal light and individual particle brightness results from Pioneer 10 suggest the presence of a dust component in the asteroid belt and a negligible concentration beyond. The penetration results show a nearly uniform concentration with no measurable contribution from the asteroid belt and no measurable decrease in concentration beyond. These results suggest different sources for the particles responsible for the penetrations and those which give rise to the individual and aggregate brightness measurements. To explain these differences, additional studies of the sources and sinks for the interplanetary dust beyond 1 AU appear warranted.

Acknowledgements

The Pioneer 10 and 11 Meteoroid Detection Experiment was directly supported by the National Aeronautics and Space Administration at the Langley Research Center; the Imaging Photopolarimetry Zodiacal Light Experiment was supported by NASA through contract NAS 2-7963 with the State University of New York at Albany; the Asteroid-Meteoroid Astronomy Experiment was supported by NASA through contract NAS 2-6559 with the General Electric Space Sciences Laboratory.

References

Alvarez, J.M. (1976), "The Cosmic Dust Environment at Earth, Jupiter and Interplanetary Space: Results from Langley Experiments on MTS, Pioneer 10 and Pioneer 11", these proceedings.

Berg, O.E., and Grün, E. "Evidence of Hyperbolic Cosmic Dust Particles" Space Research XII, Eds. M. J. Rycroft and S. K. Runcorn (Akademie-Verlag, Berlin, 1973).

Dohnanyi, J.S. "Collisional Model of Asteroids and Their Debris" Jour. Geophys. Res. 74, 2531-2554 (1969).

Hanner, M.S., Sparrow, J.G., Weinberg, J.L., Beeson, D.E. (1976), "Pioneer 10 Observations of Zodiacal Light Brightness Near the Ecliptic; Changes with Heliocentric Distances", these proceedings.

Hanner, M. S. and Weinberg, J. L. "Changes in Zodiacal Light With Heliocentric Distance: Preliminary Results from Pioneer 10" Space Research XIV, Eds. M. J. Rycroft and R. D. Reasenberg, 769-772 (Akademie-Verlag, Berlin, 1974).

Hoffmann, H. J., Fechtig, H., Grün, E., and Kissel, J. "First Results of the Micrometeoroid Experiment S215 on the HEOS 2 Satellite" Planet. Space Sci. 23, 215-224 (1975).

Humes, D. H., Alvarez, J. M., Kinard, W. H. and O'Neal, R. L. "Pioneer 11 Meteoroid Detection Experiment: Preliminary Results," Science 188, 473-474 (1975).

Neukum, G. "Micrometeoroid Erosion of Lunar Rocks" Space Research XIV, Eds. M. J. Rycroft and R. D. Reasenberg, 739-740 (Akademie-Verlag, Berlin, 1974).

Soberman, R. K., Neste, S. L. and Lichtenfeld, K. "Optical Measurement of Interplanetary Particulates from Pioneer 10" Jour. Geophys. Res. 79, 3685-3694 (1974).

Zook, H. A., and Soberman, R. K. "The Radial Dependence of the Zodiacal Light" Space Research XIV, Eds. M. J. Rycroft and R. D. Reasenberg, 763-767 (Akademie-Verlag, Berlin, 1974).

2.1.8 Sources of Interplanetary Dust: Asteroids

J.S. Dohnanyi [)]

Max-Planck-Institut für Kernphysik, Heidelberg/F.R.G.

Abstract

The asteroid belt is examined as a potential source of interplanetary dust. Using results from the Pioneer-10 experiments the relative contribution of asteroidal and cometary particles to the Zodiacal cloud is estimated using methods developed in earlier studies of meteoroidal collisions (collisional model). It is found that the contribution of asteroidal particles to dust in the asteroidal belt is small compared with the number density of cometary type particles. Similar conclusions apply to the Zodiacal cloud between the sun and the asteroid belt. When definitive criteria for differentiating between comets and asteroids become available, a reexamination of some of our conclusions may become necessary.

The distribution of asteroidal rotations is analyzed; it is found that the gross features of the distribution can be reproduced using the collisional model.

I. Introduction

The zodiacal light is thought to originate from material given off by comets (Whipple, 1951, 1955, 1967 and Dohnanyi, 1970, 1972). Because of insufficient observational material, the relative contribution of asteroidal material to the zodiacal cloud has always been difficult to estimate (Whipple, 1971, Dohnanyi, 1972 and Wetherill, 1974). It is therefore of interest to examine the significance of the asteroid belt as a potential source of zodiacal particles in the light of the new observational data obtained from Pioneer 10 and 11 satellites; this present paper summarises such a study.

II. Asteroidal Belt Particle Densities

NASA model: Our current knowledge on the distribution of dust in the asteroid belt will be briefly reviewed, in this section.

The density of large objects in the asteroidal belt has been estimated (Dohnanyi, 1971, 1972) using the collisional model of asteroids in the asteroidal belt (Anders, 1965, Bandermann, 1972, Dohnanyi, 1969, Hartmann and Hartmann, 1968, Wetherill, 1967). In this model destructive collisions between asteroids and their resulting fragmentation

---

[)] On leave from Bell Laboratories, Holmdel, N.J./USA

are found to be the dominating process in the dynamic evolution of the asteroidal population. In an attempt to estimate the density of dust in the asteroid belt one can start with extrapolating down to microparticle sizes the asteroidal number density obtained from the collisional model (Dohnanyi, 1969). The observational material (Kuiper et al., 1958, Van Houten et al., 1970) which is well explained by the collisional model consists of large objects with such a large number density that collisions with the comparatively few comets are thought to have a negligible influence on the dynamic evolution of the asteroidal population. The density of cometary microparticles may, however, be so large that destructive collisions with these cometary particles will significantly influence the dynamic evolution of the population of asteroidal microparticles (Dohnanyi, 1969). The result of such an extrapolation of the earlier results of the collisional model to microparticle size ranges therefore becomes uncertain.

Keeping in mind the possible limitations of our extrapolation, we take the steady state solution of the collisional model (Dohnanyi, 1969).

(1) $$f(m)dm = Am^{-11/6} dm$$

where $f(m)dm$ is the number density of asteroidal objects in the mass range m to $m + dm$ and where A is a constant in the region of space occupied by the asteroidal belt and is zero elsewhere.

A more detailed representation of A as a function of distance from the sun has been given by Kessler (1970). He calculated the mean number density of asteroids as a function of distance from the sun. He has carried out this calculation by computing the fraction of time spent by each catalogued asteroid at each point in its orbit and from this result he obtained a statistical estimate of the number density of asteroids as a function of distance from the sun. The mass distribution was assumed independent from the spatial distribution and of a form similar to the steady state solution of the collisional model, Eq. 1.

Figure 1, given by Roosen (1971), is a plot of the spatial variation of the particle number density for constant mass compared with the number density near earth. The curve, labelled asteroidal distribution, is a plot of the spatial dependence of the asteroidal particles obtained by Kessler (1970). The other curves represent particle distributions varying as $R^{-2.5}$, R^{-1} and a constant particle distribu-

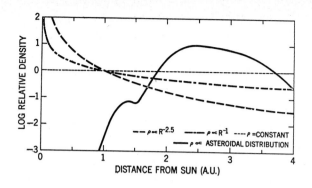

Fig. 1

Spatial distribution of particle number density relative to the number density near earth (from Roosen, 1971). R is the distance from the sun in AU and ρ is the number density at constant mass in arbitrary units.

tion, respectively. R is the distance from the sun in AU. It can be seen, from Fig. 1, that an extrapolation of the distribution of large asteroids, using the results of the collisional model, predicts a dust concentration in the asteroid belt about an order of magnitude higher than the dust density near earth.

Evidence from the Gegenschein: An upper limit to the particle number density in the asteroidal belt can be obtained from the brightness of the Gegenschein (cf. Roosen, 1971a and b). If one assumes that the Gegenschein is caused by the backscatter of light by particles along the line of sight in the anti sun direction, then reasonable assumptions for the particle albedo will enable one to estimate the upper limit to the particle density for given forms of the particle distributions in mass and space.

On this basis Whipple (1971) placed a likely upper limit on the dust density in the asteroidal belt of about 5 to 10 times the dust density near earth orbit (also cf. Kessler, 1968 and Roosen, 1971c).

Spacecraft Measurements: During the Pioneer 10 and 11 missions to Jupiter, direct measurements of the frequency of dust particles have been performed. These have been penetration measurements (Humes et al., 1974) detecting particles with sufficient kinetic energy to penetrate the walls of pressurized meteoroid detector cans ("beer cans") and optical experiments (Sisyphus) detecting particles in the field of view of an assembly of four telescopes (Soberman et al., 1974a and b). A further measurement consisted in the use of the Sisyphus telescopes for measuring the surface brightness of the zodiacal cloud (Hanner et al., 1974).

Results of the penetration experiment are given in Fig. 2 where densities of meteoroids at constant mass have been computed, using

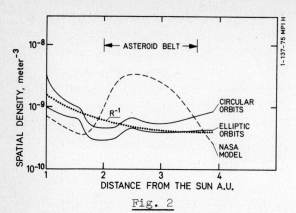

Fig. 2

Spatial distribution of the particle number density deduced from the Pioneer 10 penetration measurements. Solid lines are densities implied by the data for particles in circular and elliptic orbits, as indicated. Dotted curve indicates a density which varies inversely with the distance, R, from the sun and dashed line curve is the NASA model, discussed in the text.

two different orbital distributions for these particles (D.H. Humes, 1975, Kinard et al., 1974). The dashed line is the earlier NASA model for the distribution of cometary and asteroidal particles (Cour-Palais, 1969 and Kessler, 1970). Another curve, labelled as 1/R represents a dust density that varies inversely with the distance R from the sun.

(2) $n(R) \sim 1/R$

This inverse relationship between the number density and heliocentric distance is in good agreement with the results of zodiacal light observations from Pioneer 10 (Hanner et al., 1976). It can be seen, Fig. 2, that the number density of interplanetary dust inferred from the penetration data is a slowly decreasing function with heliocentric distance and, to within the limits of uncertainty inherent in the estimate, a constant distribution (Humes, 1974) or a distribution that varies as R^{-1} (Eq. 2) appears to fit the results quite well, to a first approximation.

The results of the Sisyphus experiment indicate a distribution from 2 AU to 3.5 AU from the sun for small particles (smaller than .15 cm in radius) that is constant with heliocentric distance to within a factor of 2 (Soberman et al., 1974). Difficulties with the calibration of this experiment (Auer, 1975) may, however, lead to a revision (Roosen, 1975) of their interpretation and we defer discussion of the Sisyphus results until this matter has been resolved.

III. Interaction between Cometary and Asteroidal Particles

It is clear, from Fig. 2, and from the results of the zodiacal light experiment (Hanner et al., 1976) that there is no abrupt change in the particle number density as we enter the asteroidal belt. To be more specific, there is no evidence for the existence of an asteroid belt concentration of micrometeoroids comparable to the concentration of large asteroids, as indicated in Fig. 1. The concentration of cometary particles therefore is likely to be about the same in the asteroid belt as it is outside of it without a sharp brake in their distribution as one enters the asteroid belt. By cometary particles we shall mean, rather loosely, those particles whose orbital distribution differs from that of the known asteroids in the asteroid belt and are therefore presumably of cometary origin (see Fechtig, 1976, for a recent review on the distribution of interplanetary microparticles).

If the number density of cometary particles is significant compared with the number density of the asteroidal particles in the asteroid belt, then, because of the relatively high relative velocity of the cometary particles, they will have a strong influence on the survival time of the asteroidal particles. Before discussing the influence of collisions, however, we turn our attention to Fig. 3.

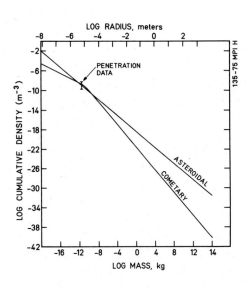

Fig. 3
Cumulative number density of cometary particles at 1 AU from the sun and the asteroidal number density obtained from the collisional model and extrapolated here into the size ranges of micrometeoroids.

Plotted in Fig. 3 is the cumulative number density of presumably cometary particles at earth's orbit and an extrapolation of the cumulative density of the asteroidal particles obtained from the collisional model. We have also indicated the results from the penetration experiment from Pioneer 10 (Humes et al., 1974).

The values used for the densities are, for the asteroidal density,

(3) $f(m)dm = 2.48 \times 10^{-19} m^{-11/6} dm$, for $2 \text{ AU} \leq R \leq 3.5 \text{ AU}$

$f(m)dm = 0$, otherwise

and, for cometary particles,

(4) $f(m)dm = 2.94 \times 10^{-22} m^{-13/6} dm$, $m \geq 10^{-10}$ kg,

$f(m)dm = 1.36 \times 10^{-15} m^{-1.5} dm$, $m \leq 10^{-10}$ kg for all relevant values of R. f(m)dm is the number density of particles, per cubic meter in the mass range m to m+dm kg. The value of f(m)dm in Eq. 4 is obtained from Dohnanyi (1973) and Eq. 3 is taken from Dohnanyi (1969). Eq. 3 is the quasi steady state solution to the collisional model representing the case when, in any given small time period, the number of asteroids in a given mass range destroyed by disruptive collisions is replenished by fragments, in the same mass range, produced by the disruption of larger objects during the same period of time. Cumulative densities are then obtained by simple integration.

It can be seen, from Fig. 3, that if the number density of cometary particles in the asteroid belt is comparable to its value at 1 AU, and this is suggested by the Pioneer 10 penetration measurements, then the number density of cometary particles is comparable to the extrapolated asteroidal number density in the asteroid belt.

We shall presently consider the influence of collisions on the population of the asteroidal particles and show that they cannot coexist in a steady state with the cometary particles in significant numbers because of the destructive influence of catastrophic collisions with cometary particles.

In order to estimate the influence of collisions we shall use a method discussed by Dohnanyi (1969, 1970 and 1972). We shall only consider the influence of catastrophic collisions; it can be shown (Dohnanyi, 1969, 1970) that erosive collisions play only a minor role for populations of the type Eq. 3 and Eq. 4.

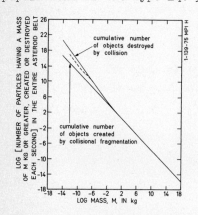

Fig. 4
Cumulative number of particles created or destroyed in the asteroid belt, as indicated, for an asteroid number density given by the collisional model and under bombardment by cometary particles, discussed in the text. Dashed line is the destruction rate implied when a refined model of cometary particle distribution is employed (see text).

The mean relative velocity between asteroidal particles in about 5 km/sec (Dohnanyi, 1969). The average relative velocity between asteroidal and cometary particles is unknown and it is therefore necessary to estimate it from first principles. If we take a "typical" short period cometary orbit with perihelion and aphelion distances of 1 AU and 4 AU, respectively, and an average inclination of the McCrosky and Posen (1961) meteors we obtain a velocity of 13.6 km/sec relative an asteroidal particle in circular orbit at 3 AU from the sun. If we use formula 7-9 given by Southworth and Sekanina (1973) for the relative velocity and an average eccentricity of .5 as obtained by these authors, then even for a zero inclination orbit we obtain a relative velocity of 16.4 km/sec for cometary particles at 3 AU from the sun. It is obvious that the relative velocity between an asteroid belt particle in near circular low inclination orbit and a cometary particle in highly eccentric and moderately inclined orbit will be much higher than the relative velocity among the asteroidal particles themselves.

For purposes of a rough, order of magnitude estimate, we shall adopt an average encounter speed of 14 km/sec between cometary and asteroidal particles and a material density of 3.5 gm/cm^3 for both cometary and asteroidal particles.

Following Dohnanyi (1973, 1969) we then estimate the amount of material crushed per unit volume and unit time in the asteroidal belt as well as the number of new fragments created by the crushing of larger objects.

Figure 4 summarizes our results: it is a double logarithmic plot of the number of objects having a mass of m kg or greater created or destroyed, as indicated, per second in the entire asteroid belt assuming an asteroidal population similar to the steady state distribution obtained from the collisional model (Dohnanyi, 1969).

It can be seen, from Fig. 4, that the creation rates are similar to the destruction rates for objects having a mass of many kg. The contribution to the collision rates by cometary objects causes the destruction rate to exceed the creation rate by an amount less than 8 % for objects larger than 100 kg and becomes negligible for even larger objects. For object with a mass of the order of 1 kg, however, the destruction rate exceeds the creation rate by about 33 % and this effect increases, rapidly, for smaller masses, where the destruction rate exceeds the creation rate by orders of magnitude.

The distribution rate, plotted in Fig. 4 as a solid line, is based on a cometary number density given by the first part of Eq. 4 and extrapolated into a mass range smaller than 10^{-10} kg without regard to the "flattening" in the distribution for those small masses. A more detailed calculation, including the change in the cometary meteoroid distribution for masses smaller than 10^{-10} kg as given by the second part of Eq. 4 and including the presence of ß-meteoroids (Dohnanyi, 1976) has been carried out. The results for the destruction rate are plotted as a broken line in Fig. 4.

The estimates for the destruction rate, plotted in Fig. 4, are probable lower limits because the particle removal rate by the Poynting-Robertson effect (Robertson, 1936) have not been included. It has been shown, (Dohnanyi, 1969) that this effect will contribute to the destruction rate by an amount equal to the creation rate plotted in Fig. 4 for particles with masses of about 10^{-13} kg; this effect increases rapidly for smaller particles.

It is therefore clear that a population of small objects, obtained by extrapolating the steady state distribution of large asteroids is not stable under the influence of bombardments by cometary particles. Destructive collisions with cometary objects will rapidly deplete the small particle of an asteroidal population given by Eq. 3 (i.e. the steady state solution of the collisional model for large asteroids).

We shall now estimate, very roughly, the likely population of small asteroidal particles. We assume steady state conditions and approximate the asteroidal distribution by a function of the form

(5) $\quad h(m)dm = H\, m^{-\sigma}\, (B + m^{ß})^{-1}\, dm$

where $h(m)dm$ is the number density of objects in the mass range m to m+dm and H, σ, B, ß are constants. $h(m)dm$ has the property that for small objects, i.e. when

(6) $\quad m^{ß} \ll B,$

we have

(7) $\quad h(m)dm \sim HB\, m^{-\sigma}\, dm,$

and for large objects

(8) $\quad m^{ß} \gg B$

we have

(9) $\quad h(m)dm \sim H\, m^{-ß-\sigma}\, dm$

We also require that for large asteroids $h(m)dm$ approach the true number density of these objects, Eq. 3. This determines the values of $\sigma + ß$:

(10) $\quad H = 2.48 \times 10^{-19} \, m^{-11/6} \, dm, \quad \sigma+\beta = 11/6$

In order to determine σ, we note that for masses much larger than $B^{1/\beta}$ we have a population given by the formula Eq. 7. This population is subject to collisions with the population of cometary objects given, as a first approximation, by

(11) $\quad f(m)dm = 2.94 \times 10^{-22} \, m^{-13/6} \, dm$

i.e. the first part of Eq. 4.

Using a method discussed by Dohnanyi (1970, cf. Eq. 47 in that paper) the steady state solution to the dynamic problem of these two interacting populations is estimated by

(12) $\quad \sigma = 1.5$

If we take, somewhat arbitrarily (cf. Eq. 6)
(13) $\quad B \sim 1 \text{ kg}$
which is the mass at which the destruction rate in Fig. 4 exceeds the creation rate by only about 25 % we have a very rough estimate of the resulting asteroidal distribution which we now can write as

(14) $\quad h(m)dm \sim 2.5 \times 10^{-19} \, m^{-11/6} \, dm, \quad m \gg 1 \text{ kg}$
$\qquad \sim 2.5 \times 10^{-19} \, m^{-1.5} \, dm, \quad m \ll 1 \text{ kg}$

In estimating $h(m)dm$ we have assumed that the density of cometary objects, Eq. 11, can be extrapolated down to masses smaller than 10^{-10} kg, as a zeroth approximation; this is not strictly correct, as can be seen from Eq. 4. The fact that the density of cometary microparticles ($m < 10^{-10}$ kg) is smaller than an extrapolation of Eq. 11 (cf. however, Dohnanyi, 1976) means that the asteroidal microparticle density is somewhat greater for very small particles than our estimate, Eq. 14, implies. It is, however clear, that the steady state density of asteroidal micrometeoroids is much smaller than that of the cometary particles. On the basis of our present results we estimate that, for masses very much smaller than 1 kg, the density of asteroidal particles is orders of magnitude smaller than is that of the cometary particles, in the asteroidal belt.

We summarize the situation in Fig. 5 which is similar to Fig. 3 but where we have sketched the estimate of the asteroidal number density, Eq. 14. The density of asteroidal dust is then somewhat underestimated as has been discussed above. The density of cometary micrometeoroids in the mass range smaller than about 10^{-14} kg is, however, also underestimated because we did not include the flux of β-meteoroids

(Dohnanyi, 1976) in our zeroth approximation treatment. Accordingly, in Fig. 5, the cometary particle density curve should start bending upwards for masses decreasing to smaller values than about 10^{-14} kg.

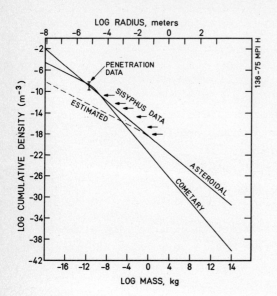

Fig. 5

Comparison of particle number densities implied by the Pioneer 10 and 11 data combined with the results of the present paper. Dashed line is a theoretical estimate of the asteroidal small particle distribution implied by the indicated cometary number density. Solid line, labelled "asteroidal" is the number density obtained from the collisional model and estimated to be valid for objects having a mass of many kg. The expected location of the Sisyphus optical data we indicated by arrows; their final calibration is in progress.

The published results of the Sisyphus (Soberman et al., 1974a and b) experiments are also plotted for comparison. These results are represented by arrows showing the direction in which the data points will probably move after final calibration of the results has been achieved.

IV. Leakage of Material out of the Asteroid Belt

So far we have considered the distribution of asteroidal dust only in the asteroidal belt and found that cometary particles appear to dominate the distribution of micrometeoroids in the asteroid belt. Since the distribution of cometary objects appears to be a slowly varying function with distance from the sun over the solar system within the orbit of Jupiter (Hanner et al., 1976) and since asteroidal objects are concentrated in the asteroid belt, it appears that the contribution of asteroidal particles to the micrometeoroid population outside the asteroid belt is much smaller than in the belt. One would therefore conclude that asteroid particles do not dominate the distribution of interplanetary dust in the solar system within the orbit of Jupiter.

In the foregoing discussion we have distinguished between objects of asteroidal and cometary origin on the basis of their orbits only.

Objects in the asteroid belt having approximately circular orbits were labelled asteroidal and all other objects in more eccentric orbits and/or outside the asteroid belt were labelled cometary. Such a distinction is somewhat arbitrary and we shall presently discuss it in greater detail.

The possible origin of earth crossing objects has extensively been discussed in the literature (cf. Anders, 1971, Marsden, 1971, Opik, 1963 and 1966, Wetherill, 1974, Whipple, 1967). Whereas some authors favor a cometary origin for most of these objects and others regard them as asteroids, the only objects that are uniformly accepted as cometary are the ones that do or have exhibited a cometary tail.

Zimmerman and Wetherill (1973) have recently suggested a mechanism by means of which a great deal of asteroidal material may escape the asteroid belt and develop eccentric earth crossing orbits. This may be accomplished as follows: belt asteroids with orbital elements near the 2:1 resonance gap with Jupiter may eject substantial quantities of collisional fragments into the gap. Those fragments may then develop somewhat eccentric orbits librating with Jupiter in such a manner that they always avoide a close encounter (at the asteroid's aphelion passage) with the major planet. Subsequent collisions may then destroy this libration relationship with Jupiter resulting in strong Jovian perturbations leading to eccentric earth crossing orbits. Through this process, Zimmerman and Wetherill estimated that enough material may leave the asteroid belt that the population of earth crossing objects (McCrosky and Ceplecha, 1970) may, to a large extent, consist of these "runaway" asteroidal objects.

It therefore appears difficult to precisely determine the relative proportion of asteroidal object in the population of earth crossing objects having a size of the order of 1 kg or larger. The dynamics of the population of small objects has been discussed by Whipple (1967) and more recently by Dohnanyi (1970). It was found that our present knowledge of the distribution of meteoroids in the mass range of less than a kg down to micrometeoroidal sizes is consistent with a cometary origin and unless it can be shown that some dynamical process (e.g. the one proposed by Zimmerman and Wetherill) can populate the Zodiacal cloud with asteroidal small particles (smaller than about a kg) more efficiently than comets are believed to do (Whipple, 1967, Dohnanyi, 1970), we conclude that the population of small meteoroids in the solar system is dominated by particles given off by short period comets (Whipple, 1967, also cf. Lovell, 1954).

V. Asteroidal Rotations

The rotation of asteroids has been discussed by McAdoo and Burns (1972) and more recently ny Napier and Dodd (1974). These authors concluded that collisional processes are definitely involved in spinning up some of the asteroids but found it difficult to explain the known distribution of asteroidal rotations in terms of a collisional origin. We shall show here that the known distribution of asteroidal rotations as given by Mc Adoo and Burns (1972) and by Gehrels (1970) can indeed be explained as having a collisional origin, thereby strengthening our confidence in the strong influence of collisions on the population dynamics of meteoroids developed in our earlier discussion. More specifically, we shall show that the gross features of the known distribution of asteroidal rotations can be reproduced from a simple random walk model.

We first consider the magnitude of the angular velocity that an asteroid may aquire over a period of time.

We assume that the number density of asteroids $f(m)dm$ is given by Eq. 1; the flux and per (m^2sec 2π sterad) is (cf. Dohnanyi, 1972)

(15)   flux = $(1/4)$ v $f(m)dm$

where v is the mean encounter speed.

The influx of particles per second into a sphere of radius r is then $(4\pi r^2) \times (\text{flux})$ and into a sphere of radius (r+dr) is $4\pi (r+dr)^2 \times (\text{flux})$. Hence the influx, per second, of particles with an impact parameter r to r+dr around a point is $(8\pi rdr) \times (\text{flux})$ which is the difference between the two previous expressions. The influx, per second, of the corresponding angular momentum $f(m,v,r)dr$ around this point using Eq. 15 is,

(16) $f(m,v,r)dr\ dm = (8\pi\ rdr)(rmv)\ \xi f(m)(1/4)vdm$

where ξ is the corresponding "momentum multiplication factor" and is generally some number greater than 1. The inclusion of ξ is necessary because the transfer of angular momentum to an asteroid by an impacting particle with momentum mv and impact parameter r is not only mvr but the momentum of the debris ejected from the impact crater will also contribute an additional angular momentum impacted to the target object. We shall attempt to include this effect with the use of the "momentum multiplication factor", ξ.

Integrating Eq. 16 over the size of the target object and over the mass of all projectile objects smaller than the critical particle size that would catastrophically disrupt the target object, using Eq. 1

and Eq. 16, we have

(17) $\int_0^R dr \int_\mu^{m/\gamma} dm' \, f(m',v,r) \simeq 2\pi \, \xi \, (R^3/3) v^2 A6 \, (m/\gamma)^{1/6}$

where the contribution of the lower mass limit at some minimum micrometeoritic mass μ has been disregarded and where m/γ is the mass of the smallest projectile object capable to catastrophically disrupt a target object whose mass is m. γ is taken here as $\gamma = 250 \, v^2$ (Dohnanyi, 1972) where the impact speed v is to be expressed in km/sec. Using Eq. 1 and an average meteoroid material density of 3.5 gram/cm^3 we now assume that the angular momentum imparted to the target adds up linearly. We can then calculate the expected value of the maximum angular momentum, H, imparted per second to our test object

(18) $H = 3 \times 10^{-20} \, \xi \, v^2 \, m^{7/6}$, MKS units.

Assuming spherical asteroids with a moment of inertia $(2/5)mR^2$ and where ω is the angular velocity, we obtain for the period T,

(19) $T = 10^{-10} \sqrt{M} \, \xi^{-1}$ hr (per 10^9 yr)

which is the period (hr) aquired during 10^9 years of bombardment by an asteroid with an average albedo of .04 (Chapman, 1975a). T is the expected value of the smallest period of rotation (hr) that may be aquired by an asteroid having a mass M (kg) during 10^9 years of exposure to an environment similar to the present asteroid belt. Since, however, the number density of asteroids was very likely greater in the past, Eq. 19 should be regarded as conservative, i.e. when past values of asteroidal number densities are considered, a smaller value for T than the one given by Eq. 19 would be obtained. In addition to the influence of collisions on asteroidal rotations, radiation forces may also contribute to the rotational state of asteroids (Paddack, 1969, Icke, 1973) causing the expected value T to be even shorter.

Napier and Dodd (1974) estimated the critical rotational period, Tcr, for asteroids; asteroids with a shorter rotational period will burst because of excessive internal tension. Their result is

(20) $Tcr \sim 10^{-6} \, M^{1/3}$ hours

where M is the asteroidal mass, in kg. A comparing of Eq. 19 with Eq. 20 shows that most asteroids "had a chance" to aquire enough angular velocity to cause rotational disruption.

We then consider the following statistical model: assume that we have

(21) $F(\omega) \, 4\pi \omega^2 d\omega$

asteroids with angular velocities in the range ω to $\omega+d\omega$ where

(22) $\quad \omega^2 = \vec{\omega} \cdot \vec{\omega}$

We further assume that these angular velocities change randomly in time. If, furthermore, any asteroid aquires an angular velocity of ω_x or greater then it disappears (= it is disrupted) and another asteroid (the largest fragment) appears i.e. is created. These new asteroids may have any angular velocity smaller than ω_x with equal statistical probability.

Our function $F(\omega)$ in Eq. 21 then satisfies the diffusion Equation

(23) $\quad \dfrac{1}{\omega^2} \dfrac{\partial}{\partial \omega} \left(\omega^2 \dfrac{\partial F(\omega)}{\partial \omega} \right) = S(\omega), \quad \omega \leq \omega_x, \quad F(\omega_x) = 0$

where the source function $S(\omega)$ is

(24) $\quad S(\omega) = $ constant $\quad \omega < \omega_x$
$\qquad \qquad = 0 \qquad \qquad \omega \geq \omega_x$

The unique solution of Eq. 23 is

(25) $\quad F(\omega) = -\dfrac{1}{\omega} \left\{ \int_0^\omega dy \int_0^y dx [xS(x)] - \dfrac{\omega}{\omega_x} \int_0^{\omega_x} dy \int_0^y dx [xS(x)] \right\}.$

Using Eq. 24 we readily obtain

(26) $\quad F(\omega) = $ constant $(\omega_x^2 - \omega^2)$

and we have for the number density

(27) $\quad F(\omega) \, 4\pi \omega^2 d\omega = c(\omega_x^2 - \omega^2) \omega^2 d\omega$

where

(28) $\quad c = (15/2) \, N \omega_x^{-5}$

where N is the total number of asteroids represented by the density function Eq. 27.

Figure 6 is a plot of the density function Eq. 27 for two values of ω_x as indicated. A sample size of N = 35 was used in the numerical plot of Eq. 27 which approximates the number of asteroids having a critical frequency ω_x within about a factor of 2 from the value used in the plot.

It can be seen, from Fig. 6, that the gross features of the distribution of asteroidal rotations can be reproduced by our simplified steady state random walk model Eq. 27. It therefore appear that most asteroids have spin rates determined by the effects of random collisions with other asteroids. It will therefore be difficult to obtain statistical information regarding the initial state of asteroidal

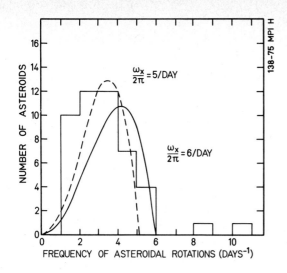

Fig. 6

Distribution of asteroidal rotations; histograms are empirical data and curves are theoretical results obtained in the text for two values of the critical bursting angular velocity, ω_x, as indicated.

rotation periods at the time of creation by only considering their present distribution of rotations. We have hereby obtained further evidence for the soundness of the collisional model and the applicability of some of the inferences one can draw from such a simple "molecular chaos" approximation.

VI. Historical Note

In this section we shall very briefly discuss the possibility that asteroids may have significantly contributed to the zodiacal cloud in past times.

Because of collisions the asteroid belt is losing mass by the production of fragments sufficiently small to be expelled from the solar system by radiation pressure (Zook and Berg, 1975) or will spiral into the inner regions of the solar system because of the Poynting Robertson effect (Wyatt and Whipple, 1950). In addition to these processes, the asteroid belt is losing some of its members because of perturbations with Jupiter (Zimmerman and Wetherill, 1975).

It therefore is clear that there was more material in the asteroid belt in the past then is there now. The total initial mass of the asteroids in the asteroid belt has been estimated from about the same order of magnitude as its present mass (Dohnanyi, 1969) to about 3000 times its present mass (Chapman and Davis, 1975). Thus, if the cometary meteoroid population has been constant at its present level, the asteroidal contribution to interplanetary dust may well have dominated the particle population of the zodiacal cloud in earlier periods of the solar system.

The situation is, however, complicated by the fact that the population of cometary objects within the orbit of Jupiter may also have been much greater in the past (Wetherill, 1975, Whipple, 1975).

Until the past distribution of comets and asteroids is better known, it appears difficult to estimate the relative contribution of asteroidal material to the zodiacal cloud during the earlier period of the history of the solar system.

VII. Conclusion

The central conclusion reached here is that the contribution of comets to the small particle population in the zodiacal cloud dominates over the asteroidal contribution to it. The distriction between comets and asteroids is in many cases, however, not yet clear. We distinguish here between comets and asteroids somewhat arbitrarily on the basis of their orbital elements. Our conclusions will have to be reexamined if many of the earth and Mars crossing (Shoemaker et al., 1975) objects turn out to be asteroids that have escaped from the asteroid belt (Zimmerman and Wetherill, 1973).

The origin of asteroidal rotations is also considered, as a corollary to our discussion of the influence of collisions on the population of asteroidal fragments. It is found that the gross features of the distribution of asteroidal rotations can be explained if one assumes that the population of asteroids (whose spins have been measured) have reached steady state conditions under the effect of mutual inelastic collisions.

VIII. Acknowledgement

I am indebted to J. Schubart for helpful discussions.

I gratefully acknowledge that my visit to Heidelberg has been made possible by an Alexander von Humboldt Senior U.S. Scientist Award.

References:

Anders, E., Fragmentation history of asteroids, Icarus $\underline{4}$, 398-408 (1965).
Anders, E., Interrelations of meteorites, asteroids and comets. "Physical Studies of Minor Planets" (T. Gehrels, ed.), NASA SP-267, 429-446 (1971).

Auer, S., The asteroid belt: doubts about the particle concentration measured with the asteroid/meteoroid detector on Pioneer 10, Science, 186, 650-652 (1974).

Bandermann, L.W., Remarks on the size distribution of colliding and fragmenting particles, Monthly Not. R. astr. Soc. 160, 321 (1972).

Chapman, C.R., The nature of asteroids, Sci. Am. 232, 24-33 (1975).

Chapman, C.R., and D.R. Davis, Asteroid collisional evolution: evidence for a much larger early population, Science 190, 553-556 (1975).

Cour-Palais, B.G., Meteoroid environment model-1969 [Near earth to lunar surface], NASA SP-8013, March, 1969.

Dohnanyi, J.S., Collisional model of asteroids and their debris. J. Geophys. Res. 74, 2531-2554 (1969).

Dohnanyi, J.S., On the origin and distribution of meteoroids. J. Geophys. Res. 75, 3468-3493 (1970).

Dohnanyi, J.S., Fragmentation and distribution of asteroids, "Physical Studies of Minor Planets" (T. Gehrels, ed.) NASA SP-267, 263-295 (1971).

Dohnanyi, J.S., Interplanetary objects in review: statistics of their masses and dynamics, Icarus 17, 1-48 (1972).

Dohnanyi, J.S., Current evolution of meteoroids, "Evolutionary and Physical Properties of Meteoroids" (C.L. Hemenway, P. Millman and A.F. Cook, ed.) NASA SP-319, 363-374 (1972).

Dohnanyi, J.S., Flux of hyperbolic meteoroids, 1976, this volume.

Fechtig, H., In-situ records of interplanetary dust particles - methods and results, 1976, this volume.

Gehrels, T., Photometry of asteroids, "Surfaces and Interiors of Planets and Satellites" (A. Dollfus, ed.) 317-375, Academic Press, London, 1970.

Hanner, M.S., J.L. Weinberg, L.M. DeShields II, B.A. Green, and G.N. Toller, Zodiacal light and the asteroid belt: the view from Pioneer 10, J. Geophys. Res. 79, 3671-3675 (1974).

Hanner, M.S., J.G. Sparrow, J.L. Weinberg, and D.E. Beeson, Pioneer 10 observations of Zodiacal light brightness near the ecliptic: changes with heliocentric distances, 1976, this volume.

Hartmann, W.K., and A.C. Hartmann, Asteroid collisions and evaluation of asteroidal mass distribution and meteoritic flux. Icarus 8, 361-381 (1968).

Humes, D.H., J.M. Alvarez, R.L. O'Neal, and W.H. Kinard, The interplanetary and near-Jupiter meteoroid environments, J. Geophys. Res. 79, 3677-3684 (1974).

Humes, D.H., J.M. Alvarez, W.H. Kinard, and R.L. O'Neal, Pioneer 11 meteoroid detection experiment: preliminary results, to be published in Science, 1975.

Icke, V., Distribution of the angular velocities of the asteroids, Astron. and Astrophys. 28, 441-445 (1973).

Kessler, D.J., Upper limit on the spatial density of asteroidal debris. J. Amer. Inst. Aeron. Astron. 6, 2450 (1968).

Kessler, D.J., Meteoroid Environment Model-1970 (Interplanetary and Planetary). NASA SP-8038, 1970.

Kessler, D.J., Estimate of particle densities and collision danger for spacecraft moving through the asteroid belt, "Physical Studies of Minor Planets" (T. Gehrels, ed.), NASA SP-267, 595-605 (1971).

Kinard, W.H., R.L. O'Neill, J.M. Alvarez, and D.H. Humes, Interplanetary and near-Jupiter meteoroid environments: preliminary results from the meteoroid detection experiment, Science 183, 321-322 (1974).

Kuiper, G.P., Y. Fujita, T. Gehrels, I. Groeneweld, J. Kent, G. Van Viesbroeck, and C.J. Van Houten, Survey of asteroids. Astrophys. J. Suppl. Ser. 3, 289-438 (1958).

Lovell, A.C.B., "Meteor Astronomy", Clarendon Press, Oxford, London, 1954.

Marsden, B.G., Evolution of comets into asteroids? "Physical Studies of Minor Planets" (T. Gehrels, ed.), NASA SP-267, 413-421 (1971).

McAdoo, D.C., and J.A. Burns, Further evidence for collisions among asteroids, Icarus 18, 285-293 (1973).

McCrosky, R.E., and Z. Ceplecha, Fireballs and the physical theory of meteors. Bull. Astron. Inst. Czechoslov. 21, 271-296 (1970).

McCrosky, R.E., and A. Posen, Orbital elements of photographic meteors. Smithson. Contrib. Astrophys. 4, 15-84 (1961).

Napier, W.McD., and R.J. Dodd, On the origin of the asteroids, Mon.Not. Roy. Astr. Soc. 166, 469-489 (1974)

Öpik, E.J., The stray bodies in the solar system Part I. Survival of cometary nuclei and the asteroids. Advanc. Astron. Astrophys. 2, 219-262 (1963).

Öpik, E.J., The stray bodies in the solar system Part II. The cometary origin of meteorites. Advanc. Astron. Astrophys. 4, 301-336 (1966).

Paddack, J.S., Rotational bursting of small celestial bodies: Effects of radiation pressure. J. Geophys. Res. 74, 4379-4381 (1969).

Robertson, H.P., Dynamical effects of radiation on the solar system. Monthly Notices Roy. Astron. Soc. 97, 423-438 (1937).

Roosen, R.B., The Gegenschein. Rev. Geophys. Space Phys. 9 (2), 275-304 (1971a).

Roosen, R.G., Spatial distribution of interplanetary dust. In "Physical Studies of Minor Planets" (T. Gehrels, ed.), NASA SP-267, 363-375 (1971b).

Roosen, R.B., 1975, private communication.

Shoemaker, E.M., E.F. Helin, and S.L. Gillett, Populations of the planet-crossing asteroids, International Colloquium of Planetary Geology, Expanded Abstracts, ed. by the Italian Consortium for Planetary Studies, Rome, 22-30, Sept. 1975.

Soberman, R.K., S.L. Neste, and K. Lichtenfeld, Particle concentration in the asteroid belt from Pioneer 10, Science 183, 320-321 (1974a).

Soberman, R.K., S.L. Neste, and K. Lichtenfeld, Optical measurement of interplanetary particulates from Pioneer 10, J. Geophys. Res. 79, 3685-3694 (1974b).

Southworth, R.B., and Z. Sekanina, Physical and dynamical studies of meteors, NASA CR-2316, October 1973.

Van Houten, D.J., I. Van Houten-Groenveld, P. Herget, and T. Gehrels, The Palomar-Leiden survey of faint minor planets. Astro. Astrophys. Suppl. $\underline{2}$, 339-448 (1970).

Wetherill, G.W., Collisions in the asteroid belt. J. Geophys. Res. $\underline{72}$, 2429-2444 (1967).

Wetherill, G.W., Solar system sources of meteorites and large meteoroids, Annual Rev. of Earth and Planet. Sci. $\underline{2}$, 303-331 (1974).

Wetherill, G.W., Pre-mare cratering and early solar system history, Proc. Sixth Lunar Science Conference, Geochemica at Cosmochemica Acta, Suppl. (1975) in press.

Whipple, F.L., A comet model, II, Physical relations for comets and meteors. Astrophys. J. $\underline{113}$, 464 (1951).

Whipple, F.L., A comet model, III, The zodiacal light. Astrophys. J. $\underline{121}$, 750-770 (1955).

Whipple, F.L., On Maintaining the Meteoritic Complex. Smithson. Astrophys. Obs. Spec. Rept. No. 239, pp. 1-46 (1967).

Whipple, F.L., A speculation about comets and the earth, Colloquium in Honor to Prof. P. Swings, Liège, March (1975).

Wyatt, S.P., and Whipple, F.L., The Poynting Robertson effect on meteor orbits. Astrophys. J. $\underline{111}$, 134-141 (1950).

Zimmerman, P.D., and G.W. Wetherill, Asteroidal source of meteorites, Science $\underline{182}$, 51-53 (1973).

Zook, H.A., and O.E. Berg, A source for hyperbolic cosmic dust particles, Planet. Space Sci. $\underline{23}$, 183-203 (1975).

2 IN SITU MEASUREMENTS OF INTERPLANETARY DUST

2.2 Lunar Studies and Simulation Experiments

LUNAR MICROCRATERS AND INTERPLANETARY DUST FLUXES

Jack B. Hartung
Department of Earth and Space Sciences
State University of New York at Stony Brook
Stony Brook, New York 11794 U. S. A.

Abstract

Conflicting data for depth-to-diameter ratios for lunar microcrater pits do not permit firm conclusions for distribution of meteoroid densities. The majority of meteoroids have equidimensional shapes. Meteoritic metal spherules have been detected in a small fraction of impact pit glasses, but contribution of meteoroidal material to most pit glasses is small to negligible. Impact pits less than 0.1 microns in diameter (impacting particle mass $\sim 10^{-16}$ grams) have been observed. Size distributions for microcrater pits less than 50 microns in diameter (particle mass $< 10^{-9}$ grams) measured on different samples differ significantly. An inflection in the cumulative size distribution curve at a diameter between 1 and 10 microns (particle masses between 10^{-14} and 10^{-11} grams) appears real, supporting the idea of a two-component model for the interplanetary dust. Data for the arrival direction of meteoroids at the moon are inconclusive.

Meteoroid flux determinations depend critically on surface exposure time measurements. Exposure time "clocks" are based on the accumulation of nuclear reaction products, etchable tracks, and sputter erosion produced by galactic cosmic ray, solar flare, and solar wind particles encountering the lunar surface. A serious problem for flux determinations is the difficulty in assuring that exposure time "clocks" measure only the time a surface is exposed to cratering. Therefore, suggestions of a lower meteoroid flux in the past must continue to be viewed with caution. Improved sample selection and cross-calibration of exposure time "clocks" should lead to better meteoroid flux and flux history measurements in the future.

Introduction

Before rocks were recovered from the surface of the moon, it was recognized by the workers in Heidelberg under the leadership of W. Gentner, J. Zähringer, and H. Fechtig that the exposed surfaces of lunar rocks would be of definite value to meteoroid science. During the last five years, the study of lunar rock surfaces in several laboratories has produced significant new data which should lead to an improved understanding of the interplanetary dust.

Reviews by Hörz et al. (1975a, 1975b) and Fechtig et al. (1975) described the progress made in this effort up to 1974. The objective of this paper is to present more recent results with particular emphasis on the problem of obtaining meteoroid flux information from lunar rock data. Also, because these authors presented results generally accepted in the field, we will emphasize those areas where conflicting interpretations and data exist.

Morphology

A description of the morphology of microcraters has been given by Hörz et al. (1971), Bloch et al. (1971), Hartung et al. (1972a), and others. A meteoroid impact origin for most microcraters has been argued by Hartung et al. (1972b).

Statistical data for the depth-to-diameter ratio for microcrater pits have been obtained. Brownlee et al. (1973, 1975) measured pits 4 to 70 microns in diameter, obtained a single peak in the depth-to-diameter distribution at a value of about 0.7, and, based on a comparison with laboratory data, concluded most interplanetary dust had a density between 2 and 4 g/cm^3. Smith et al. (1974) or Durrani et al. (1974) measured pits 1 to 8 microns in diameter and obtained three peaks in the distribution at values of about 0.3, 0.5, and 0.9 with an absence of cases at 0.7. These peaks were attributed to low density (1 to 2 g/cm^3), silicate, and iron meteoroids, respectively, with 40 to 50% of the impacts ($\sim$ 20% of the impacting mass) due to iron particles. Nagel et al. (1975) measured pits between 30 and 350 microns in diameter, obtained three peaks in the distribution, and concluded the impacts were due to particles similar to those reported by Smith et al. (1974). The discrepancies between these sets of data and the resulting interpretations have not been resolved.

Studies of the circularity of impact pits by Brownlee et al. (1973) has led to the interpretation that the majority of meteoroids have equidimensional shapes.

Chemistry

The colors of most pit glasses can be due to melting of host rock minerals. However, about 10% of the pit glasses, most often those which are dark brown to black in color, cannot consist of melted host rock alone, based on optical microscope studies by Hörz et al. (1971). For these, a contribution to the glass from impacting iron particles may have occurred.

Electron microprobe analyses of pit glasses have been made in search of contributions from impacting meteoroids. Chao et al. (1970) analyzed glasses from four pits. In general, the pit glass compositions reflected the host rock or mineral compositions. At least one pit in a microbreccia contained minute nickel-iron particles.

The presence of these particles may indicate the impact of a nickel-iron bearing meteoroid or, alternatively, may be the result of in situ reduction of iron through the action of implanted solar wind hydrogen and subsequent impact melting, as suggested by Housley et al. (1971). Bloch et al. (1971) analyzed seven pit glasses, one of which showed a nickel-iron enhancement over the host rock composition, thus indicating possibly the impact of a nickel-iron meteoroid. Brownlee et al. (1975) analyzed fifteen pit glasses, six of which contained metal spherules. Of these six pits, five were in breccia samples containing indigenous metal grains, which could account for the metal in these pits, as well as in the one studied by Chao et al. (1970). The remaining pit was in a basalt and contained only trace amounts of submicron metal grains, which could have been incorporated into the glass by any of the above described mechanisms. Nagel et al. (1975) observed metallic spherules up to 2 microns in diameter in three out of at least fifteen pits on one sample. The spherules contained iron, nickel, and sulphur, suggesting but not demonstrating conclusively a meteoroidal origin for them. Hartung et al. (1975) analyzed two metallic spherules more than 10 microns in diameter found in one mm-sized pit for iron, nickel, and cobalt. The nickel and cobalt concentrations were demonstrably meteoritic and not those expected for indigenous metal or for solar-wind-hydrogen-reduced metal. In this case, at least one out of 28 pits on the sample was formed by a meteoritic-metal-bearing meteoroid.

The origin of metal grains in any particular pit glass is not easily determined, as indicated by the different mechanisms already mentioned and other shock-related, metal-sphere-producing, processes (e.g., Gibbons et al. 1975). The data presently available suggest impacts of metallic meteoroids can be identified based on chemical analysis of additions to pit glasses. Also, the data are consistent with the idea that the proportion of metallic meteoroids impacting the moon (1 to 10%) is essentially the same as the proportion of metallic meteorites observed to fall on the earth.

Brownlee et al. (1975) also observed iron and magnesium enhancements in two pits in monomineralic plagioclase feldspar samples. On the assumption that these enhancements were due to additions of chondritic composition, up to 0.1% of the glass formed could be derived from the impacting meteoroid. This confirms the view that, at most, only very small amounts of material from impacting meteoroids may be expected to remain in the pit glasses formed.

Size Distribution

Cumulative microcrater pit size distributions measured optically on sample 12054 by Hartung et al. (1973) and on sample 60015 by Neukum et al. (1973) and Fechtig et al. (1974) deviate from one another by less than 50% over the entire pit diameter range from 20 to 500 microns.

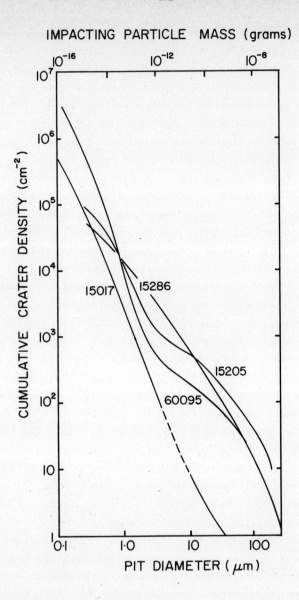

Figure 1. Cumulative microcrater pit size distributions for those samples providing data for a wide range of sizes on either side a pit diameter of 10 microns. Sources of data are for 15017, Morrison et al. (1973), 15205, Schneider et al. (1973); 15286, Brownlee et al. (1973), and 60095, Brownlee et al. (1975). Mass scale is based on calibration of Gault (1973).

Four size distributions for smaller craters have been measured where statistically significant numbers of pits greater than 20 microns in diameter were counted. These sets of data are shown in figure 1 and illustrate discrepancies which remain unresolved. Two sets of data have a definite "break" at a pit diameter between 1 and 10 microns; the other two sets do not. The curves are not normalized, so different absolute crater densities may be explained partially by differences in exposure times, exposure solid angles, or both. Relative differences between the curves remain a controversy.

The trend of present thinking is toward accepting the steeper slopes (-2 to -3) at pit diameters less than 2 microns and the existence of an inflection in the distribution. This type of bimodal size distribution is evidence for a two-component model for the interplanetary dust. The differences in the distributions may be real or may be due to pit recognition problems or the poor quality of the surfaces observed. If two components of the interplanetary dust exist, then rock surfaces with different exposure solid angles exposed for different time intervals at different times in the past might be expected to yield different microcrater pit size distributions. Knowledge of these parameters is important for future pit size distribution measurements.

Conversion of pit size distribution data to meteoroid energy, mass, or size information requires application of empirical "calibration" curves based on laboratory produced data. This aspect of the problem has been reviewed by Hörz et al. (1975a).

Impact Directions

If the arrival direction of particles impacting the moon can be determined, then it is possible to obtain information on the orbital characteristics of the particles. Blanford et al. (1974) analyzed microcrater populations in lunar-north-facing and lunar-east-facing vugs or cavities and found no difference in the size distributions observed for pit diameters between 0.1 and 1 micron. Similar studies by Hutcheon (1975) indicated no differences in pit circularity or depth-to-diameter ratio for particles arriving along and out of the ecliptic. By measuring the exposure age for each surface studied Hutcheon (1975) found up to a factor of 20 more particles arriving along the ecliptic per unit time than those approaching from out of the ecliptic. However, using the same experimental approach on different samples, Morrison and Zinner (1975) found no difference in the flux of particles arriving from the two different directions. Clearly, more experimental work is needed.

"Clocks" to Measure Surface Exposure Time

Considerable effort has been expended to obtain data on the present flux of meteoroids by making satellite-borne meteoroid-detection experiments. Because lunar rocks are exposed to space for millions of years, it is, in principle, possible to obtain values for the past flux of interplanetary dust particles by studying the microcrater populations formed on lunar rocks.

Flux is a rate parameter, which implies a measurement of time is required as a part of determining meteoroid flux. For laboratory and satellite-borne experiments the measurement of time is straightforward, if not trivial. Measurement of time over periods of thousands or millions of years is not straightforward and is probably the most difficult and uncertain aspect of lunar microcrater studies. To provide a basis for evaluating results of meteoroid flux measurements, we will describe in this section each of the time-measuring techniques or "clocks" which have been used to measure the exposure times of lunar rock meteoroid "detector" surfaces. In the next section, we will discuss how the times measured have been related to numbers of meteoroid impacts to yield flux information.

In general, time is determined by measuring the quantity of something which accumulates with time and by dividing that quantity by the rate of accumulation, which is known or independently determined and usually assumed to be constant. Basically, there are two types of things that accumulate at or near the lunar surface, fast-moving particles from outside the solar system, galactic cosmic rays, and similar but less energetic particles from the sun, solar wind and solar flare particles. Both solar and galactic particles penetrate into lunar surface material and cause nuclear reactions, which yield a variety of identifiable products, and ionization damage to crystal structures, which may be revealed as etched tracks. Because galactic particles are more energetic, they penetrate deeper and depth profiles of their nuclear reaction products are flatter. Galactic cosmic ray nuclear reaction products are dominant at depths greater than about 1 cm, and their tracks are dominant at depths greater than a few mm. Solar particle tracks and nuclear reaction products have steeper depth profiles and are dominant at shallower depths. For each "clock" there is a characteristic depth range, such that material within that range will be considered "exposed." Thus, because a rock may be eroding by meteoroid impact or slightly buried in the regolith, an exposure time measured by one "clock," in general, will not agree with an exposure time measured by another clock. Perhaps the most difficult aspect of determining meteoroid fluxes from lunar rock data is establishing the fact that the exposure time measured for a surface corresponds directly to the time interval during which meteoroids were impacting the surface presently observed.

Galactic Cosmic Ray Track "Clock"

The upper few cm of material exposed at the lunar surface is regularly penetrated by energetic charged particles or cosmic rays. Damage to crystal lattice structure results from such penetration events. Under certain conditions, by properly etching polished crystal surfaces, the tracks produced by these events may be revealed. Observable etched cosmic ray tracks are produced effectively only by very heavy (VH) nuclei ($20 \leq Z \leq 29$) (Fleischer et al., 1967a, 1967b).

Nuclei with $Z < 20$ do not produce etchable tracks because the average energy loss or primary ionization rate is too low during penetration of these particles. Nuclei with $Z > 29$ are too few in number to be statistically significant. Through a limited depth range beginning a few mm below lunar rock surfaces, tracks produced by VH galactic cosmic ray particles predominate over those from by other processes, such as spontaneous fission, spallation reactions, and solar flare particle penetrations (Crozaz et al., 1970). The production rate of VH cosmic ray tracks at a given depth depends primarily on the flux of cosmic ray particles with different energies per nucleon, the penetration ranges corresponding to those energies, and the range over which a track may be revealed by etching. Flux-vs.-energy data are obtained through experiments to analyze cosmic rays. Range-vs.-energy data are derived from laboratory penetration experiments. The etchable range parameter is difficult to evaluate because it varies with the composition and crystal structure of the host material, the etching conditions, and also the technique used to observe the etched tracks. Furthermore, the etchable range may be affected by the time-temperature history of the host material. More detailed discussions of these parameters are given by Fleischer et al. (1967a) and Lal et al. (1969).

Even if the galactic-cosmic-ray-track "clock" works effectively, the time measured is "ideal" in the sense that it corresponds to an assumed constant depth (of the order 1 cm) below an exposed surface. There is no guarantee that a point 1 cm below the surface at the time a rock was collected had been so throughout its near-surface history. In fact, the reverse situation is more often the case. From studies of Gault et al. (1972) based on expected meteoroid impact rates and laboratory cratering experiments and from actual measurement of track populations found on different sides of lunar rocks (Crozaz et al., 1970; Fleischer et al., 1970, Lal et al., 1970), we would expect that due to burial and excavation, rupture, and erosion of lunar rocks the depth history of a point within a rock would be very complicated. Therefore, in most cases the galactic-cosmic-ray-track "clock" does not record the time a particular rock surface is exposed to impacts. What is recorded is an integrated value for time at a given depth, i.e. the subdecimeter exposure age of Bhandari et al. (1971).

However, we may expect that occasionally through excavation by an impact event or rupture of a larger rock a surface will become exposed on the lunar surface without having a cosmic ray track record acquired previously. If such a rock is collected before significant erosion occurs or before superposition of one crater by another is common, then the galactic-cosmic-ray-track "clock" would record the desired time interval. This latter requirement is important both because the interval recorded by the accumulation of tracks would be unambiguous and because there would exist a one-to-one correspondence between craters and impacting meteoroids, i.e. the surface would not have approached equilibrium with respect to crater superposition.

Cosmogenic Nuclide "Clocks"

Cosmic ray particles with low atomic numbers do not leave observable tracks, but they can alter exposed materials by causing nuclear reactions which results in changing the isotopic composition of a host rock. Two types of reaction products are especially useful because their concentrations may be measured with extremely high sensitivity. These are radioactive products, which can be measured using counting methods, and rare gases, which can be measured using mass spectrometry.

The main disadvantage of using the accumulation of cosmogenic radioactive nuclides or rare gases as an exposure time measurement technique is that most nuclear reactions producing these materials occur over range of depths of up to 1 meter. Thus, for the cosmogenic nuclide "clock" also, the time recorded usually does not correspond to the time during which impacts were occurring on the rock. Because cosmogenic nuclides are formed over a greater range of depth than are cosmic ray tracks, the exposure ages derived are systematically higher (see a recent summary of exposure age data by Crozaz et al., 1974 and Hörz et al., 1975c).

However, as mentioned, a rock may be brought to the surface without having suffered cosmic-ray-produced nuclear reactions. In this case, both the galactic-cosmic-ray-track and cosmogenic-nuclide "clocks" will start from zero at the same time and give concordan exposure ages until erosion affects the track population significantly or the rock is otherwise destroyed or buried. Therefore, by using different "clocks" to measure the exposure age of the same, carefully selected sample, one "clock" may be calibrated to a second "clock." Walker and Yuhas (1973) and Behrmann et al. (1973) have measured the depth distribution of galactic cosmic ray tracks and the Kr-Kr exposure age for the same, recently exposed lunar rock. Thus, they have obtained a value for the production rate of galactic cosmic ray tracks based on a Kr-Kr exposure age.

The Kr-Kr "clock" (Marti, 1967) relies on the accumulation of a stable cosmogenic isotope of Kr. For the purposes of measurement, a ratio of the abundance of this isotope to that of cosmogenic ^{81}Kr is taken. ^{81}Kr is radioactive, and its abundance reaches a constant equilibrium value after several mean lifetimes of 0.303×10^6 yr. If the ratio of the production rates of ^{81}Kr and the stable Kr isotope is known, then the time over which accumulation occurs, i.e. the exposure age, may be determined in terms of the number of ^{81}Kr mean lifetimes. The ratio of these production rates is obtained by interpolating between values for the abundance ratios of cosmogenic ^{80}Kr to the stable isotope and ^{82}Kr to the stable isotope. These abundance ratios are equivalent to production rate ratios because all isotopes involved are stable. Thus, for this method, in addition to Kr isotopic ratios, the critical parameter requiring laboratory measurement is the mean lifetime of ^{81}Kr atoms or the disintegration rate for ^{81}Kr. Only cosmogenic Kr is used in this approach, so corrections must be applied for solar and fission-produced Kr present in the sample. Several similar exposure time "clocks" based on accumulation of cosmogenic rare gases exist, but these will not be discussed because they have not been used in connection with microcrater measurements.

The other type of cosmogenic nuclide "clock" providing a type of exposure time information relies on the accumulation of radioactive nuclides produced by low-energy interactions. Depth profiles for the nuclear reaction products for different incident particles with different energies reacting with different target elements have been calculated by Finkel et al. (1971) and Rancitelli et al. (1971). After a time interval which is long compared to the mean lifetime of a radioactive nuclear reaction product equilibrium will be reached between the production and decay of the radioactive isotope. Equilibrium values or depth profiles may be calculated for surfaces experiencing different rates of impact erosion. For isotopes with half-lives short compared to the time required for significant erosion, measured and calculated depth profiles agree without introducing a correction for losses due to erosion. For isotopes with half-lives comparable to the time required for significant erosion, for example, ^{53}Mn and ^{26}Al with half-lives of 3.7 m.y. and 0.74 m.y., measured depth profiles can be made to fit calculated profiles if 0.5 mm/m.y. (Finkel et al., 1971) or 1 to 5 mm/m.y. (Rancitelli et al., 1971) erosion rates are assumed. As shown by Hörz et al. (1971) and Gault et al. (1972), erosion rates may be transformed through laboratory impact cratering experiments and knowledge of the crater or meteoroid size distribution into estimates of meteoroid flux. Unfortunately, these measurements of erosion rates are imprecise and require analyzing the radioactivity of lunar rocks on a layer-by-layer basis, which has not been done many times.

If in a special case the integrated exposure time for a rock was less than or comparable to the mean life time for a particular cosmogenic isotope, then the radio-activity measured for the whole rock would indicate that equilibrium had not been reached between production and decay of that isotope. The growth to equilibrium is a function of time, and, in principle, this process could be used as an exposure time "clock" in this special case (Keith and Clark, 1974). Disadvantages of this "clock" are the closer equilibrium is approached, the more imprecise the time measurement becomes; the equilibrium radioactivity depends on the composition of the rock in a way that is determined empirically; and only a few rocks have such a short exposure in the lunar regolith.

Solar Flare Track "Clock"

The most widely used "clock" to measure actual surface exposure time is based on the observation of tracks produced in the upper mm of lunar surfaces by VH nuclei accelerated during solar flares (Crozaz et al., 1971, Fleisher et al., 1971a, Bhandari et al., 1971). The restriction of solar flare tracks to the upper mm of surface material is a definite advantage for this approach because impact erosion events are also dominant at this scale (Hörz et al., 1971, Hartung et al., 1972b; Gault et al., 1972).

For galactic cosmic ray tracks, exposure ages depended on knowledge of particle flux vs. energy and energy vs. particle penetration depth. Another apparent advantage for solar flare tracks resulted from the return of the glass filter from the Surveyor III camera system which was exposed to space for 2-1/2 years. Track densities and depth profiles were measured in silicate material exposed for a known length of time, thus providing a standard track production rate without the need to use laboratory-produced energy-vs.-depth data (Crozaz and Walker, 1971, Fleisher et al., 1971b, Barber et al., 1971).

However, several problems exist for this method also. The Surveyor III tracks were accumulated over a 2-1/2 year interval, which is only a fraction of the known 11-year solar cycle, and the cycle itself may have been anomalous. Thus, a question remains regarding the representivity of the standard track production rate data. Ionization tracks formed in silicate materials do not survive unchanged indefinitely. Tracks grow smaller in length and diameter under the influence of time and temperature (Storzer and Wagner, 1969). The effect is extremely nonlinear with respect to time and is different for different chemical compositions and crystal structures (Crozaz et al., 1970). Track annealing proceeds rapidly soon after formation and very slowly before the track becomes completely annealed. Tracks in glass anneal much more easily than those in crystalline material. The etchable range of a track is not a well-defined parameter (Walker and Yuhas, 1973). Values differ for different etching conditions and observing techniques. Etched tracks produced recently in laboratory experiments are significantly longer than those observed on exposed lunar surface samples.

To improve the solar flare track "clock" Hutcheon et al. (1974) measured a track density profile from a depth of 1 micron to 3.3 cm using a sample which had not suffered significantly from impact erosion. In this way, the solar flare track "clock" could be calibrated using the galactic cosmic ray "clock" instead of relying on a production rate derived from the tracks in the Surveyor III filter glass. The calibration of the galactic cosmic ray "clock" using the Kr-Kr cosmogenic nuclide "clock" was discussed previously. Solar flare track production rates using the galactic cosmic ray "clock" of Walker and Yuhas (1973) is about 50% less than those rates based on Surveyor III filter glass data, assuming an etchable range of 100 microns, not 30 microns as used before, and assuming the average particle flux during the Surveyor III exposure was one-quarter the long-term average flux. Similar work by others (Crozaz et al., 1974; Blanford et al., 1975) should lead to an improved, well cross-calibrated, solar flare track clock. At present, unfortunately, considerable uncertainty is associated with absolute time measurements made using the solar flare track "clock." Different workers using the same data could obtain ages different by as much as a factor of 10 due mainly to uncertainties in the solar flare track production rate (Morrison and Zinner, 1975). Relative ages obtained by the same worker are much more precise. We estimate for these an uncertainty of less than 50%.

Solar Wind Sputtering "Clock"

Ions, mainly of hydrogen and helium, accelerated by the solar wind encounter the lunar surface continuously and are capable of removing surface atoms by sputtering. If the amount of material removed by sputtering and the sputtering rate were known or measured, then an exposure time could be measured directly. The sputtering rate has been determined by McDonnell and Ashworth (1972) and others for lunar rock surfaces based on in situ measurements of solar wind ion fluxes and laboratory experiments to determine the sputtering yield or efficiency. Although the amount of material sputtered off lunar sample surfaces has not been measured, McDonnell and Flavill (1974) have determined the expected lifetime for microcraters of different sizes against gradual removal by sputtering. They have suggested that for craters smaller than some critical size sputtering is the dominant crater destroying mechanism. Therefore, we may expect an equilibrium to exist between production of craters by impact and destruction of craters by sputtering. Because the crater destruction rate is known (inverse of the expected crater lifetime), the crater production rate (meteoroid flux) may be calculated. The actual calculation is more complicated than described here because a range of crater sizes must be considered along with the effects of crater destruction by superposition and impact erosion (Ashworth and McDonnell, 1973, 1974).

Discussion of Meteoroid Flux Measurements

The problem of establishing a meteoroid flux may be approached from two directions. One is to measure the crater density and exposure time for a non-equilibrium surface. The ratio of these two parameters is equivalent to the flux, which can be compared to independent flux measurements. The other approach is to measure the crater density and apply an independently measured value for the flux to find an exposure time, which can be compared to other measured exposure times for the same surface. The two approaches are essentially equivalent, the one used depending on the objective of the worker. Meteoroid fluxes are of interest in meteoroid science, exposure times in lunar science.

Exposure times based on crater counts were determined by Neukum et al. (1970), but no comparison with an independent "clock" was made. Impact erosion rates were estimated by Hörz et al. (1971), Bloch et al. (1971), McDonnell and Ashworth (1972), and Gault et al. (1972), and compared to rates based on solar flare and cosmic ray track "clocks." In addition, Gault et al. (1972) considered the expected rate of lunar rock destruction. They concluded these rates were higher using present-day fluxes than those based on the track "clocks" which integrate over long times, thus the meteoroid flux must have been lower in the past. However, Crozaz et al. (1974) have shown that most cosmic ray track exposure ages give maximum values for surface exposure times. This, together with the tendency for older rocks, those surviving impact destruction, to be collected, removes part of noted discrepancy. They also point out erosion rates based on solar flare tracks refer to "microerosion" and not "mass-wastage" caused by mm-sized impacts. Other "mass-wastage" determinations, for example those based on cosmogenic ^{53}Mn and ^{26}Al "clocks" (Finkel et al. 1971; Rancitelli et al., 1971; Wahlen et al., 1972), give erosion rate values in agreement with those calculated. Thus, from erosion-rate or rock-destruction-rate arguments, no basis remains for suggesting a lower meteoroid flux in the past.

Populations of mm-sized craters on rocks for which independent exposure ages had been measured were studied by Hartung et al. (1972b, 1973) and Morrison et al. (1972, 1973). Morrison et al. (1972, 1973) assumed that the highest density of craters observed corresponded to a surface in equilibrium with respect to crater superposition and that all other surfaces having lower densities of craters had not reached equilibrium and, therefore, could yield exposure time or meteoroid flux results. Measured exposure times using the "clocks" described previously were found almost always to exceed the time required to form the observed microcrater populations using present-day meteoroid flux estimates. Consequently, they suggested a lower meteoroid flux may have existed in the past. In contrast, Hörz et al. (1971) and Hartung et al. (1972b, 1973) considered it likely that two surfaces having distinctly different densities of craters could both be in equilibrium. The differences presumably could be explained in terms of different mechanical properties of different rocks. Subsequent work by Neukum et al. (1973) and Schneider and Hörz (1974) supports this latter view, thus eliminating the need for a lower meteoroid flux in the past.

An approach to estimate rock exposure times from crater population data was used by Fechtig et al. (1975) based on the size of crater at which the transition from an equilibrium to a production population occurred. In this view, the smaller the crater size still in production, the younger the exposure age of the surface (Gault, 1970; Shoemaker, 1971). Results similar to those of Morrison et al. (1972, 1973) were suggested, but this approach also failed to demonstrate that the crater population observed developed over the same time interval measured by the "clock" being used for comparison. In addition, we may expect larger cratering events, those presumed to correspond to a production crater population, are necessarily under-represented on rock surfaces because these events contribute to rock destruction. The largest craters are also destroyed in this process. Only surviving rocks and the corresponding population of smaller craters can be observed (Hartung et al., 1973, Hörz et al., 1975c).

To estimate the meteoroid flux Hartung et al. (1972b, 1973) counted craters with pit diameters greater than 500 microns on rocks with exposure ages measured using different "clocks." Because some tracks or nuclides could have been produced before the rock was exposed to cratering, the age corresponding to the observed crater population was a maximum value. Because some of the craters may not have been observed due to superposition effects, a minimum value for the areal density of craters was obtained. Both of these factors led to a minimum value for the meteoroid flux using this approach. Because this resulted in a minimum value for the meteoroid flux which was less than that derived from satellite-borne experiments, no constraint could be placed on possible time variations of the flux.

Populations of micron-sized craters have been "dated" using the solar flare track "clock" by Neukum et al. (1972) and Schneider et al. (1973). For this work, mostly glass surfaces were used. This provided visual assurance that effects of crater superposition and erosion would be minimized. In addition, solar flare track depth profiles were measured and found to be similar to that obtained for the Surveyor III camera filter, thus confirming that crater and track production occurred over the same time interval. Because tracks in glass are severely reduced by annealing, wherever possible tracks were measured in pyroxene grains trapped in the glass. It was assumed that any track record in the grains before being trapped in the glass were annealed while the glass cooled. Meteoroid fluxes determined in this way also fell below the present-day values from satellite-borne experiments. The data for several surfaces exposed for different times could be interpreted as indicating a flux in the past lower than or equal to the present-day flux because of uncertainties involved. However, for all of these surfaces recent exposure, at the time of collection, could not be confirmed, so such an interpretation would be questionable.

Hartung and Storzer (1974) successfully observed solar flare tracks in individual sub-mm-sized glass-lined pits. Exposure-ages estimated for over fifty individual craters were concentrated toward younger ages, thus indicating a lower flux in the past. Although differential flux data may be obtained with this approach, severe annealing corrections (up to a factor of 50), which are poorly understood, were required. This result may also be questioned because it indicates a flux increasing during only the last $\sim 10^4$ years with no similar peak occurring any time during the preceding $\sim 10^5$ years.

By applying the solar wind sputtering "clock" McDonnell and Flavill (1974), McDonnell et al. (1975), and McDonnell and Carey (1975) also have concluded meteoroid fluxes have been lower in the past. Flux information, which is equivalent to rate or time information, was derived by using an equilibrium population of craters. At first glance, it would appear this approach could not succeed because time information cannot be extracted from an equilibrium situation. However, because two processes are acting to destroy craters, sputtering and "erasure" (crater superposition), each affecting different-sized craters to a different extent, the shape of the equilibrium crater size distribution becomes diagnostic. If the shape of the production crater size distribution, the size and shape of the equilibrium size distribution, and the crater destruction rate by sputtering as a function of crater size are known, the production rate of different-sized craters (the flux) may be determined.

One question related to this approach involves the assertion that both sputtering and "erasure" (superposition) are effective mechanisms for destruction of craters about 1 micron in diameter. So far no workers using scanning electron microscopes capable of resolution better than 0.1 micron have reported observing any effects attributable to sputtering. A second problem is that no measurement of an equilibrium crater size distribution for craters less than about 50 microns in diameter exists. Because mm-sized-and-larger craters dominate in the superposition of all smaller craters (a consequence indicated by the flattening of the crater size distribution in this range), the relative numbers of all smaller craters should reflect directly the relative numbers of those craters produced, i.e. the equilibrium distribution for small craters should parallel the production distribution for those craters. No attempt has been made to scan a large, representative, equilibrium, surface on a scale of 10 microns or less.

More recently, Blanford et al. (1974), Hutcheon et al. (1974), and Morrison and Zinner (1975 and this volume) have selected samples from within vugs, cavities, and crevices of lunar rocks to avoid "background" tracks. Even with this approach, Hutcheon et al. (1974) found it necessary to distinguish between acceptable and background tracks based on an evaluation of their penetration direction into the rock, a procedure requiring extreme care. The use of vug crystals for flux measurements may not satisfy entirely the requirement to avoid pre-exposure accumulation of tracks. Although cosmic rays entering from directions off the axis of the vug or cavity are effectively reduced in number by the additional overlying material, those entering along the vug axis before it is exposed have no difficulty accumulating at the base of the vug just as if it were near the surface. Nevertheless, using this approach, Morrison and Zinner (1975) obtained flux estimated for 10^{-15}-gram-and-larger meteoroids averaged over times of $\sim 10^4$ and $\sim 10^6$ years which were consistent with extrapolations of Pioneer 8 and 9 data. However, to illustrate further that not all problems are solved, data obtained using this same sample, have led Morrison and Zinner (1975) to conclude that ecliptic and out-of-ecliptic fluxes of these particles are about the same, a result which is at odds with current thinking on this question based on Pioneer 8 and 9 and Heos data (Fechtig et al., 1974; Zook and Berg, 1975).

Concluding Remarks

Up to this point, no independent approach to evaluate the flux or flux history of meteoroids is without question, and some have been shown to be invalid. To improve meteoroid flux estimates, special care must be taken in sample selection. The optimum sample should have been exposed only directly to space, with no evidence for erosion, dust covering, or residence near the surface of a larger rock before catastrophic rupture. The sample should be crystalline material which contains no tracks at the time exposure to space occurs and which collects tracks only during the time and from the direction of exposure. The ideal sample should have been excavated and exposed to space for the first time relatively recently ($\sim 10^5$ years ago) from a depth in excess of tens of cm with no subsequent tumbling until it is collected.

The possibility exists that most of the suggestions that a lower flux existed in the past stems from the fact that crater densities tend to be under estimated because craters can be lost by superposition or otherwise not accounted for and exposure times tend to be over estimated because accumulation of tracks, nuclides, or other "things" can proceed before the crater-recording surface becomes exposed and operative.

Acknowledgement

This work was supported by National Aeronautics and Space Administration grants, NSG-7036 and NSG-9013. Assistance of M. O'Dowd in preparing the manuscript is gratefully acknowledged.

References

Ashworth D.G. and McDonnell J.A.M. (1973) Lunar surface microerosion related to interplanetary dust particle distributions. In Space Research-XIII, pp. 1071-1083. Akademie-Verlag, Berlin.

Ashworth D.G. and McDonnell J.A.M. (1974) Micrometeorite influx rates on the lunar surface deduced from revised estimates of the solar wind sputter rate and lunar surface crater statistics. In Space Research-XIV, pp. 723-729. Akademie-Verlag, Berlin.

Barber D.J., Cowsik R., Hutcheon I.D., Price P.B., and Rajan R.S. (1971) Solar flares, the lunar surface, and gas-rich meteorites. Proc. Lunar Sci. Conf. 2nd, p. 2705-2714.

Behrmann C., Crozaz G., Drozd R., Hohenberg C., Ralston C., Walker R., and Yuhas D. (1973) Cosmic-ray exposure history of North Ray and South Ray material. Proc. Lunar Sci. Conf. 4th, p. 1957-1974.

Bhandari N., Bhat S., Lal D., Rajagopalan G., Tamhane A.S., and Venkatavaradan V.S. (1971) High resolution time averaged (millions of years) energy spectrum and chemical composition of iron-group cosmic ray nuclei at 1 A.U. based on fossil tracks in Apollo samples. Proc. Lunar Sci. Conf. 2nd, p. 2611-2619.

Blanford G.E., Fruland R.M., McKay D.S., and Morrison D.A. (1974) Lunar surface phenomena: Solar flare track gradients, microcraters, and accretionary particles. Proc. Lunar Sci. Conf. 5th, p. 2501-2526.

Blanford G.E., Fruland R.M., and Morrison D.A. (1975) Long term differential energy spectrum for solar flare iron-group particles. Proc. Lunar Sci. Conf. 6th, in press.

Bloch M.R., Fechtig H., Gentner W., Neukum G., and Schneider E. (1971) Meteorite impact craters, crater simulations, and the meteoroid flux in the early solar system. Proc. Lunar Sci. Conf. 2nd, p. 2639-2652.

Brownlee D.E., Hörz F., Vedder J.F., Gault D.E., and Hartung J.B. (1973) Some physical parameters of micrometeoroids. Proc. Lunar Sci. Conf. 4th, p. 3197-3212.

Brownlee D.E., Hörz F., Hartung J.B., and Gault D.E. (1975) Density, chemistry, and size distribution of interplanetary dust. Proc. Lunar Sci. Conf. 6th, in press.

Chao E.C.T., Borman J.A., Minkin J.A., James O.B., and Desborough G.A. (1970). Lunar glasses of impact origin: Physical and chemical characteristics and geologic implications. J. Geophys. Res. 75, 7445-7479.

Crozaz G. and Walker R. (1971) Solar particle tracks in glass from the Surveyor III spacecraft. Science 171, 1237-1239.

Crozaz G., Haack U., Hair M., Maurette M., Walker R., and Woolum D. (1970) Nuclear track studies of ancient solar radiations and dynamic lunar surface processes. Proc. Apollo 11 Lunar Sci. Conf. p. 2051-2080.

Crozaz G., Walker R., and Woolum D. (1971) Nuclear track studies of dynamic surface processes on the moon and the constancy of solar activity. Proc. Lunar Sci. Conf. 2nd, p. 2543-2558.

Crozaz G., Drozd R., Hohenberg C., Morgan C., Ralston C., Walker R., and Yuhas D. (1974) Lunar surface dynamics: Some general conclusions and new results from Spollo 16 and 17. Proc. Lunar Sci. Conf. 5th, p. 2475-2499.

Durrani S.A., Khan H.A., Bull R.K., Dorling G.W., and Fremlin J.H. (1974) Charged-particle and micrometeorite impacts on the lunar surface. Proc. Lunar Sci. Conf. 5th, p. 2543-2560.

Fechtig H., Hartung J.B., Nagel K., and Neukum G. (1974) Lunar microcrater studies, derived meteoroid fluxes, and comparison with satellite-borne experiments. Proc. Lunar Sci. Conf. 5th, p. 2463-2474.

Fechtig H., Gentner W., Hartung J.B., Nagel K., Neukum G., Schneider E., and Storzer D. (1975) Microcraters on lunar samples. Proc. Soviet-American Conf. on Cosmochemistry of the Moon and the Planets, June 1974, in press.

Finkel R.C., Arnold J.R., Imamura M., Reedy R. C., Fruchter J.S., Loosli H.H., Evans J.C., Delany A.C., and Shedlovsky J.P. (1971) Depth variations of cosmogenic nuclides in a lunar surface rock and lunar soil. Proc. Lunar Sci. Conf. 2nd, p. 1773-1789.

Fleischer R.L., Price P.B., Walker R.M. and Maurette M. (1967a) Origins of fossil charged particle tracks in meteorites. J. Geophys. Res. 72, 331-353.

Fleischer R.L., Price P.B., Walker R.M., Maurette M., and Morgan G. (1967b) Tracks of heavy primary cosmic rays in meteorites. J. Geophys. Res. 72, 355-366.
Fleischer R.L., Haines E.L., Hart H.R.,Jr., Woods R.T., and Comstock G.M. (1970) The particle track record of the Sea of Tranquility. Proc. Apollo 11 Lunar Sci. Conf. p. 2103-2120.
Fleischer R.L., Hart H.R.,Jr., Comstock G.M., and Evwaraye A.O. (1971a) The particle track record of the Ocean of Storms. Proc. Lunar Sci. Conf. 2nd, p. 2559-2568.
Fleischer R.L., Hart H.R.,Jr., and Comstock G.M. (1971b) Very heavy solar cosmic rays: Energy spectrum and implications for lunar erosion. Science 171, 1240-2342.
Gault D.E. (1970) Saturation and equilibrium conditions for impact cratering on the lunar surface: Criteria and implications. Radio Sci. 5, 273-291.
Gault D.E. (1973) Displaced mass, depth, diameter and effects of oblique trajectories for impact craters formed in dense crystalline rocks. The Moon 6, 32-44.
Gault D.E., Hörz F., and Hartung J.B. (1972) Effects of microcratering on the lunar surface. Proc. Lunar Sci. Conf. 3rd, p. 2713-2734.
Gibbons R.V., Morris R.V., Hörz F., and Thompson T.D. (1975) Petrographic and ferromagnetic resonance studies of experimentally shocked regolith analogues. Proc. Lunar Sci. Conf. 6th, in press.
Hartung J.B., Hörz F., McKay D.S., and Baiamonte F.L. (1972a) Surface features on glass spherules from the Luna 16 sample. The Moon 5, 436-446.
Hartung J.B., Hörz F., and Gault D.E. (1972b) Lunar microcraters and interplanetary dust. Proc. Lunar Sci. Conf. 3rd, p. 2735-2753.
Hartung J.B., Hörz F., Aitken F.K., Gault D.E., and Brownlee D.E. (1973) The development of microcrater populations on lunar rocks. Proc. Lunar Sci. Conf. 4th, p. 3213-3234.
Hartung J.B. and Storzer D. (1974) Lunar microcraters and their solar flare track record. Proc. Lunar Sci. Conf. 5th, p. 2527-2541.
Hartung J.B., Hodges F., Hörz F., and Storzer D. (1975) Microcrater investigations on lunar rock 12002. Proc. Lunar Sci. Conf. 6th, in press.
Hörz F., Hartung J.B., and Gault D.E. (1971) Micrometeorite craters on lunar rock surfaces. J. Geophys. Res. 76, 5770-5798.
Hörz F., Brownlee D.E., Fechtig H., Hartung J.B., Morrison D.A., Neukum G., Schneider E., Vedder J.F., and Gault D.E. (1975a) Lunar microcraters: Implications for the micrometeoroid complex. Planet. Space Sci. 23, 151-172.
Hörz F., Morrison D.A., Gault D.E., Oberbec V.R., Quaide W.L., Vedder J.F., Brownlee D.E., and Hartung J.B. (1975b) The micrometeoroid complex and evolution of the lunar regolith. Proc. Soviet-American Conf. on Cosmochemistry of the Moon and Planets, June 1974, in press.
Hörz F., Gibbons V.R., Gault D.E., Hartung J.B., and Brownlee D.E. (1975c) Some correlations of rock exposure ages and regolith dynamics. Proc. Lunar Sci. Conf. 6th, in press.
Housley R.M., Grant R.W., Muir A.H.,Jr., Blander M., and Abdel-Gawad M. (1971) Mössbauer studies of Apollo 12 samples. Proc. Lunar Sci. Conf. 2nd, p. 2125-2136.
Hutcheon I.D. (1975) Microcraters in oriented vugs-evidence for anisotropy in the micrometeoroid flux (abstract). In Lunar Science VI, p. 420-422. The Lunar Science Institute, Houston.
Hutcheon I.D., MacDougall D., and Price P.B. (1974) Improved determination of the long-term average Fe spectrum from ~ 1 to ~ 460 MeV/amu. Proc. Lunar Sci. Conf. 5th, p. 2561-2576.
Keith J.E. and Clark R.S. (1974) The saturated activity of ^{26}Al in lunar samples as a function of chemical composition and the exposure ages of some lunar samples. Proc. Lunar Sci. Conf. 5th, p. 2105-2119.
Lal, D. (1969) Recent advances in the study of fossil tracks in meteorites due to heavy nuclei of the cosmic radiation. Space Sci. Rev. 9, 623-650.
Lal D., MacDougall D., Wilkening L., and Arrhenius G. (1970) Mixing of the lunar regolith and cosmic ray spectra, new evidence from fossil particle-track studies. Proc. Apollo 11 Lunar Sci. Conf. p. 2295-2303.
Marti K. (1967a) Mass-spectrometric detection of cosmic-ray-produced K^{81} in meteorites and the possibility of Kr-Kr dating. Phys. Rev. Lett. 18, 264-266.
McDonnell J.A.M. and Ashworth D.G. (1972) Erosion phenomena on the lunar surface and meteorites. In Space Research-XII, p. 333-347. Akademie-Verlag, Berlin.

McDonnell J.A.M. and Flavill R.P. (1974) Solar wind sputtering on the lunar surface: Equilibrium crater densities related to past and present microparticle influx rates. Proc. Lunar Sci. Conf. 5th, p. 2441-2449.

McDonnell J.A.M. and Carey W.C. (1975) Solar wind sputter erosion of microcrater populations on the lunar surface. Proc. Lunar Sci. Conf. 6th, in press.

McDonnell J.A.M., Ashworth D.G., Flavill R.P., Bateman D.C., and Jennison R.C. (1974) Microparticle flux distributions and erosion parameters-Temporal flux variations deduced. 18th COSPAR Meeting, Sao Paulo, Brazil. To be published in Planetary and Space Research.

Morrison D.A. and Zinner E. (1975) Studies of solar flares and impact craters in partially protected crystals. Proc. Lunar Sci. Conf. 6th, in press.

Morrison D.A., McKay D.S., Moore H.J., and Heiken G.H. (1972) Microcraters on lunar rocks. Proc. Lunar Sci. Conf. 3rd, p. 2767-2791.

Morrison D.A., McKay D.S., Fruland R.M., and Moore H.V. (1973) Microcraters on Apollo 15 and 16 rocks. Proc. Lunar Sci. Conf. 4th, p. 3235-3253.

Nagel K., Neukum G., Eichhorn G., Fechtig H., Müller O., and Schneider E. (1975) Dependencies of microcrater formation on impact parameters. Proc. Lunar Sci. Conf. 6th, in press.

Neukum G., Mehl A., Fechtig H., and Zähringer J. (1970) Impact phenomena of micrometeorites on lunar surface materials. Earth Planet. Sci. Lett. 8, 31-35.

Neukum G., Schneider E., Mehl A., Storzer D., Wagner G.A., Fechtig H., and Bloch M.R. (1972) Lunar craters and exposure ages derived from crater statistics and solar flare tracks. Proc. Lunar Sci. Conf. 3rd, p. 2793-2810.

Neukum G., Hörz F., Morrison D.A., and Hartung J.B. (1973) Crater populations on lunar rocks. Proc. Lunar Sci. Conf. 4th, p. 3255-3276.

Rancitelli L.A., Perkins R.W., Felix W.D., and Wogman N.A. (1971) Erosion and mixing of the lunar surface from cosmogenic and primordial radionuclide measurements in Apollo 12 lunar samples. Proc. Lunar Sci. Conf. 2nd, p. 1757-1772.

Schneider E. and Hörz F. (1974) Microcrater populations on Apollo 17 rocks. Icarus 22, 459-473.

Schneider E., Storzer D., Mehl B., Hartung J.B., Fechtig H., and Gentner W. (1973) Microcraters on Apollo 15 and 16 samples and corresponding dust fluxes. Proc. Lunar Sci. Conf. 4th, p. 3277-3290.

Shoemaker E.M. (1971) Origin of fragmental debris on the lunar surface and the history of bombardment of the moon. Instituto de Investigaciones Geologicas de la Diputacion Provincial, v. XXV, Universidad de Barcelona.

Smith D., Adams N.G., and Khan H.A. (1974) Flux and composition of micrometeoroids in the diameter range 1-10μm. Nature 252, 101-106.

Storzer D. and Wagner G.A. (1969) Correction of thermally lowered fission track ages of tektites. Earth Planet. Sci. Lett. 5, 463-468.

Wahlen M., Honda M., Imamura M., Fruchter J.S., Finkel R.C., Kohl C.P., Arnold J.R., and Reedy R.C. (1972) Cosmogenic nuclides in football-sized rocks. Proc. Lunar Sci. Conf. 3rd, p. 1719-1732.

Walker R. and Yuhas D. (1973) Cosmic ray track production rates in lunar materials. Proc. Lunar Sci. Conf. 4th, p. 2379-2389.

Zook H.A. and Berg O.E. (1975) A source for hypervelocity cosmic dust particles. Planet. Space Sci. 23(1), 183-203.

2.2.2 THE SIZE FREQUENCY DISTRIBUTION AND RATE OF PRODUCTION OF MICROCRATERS

D. A. Morrison, NASA Johnson Space Center, Houston, TX 77058

E. Zinner, McDonnell Center for the Space Sciences, Washington University
St. Louis, MO 63130

Abstract

Crater size frequency distributions vary to a degree which probably cannot be explained by variations in lunar surface orientation of the crater detectors or changes in micrometeoroid flux. Questions of sample representativity suggest that high ratios of small to large craters of micrometeoroids (e.g., a million 1.0 micron craters for each 500 micron crater) should be the most reliable. We obtain a flux for particles producing 0.1 micron diameter craters of approximately 300 per cm^2 per steradian per year. We observe no anisotropy in the submicron particle flux between the plane of the ecliptic and the normal in the direction of lunar north. No change in flux over a 10^6 year period is indicated by our data.

Crater Size Frequency Distribution

Cumulative crater size frequency distributions developed on lunar rocks reflect the relative abundances of micrometeorites in terms of mass, velocity and density. Figure 1 shows size frequency distributions from various samples normalized (graphically) at a central pit diameter of 100 microns. The distribution of Fechtig et al. (1974) represents a compilation of normalized data from a large number of samples, one of which (15205) provides crater frequency data over nearly the whole range shown. We have plotted data from 15205 (Schneider et al., 1973) separately and included counting statistics error bars to show that the variations between distributions are beyond statistical error. Data for 15015 and 15017 (Morrison et al., 1973) range from crater diameters of 0.1 to 125 microns and 0.1 to 35 microns respectively. The densities of craters from 0.1 to 2 microns and 250 to 500 microns were measured for 76215 by Morrison and Zinner (1975). Abundances of intermediate craters were not determined. Brownlee et al. (1975) measured the distribution of craters on 60095 from 0.1 to 60 microns diameter. We have extrapolated their data to 100 microns diameter for purposes of normalization. Also included are data from 15286 for the interval 3 to 100 microns (Brownlee et al., 1973 and 1975). Data from 72315 (Hutcheon et al., 1974) are identical to 15286 over the same diameter interval and are not plotted in Figure 1. In all cases we have plotted a distribution with a slope of -2 from a diameter of 1.0 to ≤0.1 microns because a slope of -2 in this interval appears to be constant on surfaces of varying age and lunar surface orientation. The data of Hutcheon, 1975, Blanford et al., 1974, and Morrison and Zinner (1975) show that surfaces intercepting particles with orbits normal to the plane of the ecliptic and surfaces intercepting particles confined to the plane of the ecliptic have the same size frequency distribution in the submicron crater diameter range. Blanford et al., 1974 and Poupeau et al., 1975 show that a -2 slope is also characteristic of submicron crater populations on **surfaces** exposed on or before 10^9 years B.P. Both Hutcheon (1975) and Blanford et al. (1974) have extended measurements to diameters $\leq 500\text{Å}$ with no change in slope indicated. A lower limit to crater diameters has not been satisfactorily established.

The discrepancies between the various size distributions cannot easily be explained. Lunar surface orientation information is lacking for all samples except 76215 for which surfaces oriented both normal and parallel to the lunar horizontal show the same size distribution, and for our sample of 15015 which had π geometry. As we shall later discuss, the flux as well as the distribution of particles producing submicron craters does not vary sufficiently to account for the relative depletion of small craters shown, for instance, by the distribution of Fechtig et al. (1974). The exposure geometry of 76215 (Morrison and Zinner, 1975) was nearly ideal, therefore it is highly unlikely that the differences in the size distributions can be attributed to differing lunar surface orientations because an enrichment in large particle flux normal to the ecliptic would be required. A decreasing flux in the past, as postulated by Hartung and Storzer (1975), cannot account for the variations because 15205 data indicate a relative increase in large craters ($>100\mu$) compared to 76215 exposed for four times less, requiring an increasing flux in the past to explain the differences. Perhaps the most plausible explanation lies in the difficulty of selecting representative surfaces. On 15015 (Morrison et al., 1973) approximately 2 cm^2 were examined. Submicron and larger crater densities varied from 0 (over relatively large shielded areas) to approximately 10^5 1.0 micron craters per cm^2 on the most densely cratered areas. The broad band shown in Figure 1 represents variations in crater density on the most heavily cratered surfaces. 15205 data may be biased because a relatively clean half of one fragment (Schneider et al., 1973) of a group of three from which optical data were collected was selected for SEM work. Because it was more free of secondary debris it may have had a significantly smaller solid angle and therefore not directly comparable to the other samples. It is also important to consider surface renewal by depositional processes which tend to rapidly cover small craters (Morrison et al., 1973), and the possibility of dust coatings (Morrison and Zinner, 1975).

These arguments suggest that the distributions which show the largest ratio of small to large craters (i.e., the steepest) should be the most reliable.

Micrometeoroid Flux in Space and Time

Crater size frequency distributions can be converted to flux values if exposure ages of the surfaces can be determined. For this purpose the solar flare track production rate is the most useful clock, but it is necessary to select a solar flare energy spectrum which accurately predicts the long term track production rate. Standards for the solar flare spectrum have been proposed by Blanford et al. (1974, 1975), Hutcheon et al. (1974) and Yuhas (1974). These standards differ in detail (Table 1) and exposure ages derived from them differ correspondingly (Morrison and Zinner, 1975). We prefer the Blanford et al. (1975) spectrum because in this case the track production rate has been calibrated by an independent krypton isotope measurement and our track density versus depth profiles are consistent with their spectrum. The other spectra result in higher flux values.

Table 1 lists three samples which represent different situations in terms of lunar surface orientation and surface residence time and the results bear upon flux and the distribution in space of micrometeoroids. The lunar surface orientations of each are precisely known and have been described in detail by Morrison and Zinner (1975).

Submicron diameter crater densities were measured by SEM on all three surfaces and the abundances of craters greater than or equal to 250 microns diameter were measured on 76215 by binocular microscope. Solar flare track density versus depth profiles were measured in each case with the plane of the track detector normal to the plane of the micrometeoroid detector. Because of the exposure geometry of the samples, we used a modification of the Yuhas (1974) track production computer program to apply corrections to the track density observed below the surface on which the crater counts were made. We emphasize that the solid angles for the crater and track detectors are the same therefore the relative flux is insensitive to errors in solid angle estimates. Data from all three surfaces for submicron craters result in the same flux of approximately 300 0.1 micron diameter craters, produced per cm^2 per steradian per year using the Blanford et al. (1975) spectrum.

On 76215, densities of 250 to 500 micron diameter craters (Morrison and Zinner, 1975) and the track data indicate a production of 10-15 craters of 500 microns central pit diameter or larger produced per cm^2 per 10^6 years per 2π steradians. The size frequency distribution for 15015, 15017 and 76215 has been fixed to this value (10) for the production of 500 micron diameter craters (central pit), and therefore serves as a graph of crater production as indicated on the right hand ordinate of Figure 1.

Table 1 also lists the track density at 100 microns depth and the abundances of 0.1 micron diameter craters for the three samples studied. The ratio of these values is useful for interlaboratory comparisons because it avoids variations in detector geometries. In all three cases the ratio we observe is one. This value is a factor of 50 larger than the value we calculate for the data of Hutcheon (1975) i.e., we observe more 0.1 micron diameter craters per track produced than Hutcheon (1975).

As is shown in Table 1, our data indicate no differences in the flux of particles between the plane of the ecliptic and the normal to the plane in the direction of lunar north. The ratio of the density of 0.1 micron diameter craters and the solar flare track density at 100 microns is approximately one in both cases, thus eliminating fortuitous differences or similarities which may have arisen from errors in solid angle measurements. Hutcheon (1975) has observed a depletion of a factor of 7 between the south ecliptic pole and the ecliptic plane over the same crater diameters.

The data in Table 1 also indicate no substantial change in flux of particles producing submicron craters over the time sampled. 76215 was exposed for approximately

2×10^4 years whereas 76015,105 and ,40 were exposed for approximately 10^6 years, but the crater to track ratio and flux values are very similar, suggesting a constant flux.

The lack of anisotropy indicated by our data is surprising. Although there is inherent imprecision in our methods we believe it to be improbable that error could accumulate to more than a factor of five, and the consistent results between the three different samples of radically different exposure age and geometry shown in Table 1 tend to support this view. Mandeville (1975) has reported crater and track data for 15015. Referring the given track data to the Blanford et al., 1975 solar flare spectrum results in an exposure age of 8-10 x 10^4 years for 15015 (Mandeville, 1975 suggests an exposure age of 2×10^3 years) and a crater to track ratio of 1.7. These results are in accord with the relatively high flux values shown in Table 1 and Figure 1. Consistent results between three investigators and four samples lends credence to the size frequency distribution and crater production rates shown in Figure 1 for the curve 15015, 15017, 76215, and tend to corroborate our observation of no anisotropy. If this is the case, then useful constraints are placed upon the velocity regimes, spatial density, and physical processes acting upon submicron-sized particles at one A.U. over a long term.

References

Blanford G. E., Fruland R. M., McKay D. S., and Morrison D. A. (1974) Lunar surface phenomena: Solar flare track gradients, microcraters, and accretionary particles. Proc. Lunar Sci. Conf. 5th, pp. 2501-2526.

Blanford G. E., Fruland R. M., and Morrison D. A. (1975) Long term differential energy spectrum for solar flare iron-group particles. Proc. Lunar Sci. Conf. 6th.

Brownlee D. E., Hörz F., Vedder J. F., Gault D. E., Hartung J. B. (1973) Some physical properties of micrometeoroids. Proc. Lunar Sci. Conf. 4th, pp. 3197-3212.

Brownlee D. E., Hörz F., Hartung J. B., Gault D. E. (1975) Density chemistry and size distribution of interplanetary dust. Proc. Lunar Sci. Conf. 6th.

Fechtig H., Hartung J. B., Nagel K., and Neukum G. (1974) Lunar microcrater studies, derived meteoroid fluxes, and comparison with satellite-borne experiments. Proc. Lunar Sci. Conf. 5th, pp. 2463-2474.

Hartung J. B. and Storzer D. (1974) Lunar microcraters and their solar flare track record. Proc. Lunar Sci. Conf. 5th, pp. 2527-2541.

Hutcheon I. D., Macdougall D., and Price P. B. (1974) Improved determination of the long term Fe spectrum from 1 to 460 MeV/amu. Proc. Lunar Sci. Conf. 5th, pp. 2561-2576.

Hutcheon I. D. (1975) Microcraters in oriented vugs - evidence for an anisotropy in the micrometeoroid flux. In Lunar Science VI, p. 421, The Lunar Science Institute, Houston.

Mandeville F. (1975) Microcraters observed on 15015 breccia, Abs. 6th Lunar Sci. Conf. The Lunar Science Institute, Houston.

Morrison D. A., McKay D. S., Fruland R. M., and Moore H. V. (1973) Microcraters on Apollo 15 and 16 rocks. Proc. Lunar Sci. Conf. 4th, pp. 3235-3253.

Morrison D. A., Zinner E. (1975) Studies of solar flares and impact craters in partially protected crystals. Proc. Lunar Sci. Conf. 6th.

Poupeau G., Walker R. M., Zinner E., and Morrison D. (1975) Surface exposure history of individual grains. Proc. Lunar Sci. Conf. 6th.

Schneider E., Storzer D., Mehl B., Hartung J. B., Fechtig H., and Gentner W. (1973) Microcraters on Apollo 15 and 16 samples and corresponding dust fluxes. Proc. Lunar Sci. Conf. 4th, pp. 3277-3290.

Yuhas D. E. (1974) The particle track record in lunar silicates: long-term behavior of solar and galactic VH nuclei and lunar surface dynamics. Ph.D. thesis, Washington University, St. Louis.

Sample	Solar Flare Spectrum Used for Track Production	Track Density at 100 Micron Depth (Tracks/cm²)	Pit Density (Pits >1000Å/ cm²)	Exposure Age (Years)	Dust Particle Flux (>1000Å) (Particles/cm²/ Steradian/Year)
76015,105	Blanford et al. 1975	6 x 10⁷	5 x 10⁷	4.6 x 10⁵	311
76015,40		1.2 x 10⁸	10⁸	5.3 x 10⁵	505
76215	Blanford et al. 1975	3 x 10⁶	5 x 10⁶	1.6 x 10⁴	300

TABLE 1: Particle flux and data for 76015,105 (oriented parallel to ecliptic plane), 76015,40 (normal to ecliptic plane) and 76215.

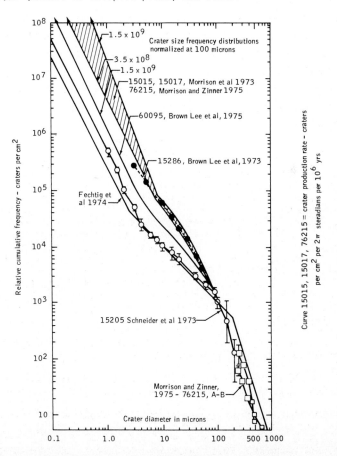

FIGURE 1: Normalized frequency distributions. Numbers for 60095; 15015, 15017, 76215 give intercept at 0.1μ. Sharp kink at 10μ in the latter is drawing artifact.

2.2.3 THE LONG TERM POPULATION OF INTERPLANETARY MICROMETEOROIDS

G. Poupeau, R.M. Walker, E. Zinner
McDonnell Center for the Space Sciences,
Washington University, St. Louis, Mo./USA

D. Morrison
NASA/Johnson Space Center, Houston, Tx./USA

Three problems will be discussed: A) The relationship between micrometeoroids and solar flare particles averaged over the recent geologic past (~ 1 my); B) the past record of this relationship as measured in lunar soils and lunar and meteoritic breccias; C) the determination of the time at which different extraterrestrial samples were exposed to free space. Data bearing on these points obtained from studies of special lunar rocks and from measurements on individual crystals removed from lunar cores will be presented. Progress in using ion-probe mass spectrometry to link measurements of micro-impact craters with the past properties of the solar wind will also be discussed. Comparing microcraters and solar flare tracks in individual crystals from lunar cores, we find no evidence of any extraordinary variations for a time span covering an interval of $\sim 10^9$ yrs. Crystals 100µ to 400µ in size in mature lunar soil samples appear to have been exposed to free space at the top of the lunar regolith for times from 10^3 to 10^4 yrs.

LUNAR SOIL MOVEMENT REGISTERED BY THE APOLLO 17 COSMIC DUST EXPERIMENT

Otto E. Berg
Goddard Space Flight Center
Greenbelt, Maryland 20771

Henry Wolf
Analytical Mechanics Associates, Inc.
Seabrook, Maryland 20801

John Rhee
Rose-Hulman Institute
Terre Haute, Indiana 47803

A. INTRODUCTION

In December, 1973, a Lunar Ejecta and Meteorites (LEAM) experiment was placed in the Taurus-Littrow area of the moon by the Apollo 17 Astronauts. Objectives of the experiment were centered around measurements of impact parameters of cosmic dust on the lunar surface. During preliminary attempts to analyze the data it became evident that the events registered by the sensors could not be attributed to cosmic dust but could only be identified with the lunar surface and the local sun angle. The nature of these data coupled with post-flight studies of instrument characteristics, have led to a conclusion that the LEAM experiment is responding primarily to a flux of highly charged, slowly moving lunar surface fines. Undoubtedly concealed in these data is the normal impact activity from cosmic dust and probably lunar ejecta, as well. This paper is based on the recognition that the bulk of events registered by the LEAM experiment are not signatures of hypervelocity cosmic dust particles, as expected, but are induced signatures of electrostatically charged and transported lunar fines.

B. THE INSTRUMENT AND FIELDS OF VIEW

The design and performance of the instrument is adequately described elsewhere in the literature (1). They are similar to those for the PIONEERS 8 and 9 cosmic dust experiments, except that the lunar experiment contains an EAST, an UP and a WEST sensor system. The front film and grid system is that portion of the basic sensor which recorded the data on charged microparticles presented here.

The LEAM experiment is located at $20°.164$ N latitude and $30°.774$ E longitude on the moon. Its WEST sensor is directed $25°$ south of west; its UP sensor is directed normal to the lunar surface; and its EAST sensor axis is directed $25°$ north of east to include in its field-of-view the solar apex direction. The field-of-view of each sensor is a square cone with a half-angle of approximately $60°$. However, the mountainous terrain blocks about 60% of the EAST and WEST sensors' fields-of-view.

C. A CASE FOR LUNAR SURFACE ACTIVITY

There are several characteristics of the data which, when considered in a process of elimination, categorically exclude extra-lunar sources as an explanation of the observed phenomena:

1) The onset of sunrise enhancement often begins as much as 60 hours before actual sunrise. This observation rules out: (a) Thermal noise from the experiment which remains at a stable, electronically ideal temperature of 250° K up till a few hours before sunrise; (b) Direct effects of electromagnetic radiation from the sun; and (c) Direct effects from the solar wind.

2) The sunrise enhancement persists for 30 to 60 hours after sunrise and wanes by two orders of magnitude while the sun is in full view. This rules out: (a) Beta meteoroids or cosmic dust ejected in a direction radially outward from the sun as seen by PIONEERS 8 and 9; (b) Again, and independently, the solar wind; and (c) Again, and independently, electromagnetic radiation, or ions, or electrons from the sun.

3) The phenomena were absent during two lunar eclipses. This observation rules out again, and independently; (a) Beta meteoroids; and (b) Solar electromagnetic radiation, electrons, and ions.

Another data characteristic which principally led to an investigation of the possibility that the instrument was detecting charged dust particles was the peculiar distribution of pulse-height -- a function of the particle's energy. Data from similar experiments in PIONEERS 8 and 9 showed a sharply decreasing distribution toward large pulse heights or the high energy particles. The LEAM data, however, showed the bulk of events represented by large pulse heights. A detailed circuit analysis of the sensors' electronics proved that the instrument would register highly charged, slowly moving particles, and would assign large pulse heights to the events as a function of the particle's speed and charge. Computer simulations of the sensors' response to charged microparticles verified the results of the circuit analysis. The precise mechanism by which the experiment responds to charged microparticles will be studied further in laboratory experiments on the spare unit. Results from that study will be reported later.

D. THE DATA

A single problem exists in the performance of the experiment on the moon. Experiment temperatures are higher than anticipated. To avoid dangerous overheating, the electronics are turned OFF during each lunar day for a period of about 8.3 earth days. To date of this publication, the instrument has been ON for 445 earth days out of 22 lunations. During that exposure it has registered a total of 7,972 events, 4900 of which are "coincidence" events requiring a simultaneous (within 1 μ sec) pulse from the front film and grid.

Figure 1 shows the number of coincident events per 3-hour period recorded by the EAST sensor for each of 6 lunations in 1973. Each lunation starts before sunset as the experiment is turned ON and continues past sunrise when the instrument is turned OFF. The time (abscissa) is in hours before and after sunrise. Of note on this figure is the consistent nature of the phenomena, relative to times of sunrise and sunset. In the 22 lunations analyzed, and processed, to date, there is no evidence of seasonal or cyclic effects. There is an interesting dearth of activity for the EAST and WEST sensors following sunset. The center of this quiet

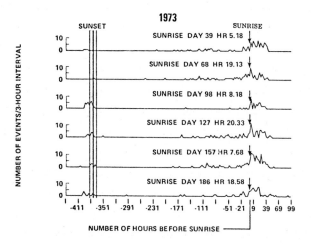

Figure 1. Number of Events per 3 Hour Interval

period occurs when the moon is preceding and aligned with the earth's orbital path. This quiet period is seen more clearly on Figure 2, which represents data from all 3 sensors over a period of 22 lunations summed into a period of one lunation. The only activity of note exhibited by the UP sensor is centered around the sunrise terminator. The EAST and WEST sensors

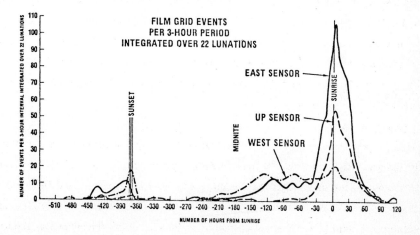

Figure 2.

show a general increase in event rates up to a first maximum near the sunset terminator, then a sharp decline to essentially no activity at 360 hours before sunrise. The quiet period lasts for 60 to 80 hours in both sensors; then the event rate slowly increases in both sensors to a quasi-plateau at about 120 hours before sunrise. The EAST sensor shows an order of magnitude increase during the passage of the sunrise terminator. The WEST sensor shows no significant sunrise terminator enhancement. The zero flux indicated after sunrise terminator enhancement is not necessarily real but represents that time when the experiment is OFF.

As a matter of interest, the average primary cosmic dust (not including lunar ejecta) at 1 AU, as measured by the PIONEERS 8 and 9 experiments, would be represented on Figure 2 as 0.4 particles/3-hour interval. However, the majority of primary particles intercepted by the PIONEER experiments were Beta meteoroids -- particles ejected by radiation pressure quasi-radially outward from the sun. Hence these should impact the UP sensor in the OFF mode. Accordingly, the average cosmic dust flux is too low to be properly represented on the scale in Figure 2.

If one assumes a mass of 10^{-12} grams for the average particle intercepted by the LEAM, the churning rate for the EAST sensor becomes 4×10^{-18} gm cm^{-2} sec^{-1}; or in 4.5 billion years it becomes 0.6 gm cm^{-2}.

Figure 3 graphically illustrates the direction and relative flux of particles intercepted by the 3 LEAM sensor systems. The arrow directions show a general direction of particle movement into the sensors. The arrow size shows the relative flux of particles intercepted.

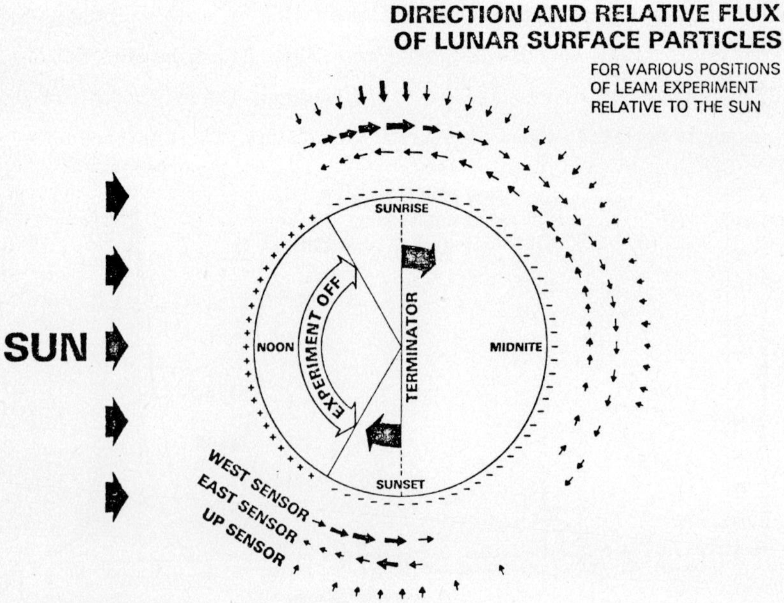

Figure 3. Direction and Relative Flux of Lunar Surface Particles

The plus and minus signs indicate polarity of the lunar surface potentials on the sunlit and dark sides. The quiet zone after sunset is shown. The fact that particles are moving both EASTWARD and WESTWARD, particularly during passage of the sunset terminator, suggests a transport of both negatively and positively charged particles.

Electrostatic transport of lunar fines has been postulated as the major mechanism for lunar soil movement for more than 2 decades (2). More recently, the effect of electrostatic forces on lunar fines is offered as the only plausible explanation for the luminous streamers and scattered light seen and described by the APOLLO 17 Astronauts (3). As for LEAM data, there remains little doubt that essentially all of the events recorded by that instrument are from lunar surface fines carrying a high surface charge. There are three mutually dependent factors which govern the feasibility of electrostatic transport: 1) The lunar surface potential; 2) adhesive forces; and 3) The surface charge on the particle. Approximate values for the first two factors have been derived from measurements on the lunar surface. Relatively accurate values for the third factor will be made available via laboratory calibrations of the LEAM spare unit to slowly moving, highly charged microspheres. When a relationship between the LEAM pulse heights and the particle charge and speed (< 1 km-sec^{-1}) is obtained and it is assumed that electrostatic transport is primarily initiated at or near the terminators, the following information may reasonably be derived: 1) The particle trajectory; 2) The lunar surface potentials and electric field strength; 3) Particle size distribution; and 4) adhesive forces.

REFERENCES

(1) Berg, O.E., Richardson, F.F., and Burton, H., "Lunar Ejecta and Meteorites Experiment", APOLLO 17 Preliminary Science Report, NASA SP-330, pp. 16-1 to 16-9; 1973.

(2) Gold, T., The Lunar Surface; Mon. Not. Royal Astron. Soc., 115, #6, 585, 1955.

(3) McCoy, J.E., Criswell, D.R., Evidence for High Altitude Distribution of Lunar Dust; Proc. 5th Lunar Sci. Conf.; Vol. 3; 2991-3005; 1974.

2.2.5 ELECTROSTATIC DISRUPTION OF LUNAR DUST PARTICLES

John W. Rhee

Rose-Hulman Institute of Technology, Terre Haute, Indiana 47803

and

Goddard Space Flight Center
Greenbelt, Maryland 20771

An investigation has been made to study a possibility that dust particles might catastrophically explode on the lunar surface due to electrostatic charging. It is shown that for the dark side along the terminator zone, dust balls and compact stony particles of micron and submicron sizes will be blown up if their surface potential is as low as a kilovolt negative. This mechanism will not operate on the sunlit side because the potential is only 3.5 ~ 20 volts positive. Some of these fragments may possibly levitate in the vicinity of the terminator.

One aspect of the lunar environment which has created much interest and discussion is the method of formation and transportation of dust particles. Several investigators studied various mechanisms (Gold, 1955; Singer and Walker, 1962; Gold, 1962; Rennilson and Criswell, 1974). The lunar surface is covered with dust. It is expected that surface dust particles do have certain electric charge due to the solar wind and solar radiation. Accordingly, the electromagnetic field carried in the solar wind will exert a Lorentz force on the charged dust particles. Very likely the moon as a solid body is charged positively on the sunlit side (Singer and Walker, 1962; Gold, 1962; Rhee, 1972). The magnitude of this field is rather difficult to estimate, but the currently accepted value seems to range from +3.5 to +20 volts. On the dark side of the moon Knott (1973) recently predicted an equilibrium surface potential of a few kilovolts negative. It is also demonstrated that for the dark side along the terminator the potential can be as low as -1.2×10^3 volts (Rhee, 1974). On the basis of the Apollo ALSEP/SIDE experiment Freeman et al. (1974) have determined that the surface potential at the lunar nightside ranges from -250 to -750 volts in the solar wind.

It is the objective of this note to show that in this range of surface potential micron and submicron size surface dust particles might break up due to the large electrostatic stress which is greater than the tensile strength of the material.

The internal disrupting stress F due to electrostatic repulsive force acting on a sphere of radius r and surface potential ϕ (Öpik, 1956) is given by

$$F = \epsilon_0 \phi^2 / r^2 \tag{1}$$

where ϵ_0 is the permittivity. Equation (1) can be rewritten as

$$F = 8.85 \times 10^{-7} \phi^2 / r^2 \qquad (2)$$

where F is the tensile strength in dyne/cm^2, ϕ is in volts and r is in cm.

Table 1 shows diameter of dust particles for various materials of different tensile strengths computed as a function of surface potential. Loosely bound and fluffy dust balls have a low strength of about 10^4 dyne/cm^2 and $10^{-3} \sim 10^{-2}$ cm size dustballs will fragment due to electrostatic tension if the surface potential ranges from -10^2 to about -10^3 volts. A typical tektite glass requires a much higher potential to be electrostatically broken up. For a compact stone charged to about -10^3 volts, the break-up diameter is about two microns.

Table 1

Relation between material strength and break-up diameter

Material	Tensile strength (dyne/cm^2)	particle diameter (micron)			
		-100 volts	-300 volts	-500 volts	-1000 volts
dust ball	10^4	19	57	95	190
compact stone	10^8	0.19	0.57	0.95	1.90
tektite	6.9×10^9	0.02	0.06	0.10	0.20
iron	2×10^{10}	0.01	0.03	0.05	0.10

In the absence of any pertinent laboratory data on the electrostatic explosion of dust particles, it is very difficult to explain the dynamics of these resulting fragments. Nevertheless one might consider a specific case in order to find out just how far and high these fragments might travel on the lunar surface. A compact stone of radius 1 micron and charged to -10^3 volts carries an electrostatic energy of 1.1×10^{-3} erg. In the event that an electrostatic explosion takes place and creates a fragment of mass 10^{-14} gram and if about 10% of this energy is imparted to this fragment, it will have an ejection speed of about 1.05 km/sec. Assuming an ejection angle of 45° and considering lunar gravity only, it can travel a distance of 728 km and reach a maximum altitude of 208 km on the lunar surface.

Evidence for the transportation of dust on the lunar surface has been reported from Surveyor 3 observations. LUNOKHOD-II observation of a post-sunset brightness of the lunar sky, astronaut observations of streamers associated with sunrise and sunset as seen from lunar orbit, and results from the Apollo 17 LEAM experiment (Berg et al., 1974) indicate strong evidence of lunar soil transport associated with the terminators. The impact data from the Apollo 17 LEAM experiment strongly suggest electrostatic levitation and horizontal soil transport as postulated by Gold (1955); Singer and Walker (1962), and others. More recently Rennilson and Criswell (1974) have proposed electrostatic levitation as an explanation for the sunset horizon glow observed from the Surveyor missions.

The mechanism of electrostatic disruption proposed here is effective only in the vicinity of the terminator zone on the dark side. At the lunar terminator and on the dark side adjacent to it, dust particles are expected to assume a very low negative potential and, depending on their material strength, they may explode if the electrostatic stress exceeds the tensile strength.

It is impossible at the present time to invoke dust charging as a possible mechanism to explain the differences between the near and far side of the moon because of the uncertainties in the physical parameters on the dark side.

In summary, an attempt has been made to investigate the possibility that dust particles might catastrophically explode on the lunar surface, even though it is extremely difficult to attack the problem satisfactorily. It is pointed out that for the dark side along the terminator zone, dust balls and stony particles of micron and submicron sizes will explode if the surface potential is as low as a kilovolt negative. Some of these fragments may conceivably levitate in the vicinity of the terminator. It is unlikely that the entire dark side of the moon would assume a negative potential as low as a few kilovolts because of the presence of the plasma cavity behind the moon.

REFERENCES

Berg, O. E., F. F. Richardson, John W. Rhee, and S. Auer, Preliminary Results of a Cosmic Dust Experiment on the Moon, Geophys. Res. Letters, 1, 289-290, 1974.
Freeman, J. W., and H. K. Hills, The lunar surface potential and plasma sheath effects, A paper to be presented at the Fall Meeting, AGU, San Francisco, December, 1974.
Gold, T., The lunar surface, Mon. Not. Royal Astron. Soc., 115, #6, 585, 1955.
Gold, T., Processes on the Lunar Surface, in the Moon, edited by Z. Kopal and Z. K. Mikhailov, 433-439, Academic Press, New York, 1962.
Knott, K., Electrostatic Charging of the Lunar Surface and Possible Consequences, J. Geophys. Res., 78, 3172-3175, 1973.
Öpik, E. J., Interplanetary Dust and Terrestrial Accretion of Meteoric Matter, Irish Astron. J., 4, 84-135, 1956.
Rennilson, J. J., and D. R. Criswell, Surveyor observations of lunar horizon-glow, The Moon, 10, #2, June, 1974.
Rhee, J. W., Lunar dust potential, Space Res. XI, 275-277, 1972.
Rhee, J. W., Electrostatic disruption of lunar dust particles, Univ. of Md. Technical Report No. 75-030, October, 1974.
Singer, S. F., and E. H. Walker, Electrostatic Dust Transport on the Lunar Surface, Icarus, 1, 112-120, 1962.

2.2.6 Microcraters Produced by Oblique Incidence of Projectiles

Stähle, V., Nagel, K. and Schneider, E.
Max-Planck-Institut für Kernphysik, Heidelberg/F.R.G.

Using a Van de Graaff dust accelerator an experimental program has been carried out in order to study crater parameters as a function of projectile incidence angle. Iron particles were shot into quartz glass targets. The angle of incidence which is the angle between target normal and the impact direction of the projectile has been varied from $0°$ to $70°$ in steps of $10°$. Projectile masses ranged from 10^{-11} to 10^{-13} g with velocities between 3 and 20 km/sec and projectile masses at 10^{-2} g with 4 km/sec impact velocity using a light gas gun at the Ernst-Mach-Institut, Freiburg i.Br. The so called circularity index which is the ratio of the crater area to the area of the smallest circle around the crater is a measure of the asymmetry of a crater. The circularity index decreases linearly with increasing angle of incidence. Also a small increase of the circularity index with increasing projectile velocity has been found i.e. the craters have a rounder shape with increasing velocity at the same angle. The circularity index appears to be independent from the projectile mass in the mass range from 10^{-11} to 10^{-2} g for stainless steel targets.

Over the entire range of angles of incidence (m,v constant) the ratio of the width of the crater to the diameter of the projectile (D/d) is constant. From this ratio one can infer the diameter of the projectile.

Craters on stainless steel targets were used for comparing simulated craters with those on exposure areas on board of satellites. Craters on moon-like targets (feldspar) were used for comparing simulated craters with those on silicate lunar material.

The small spherules inside the craters were analysed by means of energy dispersive X-ray spectrometry. They consist mostly of projectile material.

Full paper: to be published in "Planetary and Space Sciences".

2.2.7 **Measurements of Impact Ejecta Parameters in Crater Simulation Experiments**

E. Schneider
Max-Planck-Institut für Kernphysik, Heidelberg/F.R.G.

Ejecta parameters such as masses and velocities as a function of ejection angle have been determined in impact experiments performed with a light gas gun. Impact plasma detectors and secondary targets mounted around the primary impact site served for the detection of secondary particles.

The fraction of mass ejected with velocities $\gtrsim$ 3 km/s is only a factor of 10^{-4} times the mass of the primary projectile.

A conservative calculation shows that the contribution of secondary microcraters (caused by fast ejecta) to primary crater number densities (caused by interplanetary particles) on lunar rock surface is on the statistical average below 1 % for any lunar surface orientation. Therefore lunar samples showing favorable exposure conditions can be considered as good detectors for interplanetary particles exposed over geological periods of time.

Calculation of the dust flux enhancement due to moon ejecta turned out to be in good agreement with Lunar Explorer 35 in situ measurements.

Full paper: Schneider, E. (1975), "Impact Ejecta Exceeding Lunar Escape Velocity", published in The Moon **13**, 173.

2.2.8

Impact light flash studies: Temperature, Ejecta, Vaporization

G. Eichhorn

Max-Planck-Institut für Kernphysik, Heidelberg/F.R.G.

Abstract

Impact experiments have been performed with the 2 MV dust accelerator; the dependence of the maximum light flash energy and intensity on the projectile mass and velocity has been determined experimentally. The temperature of the radiating gas and plasma was estimated to be in the range from 2500K to 5000K, depending on the impact velocity. The distribution of the maximum ejecta speed as well as the normalized distribution of ejected mass have been determined as a function of the ejection angle. A rough estimate of the degree of vaporization of the displaced mass was obtained.

In this paper we discuss measurements we have performed on the light flash produced during impacts of very high velocity projectiles. We also discuss information we have obtained in these experiments such as the temperature of the luminous material, distribution of secondary particles and the degree of vaporization of the displaced material. The measurements were performed with an electrostatic dust accelerator (particle masses between 10^{-15} g and 10^{-9} g) and a light gas gun (projectile mass 1.5×10^{-2} g).

In agreement with earlier studies (Früchtenicht (1965), Rollins and Jean (1968), and Eichhorn (1975)), the maximum impact flash intensity I and the total energy E were found to depend upon the projectile mass m and velocity v in the form $I = c_1 m^{\alpha_1} v^{\beta_1}$ and $E = c_2 m^{\alpha_2} v^{\beta_2}$. The parameters α_1, α_2, β_1, and β_2 were measured for different projectile-target combinations (projectile materials: Fe, Al, C, W, glass; target materials: Au, W, Fe, Rh, Al, glass). α_1 and α_2 were found to be $\alpha_1 = \alpha_2 = 1$ for all projectile-target combinations measured. A comparison of experiments with the light gas gun with those performed with the dust accelerator showed that the light intensity is proportional to the projectile mass over a mass range of ten orders of magnitude.

Values between 3.8 and 4.6 with an average of 4.1 were found for β_1, while β_2 ranges from 3.0 to 3.6 with an average of 3.2. These results yield the functions $I = c_1 mv^{4.1}$ and $E = c_2 mv^{3.2}$.

The absolute light energy of the impact flash was determined by measuring the spectral distribution of the light flash. Fig. 1 shows such a spectral distribution. The light energy measured with calibrated photomultipliers in different spectral ranges defined by interference

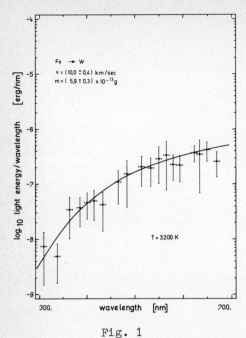

Fig. 1
The spectral distribution of the light flash. The curve is obtained from theoretical calculations on the basis of black body radiation with an initial temperature of the luminous material of 3200K.

filters is plotted there as a function of the wavelength. Integrating this spectral distribution over the wavelength yields the total light energy emitted in the range of the visible spectrum. Depending upon projectile and target material, the light energy varies from 2×10^{-6} (for iron impacting gold at 4 km/sec) to 10^{-2} (for carbon impacting gold at 7.5 km/sec) in units of the projectile energy.

Assuming, that the light is black body radiation, we calculate theoretical spectral distributions and fit them to the measured ones by adjusting the initial temperature of the radiating material. The curve in Fig. 1 is such a spectral distribution calculated with an initial temperature of 3200K.

The temperatures that have been estimated in this way for various measured spectral distributions are plotted versus the particle velocity in Fig. 2. The temperature increases with increasing particle velocity. The two curves represent measurements by Friichtenicht (1965) and serve as a comparison of the estimated temperatures at two different wavelengths of light.

In experiments with iron particles and a gold target, secondary particles with velocities greater than 1 km/sec were detected from the light they produce when they impact the entrance windows of the photomultiplier. Their velocity is calculated from data on the time between the primary and the secondary impacts and their locations. Fig. 3 shows the velocity of the fastest ejecta in different angular intervals plotted versus the ejection angle with respect to the target surface. At angles smaller than 20° the ejecta have velocities of up to 30 km/sec at an impact velocity of about 5 km/sec. The dashed curve represents measurements with high speed framing cameras of

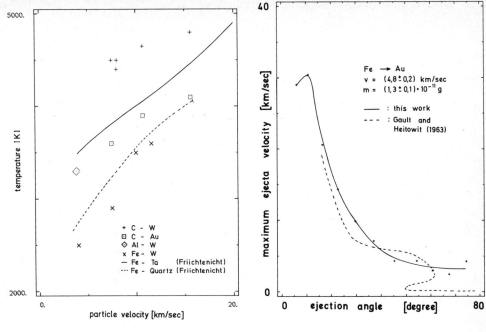

Fig. 2

The initial temperature of the luminous material estimated from the measured spectral distribution as a function of the impact velocity. The curves represent measurements by Friichtenicht (1965).

Fig. 3

The maximum ejecta velocity as a function of the ejection angle. The dashed curve represents measurements of Gault and Heitowit (1963) for iron impacting basalt.

Gault and Heitowit (1963) for iron impacting basalt. The light intensity produced by these ejecta is a measure of the mass ejected into a specific angular interval. This mass distribution is shown in Fig. 4. About 90 % of the ejecta are seen to be ejected at angles between $50°$ and $70°$. The fastest secondary particles are ejected at angles below $20°$ and comprise only about 0.01 % of the total ejected mass.

Since the light emission is due to radiating gas and plasma (Jean and Rollins (1970)), information about the luminous gas can be gained by varying the pressure in the target chamber. The light intensity as a function of the pressure is shown in Fig. 5. Between 10^{-6} torr and 10^{-3} torr, the light intensity does not depend on the pressure. Increasing the pressure further, the light intensity also increases and reaches a maximum at about 0.1 torr. This increase is caused by collision of the vaporized material with the restgas. The number of atoms in the vapour can be calculated from the thermal energy of the

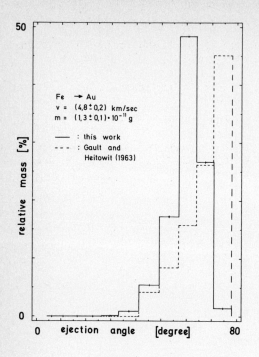

gas when converted into light energy, and combined with the above estimated temperatures. Assuming, that all of the thermal energy is converted into light, 0.4 % and 1.6 % of the displaced mass are vaporized at impact velocities of 5 km/sec and 7.4 km/sec respectively. Since nothing is known about the degree of energy conversion, these values are to be considered as lower limit estimates.

Further experiments are expected to provide more informations about the partition of energy during the impact process.

Fig. 4
The relative mass distribution of the ejecta.

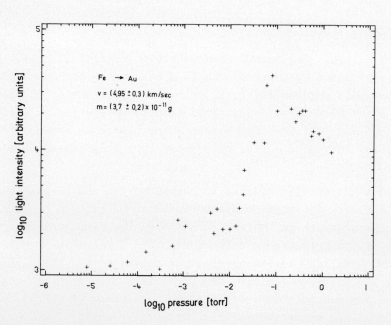

Fig. 5: The light flash intensity as a function of pressure in the target chamber.

References

Eichhorn, G. (1975), "Measurements of the Light Flash Produced by High Velocity Particle Impact", accepted for publication in Planet. Space Sci.

Friichtenicht, J.F. (1965), "Investigation of High-Speed Impact Phenomena", NASA Contr. No. NA Sw-936.

Jean, B., and Rollins, T.L. (1970), "Radiation from Hypervelocity Generated Plasma", AIAA Journal, Vol. 8, No. 10, p. 1742.

Rollins, T.L., and Jean, B. (1968), "Impact Flash for Micrometeoroid Detection", Com. Dev. Report 9899/FR1, NASA Contr. No. NAS 9-6790.

2 IN SITU MEASUREMENTS OF INTERPLANETARY DUST

2.3 Particle Collection Experiments and Their Interpretation

2.3.1   Submicron Particles from the Sun

by

Curtis L. Hemenway
Max-Planck-Institut für Kernphysik*

Abstract

A review is given which suggests that cosmic dust theoretical and experimental studies are still beset with uncertainty and inaccuracy. A significant body of interrelated evidence exists which indicates that the solar system has two populations of dust particles, a submicron population generated and emitted by the sun and a larger size population spiraling inward toward the sun. The submicron component may provide the missing coupling mechanism between solar sunspot activity and meteorological activity in the earth's atmosphere.

Troubles with Measurements and Theories

In the attempts to infer the nature of the interplanetary dust particles from Zodiacal Light studies, a single spectral index has been assumed in the size distribution (dn $\propto r^{-c} dr$), even though it is clear from the collection studies, lunar microcrater studies and the interplanetary detection studies that such an assumption is unreasonable. The collection techniques have been questioned from a contamination viewpoint and when inflight-shadowing was introduced to obtain greater reliability of particle identification, a loss in small particle sensitivity resulted. The coincidence techniques currently employed in detection experiments are always limited by the sensitivity of the least sensitive element and thus also sacrifice sensitivity for reliability. The penetration studies, which use foil thicknesses 25 to 50 μ thick, have to be calibrated by extrapolation since electrostatic accelerators cannot accelerate particles larger than

---

*On leave from the Dudley Observatory and the State University of N.Y. at Albany, U.S.A

$1\,\mu$ to typical interplanetary impact velocities. When we calibrated our S-149 experiment at both the Goddard Space Flight Center and the Max-Planck-Institut für Kernphysik we found that to produce the same microcrater size in the same materials with the same kind of particles, that unexpected factors of 8 in momentum and of 3 in energy appeared. Thus it seems there are problems and limitations with all phases of cosmic dust work and this paper will be no exception. The purpose of this paper is to suggest that some of the observations in the cosmic dust field become more understandable if one views the dust population of the solar system as having two components, a component of relatively large particles ($\gtrsim 10^{-13}$gms) spiraling inward towards the sun and a component of submicron particles ($\lesssim 10^{-13}$gms) flowing outward from the sun, having its origin in sunspots.

Evidence from Collection Studies suggesting a Solar Origin for Submicron Cosmic Dust

Figure 1 shows a core-mantle particle collected during a noctilucent cloud (NLC) display. It will be noted that the small core appears black, which in a transmission electron microscope indicates a large number of electrons per unit volume which in turn requires a high Z material for the core. The mantle is very fragile and is only an ash of its former self. The mantle is consistent with a hydrogen-soaked, carbon mantle which has oxidized[1] in transit through the ionosphere. The reality of these core-mantle particles appears to be confirmed by the existence of submicron penetration holes ("evil eyes") at 230km altitude[1] (Gemini S-012) and at 430km altitude[2] (Skylab S-149). The existence of submicron particles in the solar system has also been shown by the presence of submicron craters in the Skylab S-149 experiment[2].

Figure 2 shows a NLC particle which has experienced a violent thermal experience, probably more violent than entry into the earth's atmosphere. The particle appears to have a high-Z refractory nature and may have encountered a high temperature environment during ejection from the sun's atmosphere. Notice that the particle is curved only on one side and has only a thin mantle. Generally particles smaller than $.1\,\mu$ have no mantles. The presence of carbon mantles in the larger submicron particles may be nature's way of reducing the average density of the larger particles to permit solar light pressure to approximately balance solar gravity to enable the particles energetically to escape from the sun.

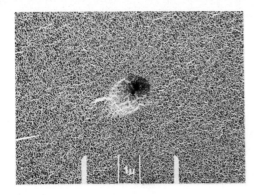

Noctilucent Cloud Particle
Figure 1

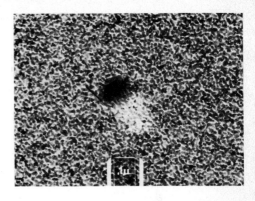

Noctilucent Cloud Particle
Figure 2

Figure 3 shows the sort of evidence which suggested to us[3] the presence of high Z elements in the cores of the particles. The top scan is a portion of an X-ray probe analysis of the particle shown and the bottom scan of a region about 30μ away. The extra peak best fits hafnium and two additional peaks also fitting hafnium were also found on the same trace. Similar X-ray peaks in other particles have given evidence of Ta, La, W, Th, etc., all high temperature, high Z elements found in NLC collections. Three of the four successful NLC collections found significant particle flux enhancements at times of NLC displays. One did not. Probably both the enhanced particle flux and the hydroxyl-ion mechanism are playing roles in NLC formation.

Figure 4 shows a fission track cluster which shows the presence of uranium in particulate form in a sample of lexan exposed facing the sun during the S-149 experiment[4]. Significantly more such events appear to have been found in the solar facing samples than in the anti solar exposure. More work needs to be done before flux numbers can be given. Their presence does suggest the presence of uranium (a high Z, high temperature refractory material) in significant amounts in the submicron particles of the solar system.

Next we note that the times of NLC displays do not appear to be related to meteor showers and that significant flux enhancements of collected submicron particles have been observed in rocket collection experiments on four occasions. Table I summarizes these data. During these particle enhancements, a constancy of observed particle flux with altitude has been found, suggesting an influx from outside the earth's atmosphere. Also Fechtig[5] has found all particles measured in rocket experiments in the

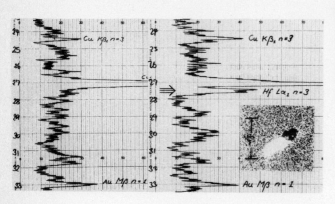

X-ray Probe of
Noctilucent Cloud Particle
Figure 3

Fission Track Clusters
in Lexax S-149 Sample
Figure 4

Location	Date	Enhanced Particle Flux	
Kronogård, Sweden	Aug. 11, 1962	$10^{5.5}$	part/m²s
White Sands, N.M., USA	Nov. 18, 1965	10^{4}	"
Kiruna, Sweden	Aug. 8, 1970	10^{4*}	"
Kiruna, Sweden	July 31, 1971	10^{3*}	"

Table I

atmosphere to be falling. These enhancements may be the result of concentrations of particles ejected from the vicinity of sunspots by prominence outbursts in the solar atmosphere.

Table II shows the melting and boiling points of some of the elements or molecules which appear to exist in the NLC particles. It is to be noted that these materials have such low vapor pressures that their boiling points (STP) are so high that their values are not well known. In some cases the oxides or carbides appear to be more stable and have lower vapor pressures than the elements alone. Thus there does appear to exist materials in the NLC particles which could nucleate and grow in the vicinity of sunspots. Furthermore, the composition discrepancy[6] between the solar abundances and, say, the moon might well be the result of material of the refractory elements existing in particulate form in the atmosphere of the sun.

---
*Estimated flux may be low due to use of inflight shadowing.

Element or Molecule	Melting Point	Boiling Point
W	3683°K	6200°K
Hf	2248 "	5673 "
HfC	4163 "	----
Ta	3500 "	5698 "
TaC	4153 "	5773 "

Table II

The nose cone collections which were made during the first NLC sampling expedition and the corresponding control collections in the absence of a NLC both showed a particle size-cutoff at about 0.03 μ in radius. Figure 5 shows the size distribution[7] of this NLC collection in 1962. Since inflight shadowing was not used and the fluxes were high, it was possible to detect the small particle cutoff in this experiment. Smaller particles have been observed in the rocket collection experiments but always in association with the larger particles. The largest particles observed in most rocket experiments are about .3 μ in radius. The integrated size distribution shown in Figure 5 has a slope of -3.2 (differential size distribution, $dn \propto r^{-4.2} dn$). An important question to be explained is why over such a relatively narrow range of sizes is there such an enhancement of the small particle fluxes over the larger sizes. It may be because the particles in this size range are the only ones which can approximately balance solar light pressure and gravity and thus can energetically escape from the sun in relatively large numbers. Under these conditions then

$$\rho \cdot r = .58 \frac{gm}{cm^3} \cdot \mu \qquad (1)$$

would be expected to hold where ρ and r are the particle density and radius[8]. The minimum particle size observed would represent the point where nature runs out of materials of sufficient mass density ($\sim 20 gm/cm^3$) and the maximum size would be determined by the smallest effective density that the materials can have in particulate form.

Intercomparisons Between Other Techniques to

Study Cosmic Dust

Figure 6 shows the work of Blackwell and Ingham[9] which indicated that the brightness of the Zodiacal Light (ZL) as a function of elongation angle matches up with the brightness of the outer Corona. The standard explanation of this matching is by forward scattering of sunlight by large dust particles. However, this explanation encounters difficulties in the inability of the large dust grains to provide sufficient polarization for the ZL. An alternate possibility is that a significant part of the brightness of the ZL is due to scattering by submicron particles flowing outward from the sun.

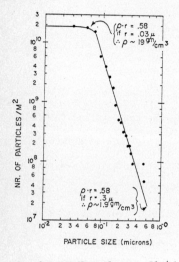

Cumulative Noctilucent Cloud
Particle Size Distribution
Figure 5

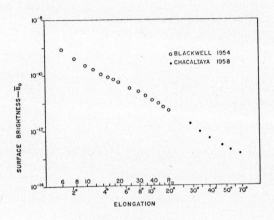

Zodiacal Light and
Coronal Brightness
Figure 6

When the samples were brought back from the moon, the initial conclusion was that there were no submicron particles in the solar system since no small impact craters were found. With more careful work on smoother samples at higher magnification, relatively large numbers of small impact craters were found. Figure 7 shows an example of recent studies[10]. It will be noticed that there is about 2.5 to 3.5 orders of magnitude enhancement of the relatively steep small crater data. This suggests a flux of submicron particles hitting the moon (omitting corrections for particle directivity and small particles impact craters missed) of about 1 particle $/m^2 s$, since these numbers must be added to the relatively flat flux curve of larger particles with masses greater than 10^{-13}gm (3×10^{-4}part/$m^2 s$). It will also be noticed that the slopes of the lunar microcrater data are relatively steep. For example, a slope of -4.2 fits the data of sample (15301.79) better than the slope of -3.2, suggesting (assuming a constant impact velocity and that the particle size distribution is that given in Figure 5) that the particle density may increase as the particle size decreases according to equation (1). There is also a suggestion of a small particle cutoff in the data despite the difficulties of finding very small impact craters.

In Figure 8 the number density as a function of particle size of the particles scattering solar photons to form the Zodiacal Light (ZL) is shown as computed from the work of Powell[11]. If one performs a conversion from the NLC particle size distribution shown in Figure 5 for the constant mass density case and the variable mass density case,

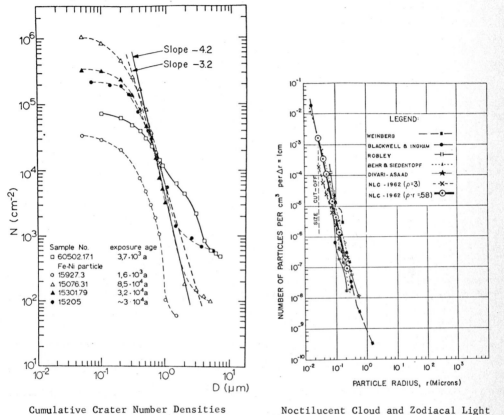

Cumulative Crater Number Densities
N vs. crater pit diameter D
Figure 7

Noctilucent Cloud and Zodiacal Light
Particle Size Distributions
Figure 8

equation (1), one can see that the variable mass density case fits the predictions of Powell better than the constant density case. This fit uses a flux of about 10 particles/m^2s and assumes a constant velocity of outflowing particles from the sun of 120/km/s. At 60km/s the flux to match the ZL brightness is about 5 particles/m^2s. It is also to be noted that the work of Powell provides an explanation of the polarization of the ZL and thus the size distribution of Figure 5 does also. In Figure 9 a high resolution plot of one of the ZL spectral lines given by Ring's group[12] is shown. If the satellite line represents the brightness contribution of an outward flowing submicron flux from the sun, it suggests that about 1/5 of the ZL brightness is due to submicron particles and thus also suggests that about 1 to 2 particles/$m^2 s$ should represent the average submicron flux from the sun.

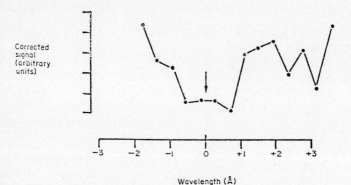

High Resolution Zodiacal Light Spectral Line
Figure 9

We next attempt to intercompare the optical characteristics of the NLC particles and interstellar grains. If one uses a plane slab model and a NLC differential size distribution of $dn \propto r^{-4.2} dr$ and assume constant absorption and scattering efficiencies, one gets an extinction approximately inversely proportional to wavelength. In Figure 10 one can see a comparison with the interstellar extinction measurements of Whitford, Boggess and the theoretical estimates of Greenberg[13]. It is to be noted that the prediction from the NLC size distribution is in fair agreement with the interstellar extinction values when converted to the same ratio of magnitude difference units. Furthermore the average size of the NLC particle cores at $.044 \mu$ obtained from Figure 5 agrees quite well with the $.05 \mu$ given by Greenberg for the average size of the cores of interstellar grains. This suggests that the process of stars generating submicron particles may be a common process and the difficulties of nucleating interstellar grains in interstellar space may disappear.

In Figure 10 one also sees the wavelength dependence of solar limb darkening[14] converted to the same units and compared with Greenberg's[13] work. It is to be noted that the data also agree remarkably well and a second interesting coincidence has appeared. It may also be shown that the limb darkening data converted to the ratio of magnitude difference units as in Figure 10 is independent of the angle with respect to the center of the sun at which the limb darkening data is taken. This limb darkening agreement

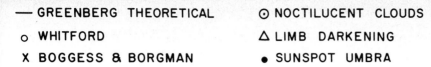

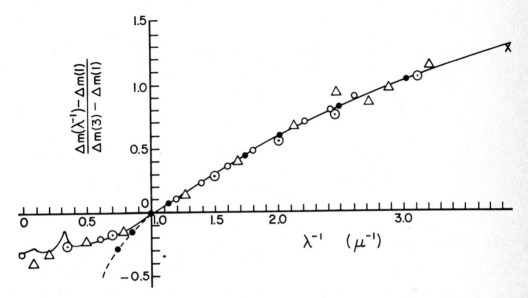

Interstellar Grains, Noctilucent Cloud Particles, Limb Darkening, and Umbra
Figure 10

suggests that high temperature refractory particles having the size distribution given in Figure 5 extending down into the Angstrom range, located in the region of the temperature minimum between the photosphere and the chromosphere could provide an alternate explanation of limb darkening. Chandresekar[15] gives wavelength dependence of solar extinction at the center of the sun and finds significant departures below 4500Å and around 9000Å. Pottasch[16] in a later review article says that an unexplained opacity in the solar brightness still exists in the UV.

Also if one converts the wavelength dependence of the umbral brightness of a sunspot[17] to these ratio of magnitude difference units, one finds fair agreement in the visual but significant departures in the infrared as shown in Figure 10. This suggests that we are dealing with the same particles except that they have not had time to grow to a sufficient size to be significant absorbers in the infrared. Thus there appears to be a four-fold coincidence between interstellar grains, NLC particles, particles sufficient to explain limb darkening and the particles to explain the wavelength dependence of the sunspot umbral brightness. If sunspots are particle generators such a coincidence is not surprising since most of the particles which escape from the sun escape from the solar system. If our sun generates submicron particles which can escape to interstellar space, it is likely that most stars also do.

The temperature of the corona may be estimated by comparing Parker's theory of the solar wind with Vela satellite data to be about 750,000°K. Similar values result from radio and X-ray brightness studies[18]. If one determines the coronal brightness by the width of the Fraunhofer lines, the temperature of the corona becomes 2 to 3×10^6°K, a significant difference from 750,000°K. Let us assume that the high coronal "temperature" is due to doppler shifted photons scattered by submicron particles moving radially from the sun in different directions at the velocity equivalent of 2.3×10^6°K and estimate the flux of submicron particles to be expected at the earth, ϕ_e, to provide the total coronal brightness (Baumbach). We use Mie theory and assume m=2, $\lambda = .5\mu$ and all particles to have the average size of the NLC distribution 0.044 μ (Figure 5). The results are given in Table III, where r_m is the minimum distance to the sun along the straight line integration paths. It is interesting that

r_m ($r_\odot$)	ϕ_e (Part/m^{2}s)
2.0	10.7
3.0	7.4
4.0	7.6
5.0	8.3
6.0	8.9

Table III

the values of ϕ_e are about the same as needed to provide a match between the NLC distribution and Powell's work as might be expected from Blackwell and Ingham's work (Figure 6). Perhaps the flow of submicron particles from the sun can provide an explanation of the temperature inconsistency of the solar corona and the coronal temperature is really only about 750,000°K. It is interesting that the outward velocity of the particle reaches a maximum at $r_m=3.5(r_\odot)$ of about 200km/sec and appears to decrease linearly with $1/\sqrt{r_m}$ to about 80km/sec at the earth's distance from the sun, suggesting that the balance between solar light pressure and gravity is lost as the particles cool as they move away from the sun and that drag forces are important for small particles in the solar corona. The fluxes are reduced somewhat if one intercompares with the F coronal brightness and the extrapolated velocity at the earth drops to about 60km/sec.

If one intercompares the steep lunar microcrater data given in Figure 7 with the relatively flat data[10] for larger particles ($\gtrsim 10^{-13}$gm) and inquires how do the submicron solar particle flux enhancements, velocities, fluxes and maximum particle masses vary, the data of Table III result.

Submicron Radial Velocity	Enhancement	Solar Part. Flux (ϕ_e)	Max. Part. Mass
20 km/s	8×10^4	32 part/m²s	2×10^{-12} gm
60 "	7×10^3	3 "	2×10^{-13} "
→ 80 "	3×10^3	1.5 "	3×10^{-13} " ←
100 "	1×10^3	.4 "	1×10^{-13} "

Table III

These estimates assume that the NLC size distribution data of Figure 5 is applicable for the submicron solar particles with its small and large particle size cutoffs (.03μ to .3μ radius) that equation (1) applies and that the large inwardly spiraling particles $\gtrsim 10^{-13}$ gms hit the moon with an average velocity of about 8km/s. It is to be noted that 3×10^{-13} gm is the mass of a spherical particle of radius .3μ having the density of carbon. The intercomparison data of Table III with that of Figure 7 suggests an outward flowing velocity of submicron particles of about 80km/s at the earth's distance from the sun and a flux of about 1 particle/m²s.

The velocity of the outward bound solar submicron particles may be estimated by direct integration, assuming an injection velocity of about 20km/s, that light pressure and solar gravity balance at the solar surface, and neglecting coronal gas drag. Table IV shows the results of such an estimate. This estimate suggests that if solar light pressure and gravity are in balance at the solar surface, then the imbalance which results as the particles move away from the sun and are struck by photons in a more radial manner, can accelerate particles to velocities of the order of 100 km/s. If one were to take into account coronal gas drag and the change in the index of refraction of the particles as they cool whyle moving away from the sun, then lower velocities would be expected for the solar submicron particles at the earth's distance from the sun. It is interesting that this estimate is independent of particle size (0.03μ to .3μ particle radii) if equation (1) holds.

An additional way of estimating the average particle radial velocity (v_r) at the

Height Above the Sun (r_o)	(Solar Light Pressure-Solar Gravity)/Solar Gravity	Velocity
.025	0.000	20.0 Km/s
.055	.000	20.0 "
.125	.001	20.4 "
.25	.006	25.1 "
.50	.016	37.4 "
1.5	.038	69.6 "
2.5	.057	81.4 "
4.5	.075	94.2 "
8.0	.086	105.9 "
35	.096	113 "
75	.098	115 "
150	.103	116 "

Table IV

earth is to analyze the frequency of NLC data[19] as a function of time of year (Figure 11) It will be noted that the peak frequency occurs approximately 18 days after the summer solstice, which is the time that the earth's pole tilts in the plane of the ecliptic about $17.7°$ away from the sun. If one assumes the peak frequency time is the time that

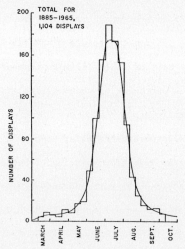

Time Distribution of Noctilucent Cloud Displays
Figure 11

the greatest number of submicron particles from the sun are intercepted by the high latitude, northern hemisphere region of the earth, then because of the earth's motion

$$v_r = \frac{29.8 \text{km/sec}}{\tan 17.7°} \approx 90 \text{km/sec}. \tag{2}$$

Thus this crude parallax estimate appears to be consistent with the earlier velocity estimates. It is to be noted that NLC displays peak in the southern hemisphere six months later than in the northern hemisphere.

Sunspots and Submicron Particles

An important question is can the cool regions above the observable sunspot umbra nucleate particles? Let us estimate the magnitude of the cooling to be expected by computing the adiabatic lapse rate; we note that the partially conducting gas is carried to greater heights over sunspots (by virtue of the strong magnetic fields in the vicinity of sunspots by the Evershed effect-300 meters/sec outward velocity), than the heights the photospheric gases can reach by convective mechanisms over the normal photosphere. If one uses the solar gravity at the surface of the photosphere and assumes the gas to be mostly hydrogen, then

$$\frac{\Delta T}{\Delta z} = \frac{g}{C_p} = \frac{2.7 \times 10^4 \text{cm/s}^2}{4.08 \text{ cal/gm} \times 4.2 \times 10^7 \text{ergs/cal}} = 1.6 \times 10^{-4} \text{°K/cm} = 16 \text{°K/km}, \quad (3)$$

where g is the solar gravity and C_p the specific heat.

Thus in equation (3) a cooling of about 16°K/km is estimated. This means that one need only to elevate the umbral gas through a distance of 100 km (a distance small compared to the diameters of typical sunspot umbra) to drop the temperature of the umbral gas from a temperature of 3600°K to a temperature of about 2000°K. At such temperatures Wickramasinghe[20] and others have suggested that carbon can nucleate in the atmosphere of M stars provided the nucleation occurs on ions. It is to be noted that, despite their much smaller abundances, high Z refractory elements or molecules might nucleate at higher temperatures since they are much more easily ionized and have very much smaller evaporation rates from solid surfaces. It is to be noted that the abundance of high Z refractory elements determined in the solar atmosphere by atomic spectral line amplitude estimates are relatively low compared to other abundance evidence, such as from lunar abundances[6]. The relatively low solar abundances might result from the high Z refractory elements existing in the solar atmosphere in molecular form and the resulting bands being too weak to be detected in the noisy solar spectrum. In addition, the ease with which high Z elements can be ionized might lead to some concentration of the high Z refractory elements in the vicinity uf sunspots. It can be shown that the thermal conductivity is negligable for the temperature lapse rate given by equation (3) if an Evershed velocity of 300 meters/sec is used for the outflowing of solar gas from the vicinity from the umbra of a sunspot.

Figure 12 displays a diagram showing the presence of faculae[18] (white light,

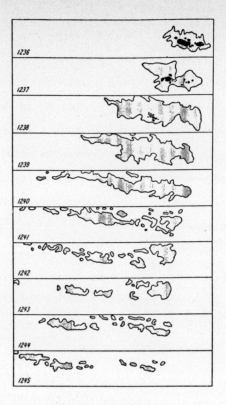

Faculae Duration
Figure 12

higher brightness areas) around a sunspot group. These faculae were observed to persist around a sunspot group for about 9 solar rotations. It is to be noted that all sunspots have faculae and that faculae are best observed near the solar limb. It is suggested that faculae are regions where clouds of high Z refractory particles nucleated above sunspot umbra have grown to a size sufficient to scatter light and that the brightness observed in faculae is the result of increased photon scattering toward the direction of the observer. The average life of a faculae is given by Allen[14] to be 15 days, De Jager[1] says 80 days.

If one takes the high temperature region of the corona to be located within a distance of two solar radii from the solar surface and assumes a constant velocity of 50km/sec for the velocity of the submicron particles traveling radially outward from the solar surface, then a time of about .3 day is required for transit, as shown below.

$$\frac{1.4 \times 10^6 \text{km}}{50 \text{km/sec}} = 2.8 \times 10^4 \text{sec} \approx .3 \text{ days} \qquad (4)$$

Since the particle is cooler in the solar corona than in the interface region between the photosphere and chromosphere (since the temperature of the particle in the corona is determined mostly by photon absorption from the sun and reemission, the high temperature gas pressure being very low), the evaporation rate of the material from the particle in the corona is much reduced and it appears that for transit times of the order of a day the solar particles can readily survive transit through the sun's corona, particularly if the faculae are clouds of high Z, refractory, low-vapor pressure particles.

It is interesting to estimate the evaporation rate ($gm/cm^2 s$) for the particles which we assume are responsible for faculae. We let

$$dm = 4\pi r^2 \rho\, dr = \omega\, 4\pi r^2 dt \tag{5}$$

where dm is the mass of a spherical shell, r the particle radius, ρ the mass density of the particle, ω the evaporation rate from the particle and dt the time for a spherical shell to grow or disappear. Integrating equation (5) gives

$$t = \frac{\rho r}{\omega} = k_w \frac{\rho r}{\omega_w} \tag{6}$$

where ω_w is the evaporation rate for tungsten and k_w the constant necessary to convert ω_w to the evaporation rate of the real particle. If the temperature of the interface between the photosphere and the chromosphere is taken to be $4300°K$, $\omega_w = 7 \times 10^{-2} gm/cm^2 s$ and taking 15 days for the average life time of a faculae gives in equation (6)

$$15 \text{ days} \times 86400 \text{sec/day} = k_w \frac{19 gm/cm^3 \times 0.03\,\mu \times 10^{-4} cm/\mu}{7 \times 10^{-2} gm/cm^2 s} \tag{7}$$

$$k_w = 1.6 \times 10^7$$

This says that the effective evaporation rate of the particle is a factor of about 1.6×10^7 lower than the evaporation rate a $.03\,\mu$ radius tungsten particle would have at a temperature of $4300°K$. There are several possible reasons for this large factor: (1) the particle may have a lower temperature if it is not able to absorb all of the photons incident upon it, (2) the particle is probably charged negatively by excess electron accretion and the easily ionized high Z atoms evaporated from the particle are returned to it in significant numbers electrostatically, (3) the particle materials may well have lower evaporation rates than tungsten and (4) the particle is accreting atoms from the solar photosphere and the effective k_w is the result of the difference between the accretion rate and the loss rate of the atoms. The problem of how particles accrete and lose atoms or molecules in a stellar atmosphere needs much further work.

Possible Meteorological Implications of Solar Submicron Particle Enhancements

Many people have attempted to correlate weather phenomena with sunspot activity. Bowen[21] has suggested that the average latitude of "high" tracks across Australia varies with the sunspot cycle. Roberts[22] has suggested that droughts in the midwest of the USA are related to sunspot cycles. Douglas[23] has related tree ring widths and hence tree growth to sunspot activity. The problem has been how to relate sunspot activity physically to meteorological phenomena. We note that if greater numbers of sunspots are present, that a greater area of the sun is covered with faculae particles and the numbers of submicron particles ejected from the sun of a type suitable for NLC formation could be increased. Figure 13 shows how NLC numbers are correlated with sunspot numbers if one averages over 4 year periods. If greater numbers of NLC are formed during high sunspot activity it suggests that the number of particles entering the earth's atmosphere suitable for serving as centers for ice crystal growth in the mesopause vary with sunspot numbers. These particles must fall through the earth's atmosphere and eventually be swept out by rain. It can be shown that the average submicron particle flux is approximately constant

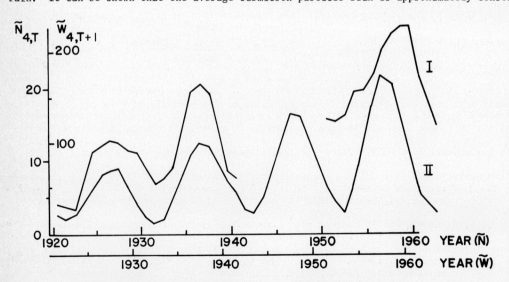

Sliding 4-year curves of the number of nights with noctilucent cloud displays (according to observations in the USSR in 1921 - 1940 and 1953 - 1963) (I) and of Wolf numbers (II).
Figure 13

from 430km to 25km (alt.) and that the fall-time through the earth's atmosphere is independent of particle size if equation (1) holds (under free molecular conditions).

Rosinski[25] has measured the particle elemental compositions in clouds at an altitude of 9km. A preliminary look at his data suggests the presence of significant excess abundances of Hf, Ta and Th with respect to Fe whereas Ni appears to have normal abundance with respect to Fe.

If the earth encounters enhancements of submicron particle fluxes from the sun as the NLC data suggest, then these particles entering the troposphere in variable amounts might well be the missing variable for long and short range weather forecasting.

Conclusions

There exists a substantial body of interrelated data concerning studies of the Zodiacal Light, the collection and detection of cosmic dust, solar coronal, photospheric and sunspot studies consistent with the view that the sun generates and emits submicron particles. However, much more theoretical and experimental work needs to be done before the sun and most stars can be considered to be submicron particle generators and the source of interstellar grains. It will be particularly interesting to study the polarization characteristics of the photospheric white-light faculae near the solar limb and also to look for polarization in Waldmeier's blue rings[26] around sunspots. Better evidence is needed for the existence of high temperature, refractory materials in the submicron particle complex and can be obtained from rocket collection studies. The meteorological implications of flux enhancements of submicron particles of solar origin may become important for both long and short range weather forecasting.

References

1. H. Fechtig and C. Hemenway, "Near-Earth Fragmentation of Cosmic Dust", IAU Colloquium 31, 1976, this volume.
2. C. Hemenway, et al "Near-Earth Cosmic Dust Results from S 149", AIAA/AGU Conference on the Scientific Experiments of SKYLAB, Huntsville, Ala. 1974.
3. D. Hallgren. et al "Noctilucent Cloud Sampling by a Multi-Experiment Payload", Space Research XIII, p. 1105-1112, 1973.
4. E. Fullam, Private Communication.
5. P. Rauser, H. Fechtig, "Combined Dust Collection and Detection Experiment During a

Noctilucent Cloud Display Above Kiruna, Sweden", Space Research XII, p. 391-402, 1972.

6. G. Morrison, et al "Elemental Abundances of Lunar Soil and Rocks", Proceedings of the Apollo II Lunar Science Conference, Houston, 1970, V. 2, Geochimica et Cosmochimica Acta Supp. 1, p. 1383-1392, 1970.

7. C. Hemenway, et al "Electron Microscope Studies of Noctilucent Cloud Particles", Tellus, V. 16, p. 96-102, 1964.

8. L. Standeford, "The Dynamics of Charged Interplanetary Grains", Thesis, University of Illinois, 1968.

9. D. Blackwell and M. Ingham, "Observations of Zodiacal Light from a Very High Altitude Station. I The Average Zodiacal Light", Monthly Notices of the Royal Astronomical Society, V. 122, p.113-127, 1961.

10. E. Schneider, et al "Microcraters on Apollo 15 and 16 Samples and Corresponding Cosmic Dust Fluxes", Proceedings of the Fourth Lunar Science Conference, Houston, 1973, V. 3, Geochimica et Cosmochimica Acta Supp. 4, p. 3277-3290, 1973.

11. R. Powell, et al "Analysis of All Available Zodiacal-Light Observations", The Zodiacal Light and the Interplanetary Medium, National Aeronautical and Space Administration, Washington, NASA SP-150, p. 225-241, 1967.

12. T. Hicks, et al "An Investigation of the Motion of Zodiacal Light Particles I", Monthly Notices of the Royal Astronomical Society, V. 166, p. 439-448, 1974.

13. J. Greenberg, "Interstellar Grains", Nebulae and Interstellar Matter, Stars and Stellar Systems, University of Chicago Press, V. 7, p. 221-361, 1968.

14. C. Allen. "Astrophysical Quantities", 2nd. ed., University of London, The Athlone Press, p. 170, 185, 1963.

15. M. Minnaert, "The Photosphere", The Sun, The Solar System, University of Chicago Press, V. 1, p. 88-185, 1953.

16. S. Pottasch, "Review of Astrophysical Conclusions from the UV Solar Spectra", International Astronomical Union Symposium. 36th. Lunteren, 1969. Dordrecht: D. Reidel, 1970, p. 241-249.

17. E. Pettit and S. Nicholson, "Spectral Energy-Curve of Sun-Spots", Astrophysical Journal, V. 71, p. 153-162, 1930.

18. C. De Jager, "Structure and Dynamics of the Solar Atmosphere", Handbuch der Physik, V. 52, Berlin: Springer-Verlag, 1959, p. 174.

19. C. Villmann, "Space-Time Regularities of Noctilucent Cloud Displays", Physics of Mesopheric (Noctilucent) Clouds, Proceedings of the Conference on Mesopheric Clouds, Riga, 1968. Jerusalem: Israel Program for Scientific Translations, 1973, p. 86-95.

20. N. Wickramasinghe, "Interstellar Grains", London: Chapman and Hall, 1967.

21. E. Bowen, "Kidson's Relation Between Sunspot Number and the Movement of High Pressure Systems in Australia", Possible Relationships Between Solar Activity and Meteorological Phenomena, Proceedings of a Symposium held at NASA Goddard Space Flight Center, 1973. NASA Goddard SFC Preprint X-901-74-156, p. 56-59 (To be published subsequently as a NASA Special Publication).

22. W. Roberts, "Relationships Between Solar Activity and Climate Change", Possible Relationships Between Solar Activity and Meteorological Phenomena, Proceedings of

a Symposium held at NASA Goddard Space Flight Center, 1973. NASA Goddard SFC Preprint X-901-74-156, p. 3-23 (To be published subsequently as a NASA Special Publication).

23. A. Douglass, "Tree Rings and Their Relation to Solar Variations and Chronology", Annual Report of the Smithsonian Institution, p. 304, 1931.

24. O. Vasil'ev, "Frequency Spectrum of Noctilucent Cloud Displays and Their Connection with Solar Activity", Physics of Mesopheric (Noctilucent) Clouds, Proceedings of the Conference on Mesopheric Clouds, Riga, 1968, Jerusalem: Israel Program for Scientific Translations, 1973, p. 100-113.

25. J. Rosinski, Private Communication.

26. M. Waldmeier, M. 1939 "Über die Struktur der Sonnenflecken", Astron. Mitt. Zürich, No. 138, 439.

2.3.2 ANALYSIS OF IMPACT CRATERS FROM THE S-149 SKYLAB EXPERIMENT

D.S. Hallgren and C.L. Hemenway
Dudley Observatory and State University of New York at Albany,
Albany, N.Y. USA

Abstract

Analysis of craters found on polished plates exposed during Skylab has provided data for a flux measurement over the mass range 10^{-15} to 10^{-7} gms. Chemical analysis of residues in the craters shows a high incidence of aluminium. A variety of morphological forms are described.

A major part of the S-149 experiment[1] on Skylab involved the study of impact craters on highly polished metal surfaces. Samples of polished copper, stainless steel, and silver were exposed in sets of four cassettes, each set having a polished plate sample area of 0.06 m^2. Each cassette is in two parts, a stationary half called the "pan" and a movable half called the "cover". During exposure the covers face in either the solar or anti-solar direction. Figure 1 illustrates the deployment of the S-149 experiment in the solar and anti-solar modes. The Z axis direction was highly stabilized with respect to the sun but the spacecraft underwent slow oscillation about Z. The pans, therefore, being parallel to the Z axis were not directed toward a fixed direction in space. Table 1 shows the exposure data for each of the three sets of cassettes which were returned. A fourth set of samples is currently being exposed awaiting a possible future return to Skylab.

Optical scanning is the principal method used for the detection of craters. All of the samples have been scanned at 200 X, and about 25 % have been re-examined at 500 X. In addition a small area (approx. 63 mm^2) has been scanned in a scanning electron microscope at 5000 X. The optical scanning to date has revealed a total of 78 craters ranging in size from 1.9 μ inside diameter to 135 μ inside diameter. The small area studies in scanning microscope have shown six classic submicron craters as small as 0.3 μ. The fluxes computed from the data are shown in Figure 2.

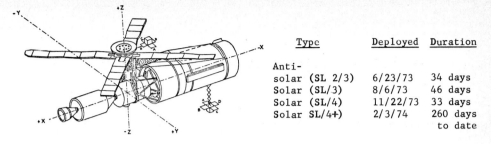

Figure 1
Schematic of S-149 Deployment

Table 1
Exposure Times

Type	Deployed	Duration
Anti-solar (SL 2/3)	6/23/73	34 days
Solar (SL/3)	8/6/73	46 days
Solar (SL/4)	11/22/73	33 days
Solar SL/4+)	2/3/74	260 days to date

Due to the orientation of the collector, the region of space seen by the pans is the same for both the solar and anti-solar exposures while the covers point along a single axis. In each case the fluxes for the pans exceed those for the covers by a factor of 5 to 10. Note in Figure 2 that the fluxes computed from the submicron craters found in the scanning electron microscope are higher than those for the larger craters by a factor of about 10^3.

All fluxes were computed using a crater-to-projectile diameter ratio of 3 which presumes all particles to have the same densities and impacting velocity. Shielding factors to correct for the field of view for each pan and cover were applied in the flux calculation; these are listed in Table 2.

In addition to the flux determination the craters have been examined in the

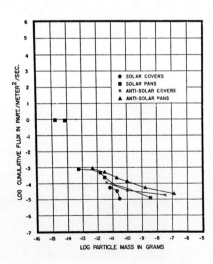

Cassette Half	Solar	Anti-solar
Cover	52.8	56.3
Pan	40.9	43.6

Figure 2
Flux from Crater Data

Table 2
Combined Shielding Factors (%)

scanning electron microscope to study the variability in morphology of the crater interiors. This data has been computed with chemical analyses of the residues seen in the crater interior.

Figures 3 - 6 show representative examples of four general types of crater morphology observed on copper plates. In Figure 3 the crater interior shows distinct signs of melting. This type of structure was limited to the larger craters i.e. greater than 20 μ inside diameter.

Figure 4 shows an example of the most prevalent structure. The term "textured" has been used to describe the generally rough interior.

Figure 5 shows one of three craters in which lumps of residue could be seen within the crater. The last type, shown in Figure 6, has no internal structural detail.

While it is apparent that the craters in Figure 5 and 6 show distinctly different interior structure we emphasize that the structures seen in Figures 3 and 4 are also

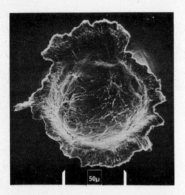

Figure 3
Crater with Melted Interior

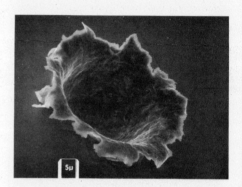

Figure 4
Crater with Textured Interior

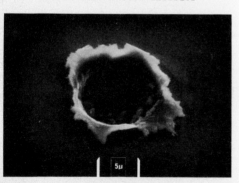

Figure 5
Crater with Lumpy Structure
or Residue

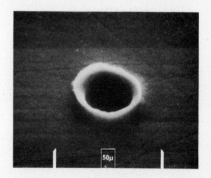

Figure 6
Smooth Structureless Crater

sufficiently distinct to define crater types. Because of what looks like a sequence of potential physical relevance in these micrographs, attempts were made to correlate morphology with observed parameters such as crater diameter, crater depth and diameter to depth ratio; no systematic relationship could be established. It would appear that crater morphology may be related to other physical parameters involved in the collision such as the strength and structure of the original particle.

Chemical analysis of the interior walls and lips of the craters was carried out using an energy dispersive x-ray spectrometer in the scanning electron microscope for counting times of 100 - 1000 seconds. Even though there appear to be distinct types of craters as seen from the morphology no classification can be made on the basis of the residual elements from the impacting particle detected within the craters. Size alone does not determine the ability to detect residues: some craters one micron or less in diameter show residues while the largest crater analyzed showed no detectable residue even with long counting times (1000 sec.). As seen in Table 3, the element most frequently detected was aluminum and, in all, 10 different elements have been detected. Residues may be found exclusively on the lips of a crater, exclusively in the interior, or distributed over either but it is not possible to predict from the appearance of a crater alone where the residue will be found. About 55% of the impact craters yielded

Type	Dia (μ)	Dia/depth	Elements
Smooth			
B-1-12-2	5.2	4.3	Al
C-2-9-9-2	1.8		Cr Fe Ni
Melted			
B-1-12-1	47	2.1	Fe
A-3-15-2	21	1.8	Al
A-3-16-2	20	1.9	Al
A-3-16-3	31	1.9	Al
Lumpy			
B-1-12-4	15	3.9	Mg Al Si S K Fe Ca
B-2-9-3-1	1.8	-	Si S Zn
B-3-15-9-1	12	2.6	Al
Textured			
B-1-12-3	7.8	2.8	Mg Si
D-2-16-8-1	2.5	-	Al Si
A-3-15-1	7.8	3.0	Al
B-3-13-1	6.9	2.5	Si S Fe
C-3-14-1	5.6	3.1	Al
C-3-14-2	5.4	3.6	Al
C-3-14-4	4.9	4.1	Al
C-3-14-5	6.5	3.3	Al
C-3-16-1	2.1	-	Al Si

Table 3
Elements Detected Within Craters

detectable residual elements.

It might be supposed that detection of many craters with aluminum residues indicates that the craters were produced by secondary impacts resulting from micrometeorites striking the spacecraft. There is a possibility of this happening during the solar exposure[2] but on the anti-solar exposure, where most of the aluminum containing craters were found, this is highly unlikely because the S-149 samples were out of the field of view of most of the spacecraft.

In summary the S-149 experiment has provided a long duration measurement of the micrometeorite flux in near-earth vicinity over the mass range 10^{-15} to 10^{-7}gm. Over this mass range the flux is observed to be greater by about an order of magnitude than those measured on the lunar surface. A sharp discontinuity is suggested in the flux at a mass of about 10^{-13}gm. The fact that 70 out of the 78 craters found were on the pans indicates that most of the detected particles were in near circular heliocentric orbits. Examination of the crater interiors as seen in the scanning electron microscope indicates that there are considerable variations in the interior structures of the micrometeorite impact craters. Detection of residual elements from micrometeorites within the craters has been successful for craters even those as small as one micron. The most frequently detected element is aluminum which, considering meteoritic abundances, is quite surprising. Evidence for parcicle clustering has also been found.

References

1. C. L. Hemenway, D. S. Hallgren and C. D. Tackett, "Near-Earth Cosmic Dust Results from S-149", AIAA/AGU Conference on Scientific Experiments of Skylab, Huntsville, Alabama, 1974, AIAA Paper 74-1226, p. 7.

2. D. S. Hallgren, C. L. Hemenway and W. Radigan, "Micrometeorite Penetration Effects in Gold Foil", COSPAR, Varna, Bulgaria, 1975, Space Research XVI, in press.

2.3.3 Micrometeorite Impact Craters on Skylab Experiment S 149

K. Nagel, H. Fechtig, E. Schneider, G. Neukum
Max-Planck-Institut für Kernphysik, Heidelberg, Germany

Abstract

During the Skylab experiment S 149 three different sets of areas were exposed. 71.5 cm^2 were facing the sun for 46 days, and 36 cm^2 for 33 days, whereas 77.5 cm^2 were exposed in anti-solar direction for 34 days. A fourth set is currently being exposed with the hope of future recovery. The exposed surfaces consisted of stainless steel, aluminium, platinum, glass, and pyroxene. The recovered targets have been investigated with a light microscope and a scanning electron microscope. We found two groups of possible impact structures:

1.) Five craters between 1 and 30 µm. These craters show clear signs of hypervelocity impact. Measurements yielded diameter to depth ratios between 2 and 3. Chemical investigations in the craters yielded an enhancement in aluminium in one case.

2.) 44 crater-like structures between 1 and 4 µm in diameter. These features have been found on 4 cm^2 of pyroxene exposed in solar direction. They show diameter to depth ratios between 5 and 8. Chemical measurements of the interior of these structures indicate the elements of the pyroxene composition.

The five impacts of the first group correspond to a cumulative flux of the order of 10^{-4} (m^{-2}s^{-1}) for masses of about 10^{-12} g. The second group may indicate a fragmentation process at altitudes around 450 km. Considering these 44 crater-like structures having been produced by fragments of one projectile, the impact rate could be comparable to that calculated for the first group. If individual projectiles had produced these structures, the corresponding flux could be 2 orders of magnitude higher.

Introduction

The Skylab experiment S 149 was designed to measure the interplanetary dust flux near the earth (altitude approximately 430 km). As a result of a kind invitation by Dr. Hemenway, Dudley Observatory, Albany, N.Y., surfaces were exposed in three missions. The exposure areas consisted of stainless steel, pyroxene, phosphate glass, and foils of aluminium, and platinum. One set of 77.5 cm^2 was looking for 34 days in the antisolar direction; in solar direction there has been one set of 71.5 cm^2 exposed for 46 days, and another one of 36 cm^2 for 33 days. Fig. 1 shows a model of the spacecraft and the position of the solar and antisolar positions.

Methods

The exposed surfaces have been scanned for micron-sized impact craters or penetration holes, respectively. All areas have been investigated

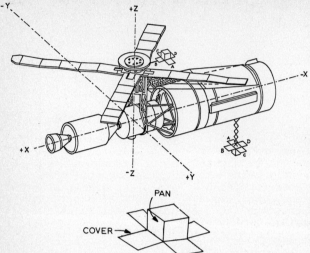

Fig. 1: Model of Skylab with the exposure positions of the collection surfaces.

with a light microscope (detectable crater size 5 μm diameter), and a scanning electron microscope to detect craters .5 μm in diameter. In addition, depth measurements of the individual craters have been performed. Chemical investigations in the interior of the detected craters have been done using a solid state detector Si(Li) and X-ray spectrometers in order to detect projectile residues.

Results

A total area of 52 cm^2 has been investigated with a scanning electron microscope. Five craters have been found between 1 μm and 30 μm in diameter and they all show clear signs of hypervelocity impact. In Table 1, data from the five craters are listed. In addition 44 crater-like features were found on 4 cm^2 of spodumen surface which are interpreted as possible fragmentation products of a low-density projectile near the collection surfaces.

Table 1: List of Impact Craters

Material	Number of craters	Diameter (μm)	Diameter to depth ratio	Exposure direction	Exposure position
stainless steel	1	3.1	2	solar	pan
pyroxene	1	1.0		solar	pan
phosphate glass	1	2.0	to	solar	pan
aluminium foil	1	30.0		solar	cover
stainless steel	1	16.0x22.0	3	antisolar	pan
pyroxene	44	1.0-4.0	5 to 8	solar	cover

Fig. 2 shows the larger crater on stainless steel, the penetration hole in aluminium foil, and one of the 44 craters on pyroxene.

The ratios of crater diameter to crater depth of some of the 44 craters vary between 5 and 8 (Table 1). A comparison with simulated craters suggests low density projectiles (Nagel et al., 1975). If we assume that these 44 craters were produced by fragments of one projectile (total of 6 projectiles) the calculated particle flux is $10^{-4} m^{-2} s^{-1}$ for particle masses of about 10^{-12} g. If we assume that each crater was produced by a primary projectile (total of 49 projectiles) the corresponding flux is two orders of magnitude higher.

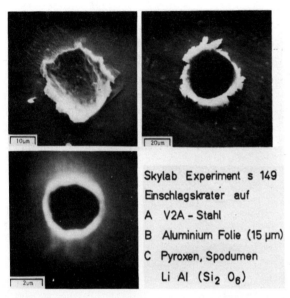

Skylab Experiment s 149
Einschlagskrater auf
A V2A - Stahl
B Aluminium Folie (15 μm)
C Pyroxen, Spodumen
    Li Al ($Si_2 O_6$)

Fig. 2: Scanning electron micrographs of three craters detected on exposed areas.

Chemical measurements in some of the 44 craters yielded the pyroxene composition and no indication of other materials. Investigations of the interior of the larger crater in stainless steel showed an enhancement in aluminium. Chemical analysis of the crater in phosphate glass revealed Al, Si, P, Ca, and Fe, but since all these elements are present in the target it is difficult to measure the residues of the projectile if it had stony meteorite composition. However, we can still exclude the possibility that the projectile was composed of FeNi since these elements were not recorded in the crater.

Discussion of results

The results given here are in a good agreement with the results of the experiment S 149 of Hemenway at al. (1974).

The evidence for fragmentation processes near the collection surfaces as reported in the previous section has also been found by Hemenway et al. (1975). These fragmentation effects have been observed by the satellite experiments Prospero (Bedford et al., 1975) and HEOS 2

(Hoffmann et al., 1975b). In a separate paper (Fechtig and Hemenway, 1976) possible fragmentation mechanisms are discussed.

The location of 4 (out of 5) craters on the pans (cf. Fig. 1, Table 1) suggests that they represent the "apex"-population of the meteoroids, since most of them are found in apex direction (Hoffmann et al., 1975a).

The results for the fluxes are also in good agreement with the results reported from the HEOS 2 dust experiment S 215. The reported fluxes are close to the so-called "apex"-fluxes from the HEOS 2 results.

Acknowledgments

We like to thank Dr. C.L. Hemenway and Dr. D.S. Hallgren for the invitation to participate in their experiment S 149 and for their efforts in mounting and in handling our samples. Mrs. Papp has done a great part of the scanning work.

References:

Bedford, D.K., Adams, N.G., and Smith, D. (1975), "The Flux and Spatial Distribution of Micrometeoroids in the Near-Earth Environment", Planet. Space Sci. 23, 1451.

Fechtig, H., and Hemenway, C.L. (1976), "Near Earth Fragmentation of Cosmic Dust", this Volume.

Hemenway, C.L., Hallgren, D.S., and Tackett, C.D. (1974), "Near Earth Cosmic Dust Results from S-149", AIAA/AGU Conference on Scientific Experiments of Skylab, Huntsville, Alabama.

Hoffmann, H.-J., Fechtig, H., Grün, E., and Kissel, J. (1975a), "First Results of the Micrometeoroid Experiment S 215 on the HEOS 2 Satellite", Planet. Space Sci. 23, 215.

Hoffmann, H.-J., Fechtig, H., Grün, E., and Kissel, J. (1975b), "Temporal Fluctuations and Anisotropy of the Micrometeoroid Flux in the Earth-Moon System Measured by HEOS 2", Planet. Space Sci. 23, 981.

Nagel, K., Neukum, G., Eichhorn, G., Fechtig, H., Müller, O., and Schneider, E. (1975), "Dependencies of Microcrater Formation on Impact Parameters", VIth Lunar Science Conference, Houston, 1975. To be published in the Proceedings of the VIth Lunar Science Conference, Vol. 3, p. 3417.

2.3.4   EXTRATERRESTRIAL PARTICLES IN THE STRATOSPHERE

D.E. Brownlee, D.A. Tomandl, and P.W. Hodge
Dept. of Astronomy, University of Washington, Seattle, WA

Over the past several years we have collected 2μm to 30μm particles from the stratosphere using high volume air sampling techniques. In 1970 and 1971 we flew balloon experiments to 34 km, sampling particles from 1.1×10^4 m^3 of ambient air. Beginning in March 1974 we have flown 100 hours of sampling time on a NASA U-2 aircraft yielding a sampling volume of 9.3×10^4 m^3. In both programs particles are collected by inertial deposition from a 200 ms^{-1} airstream on to clean surfaces coated with thick films of 500,000 centistokes silicone oil.

Collected particles are analyzed by individually removing them from collection surfaces, washing them in zylene and mounting on stubs for SEM analysis. The major data output from the analysis program is morphology and relative elemental abundances as determined in the SEM using a solid state X-ray detector. By raster scanning on the portion of the particle facing the X-ray detector and calibration with mineral standards (similar to the unknowns), element ratios are routinely determined to an accuracy better than a factor of two.

On all the U-2 and balloon flights the same types and spatial densities of particles have been collected. In all collections the most common particles have been pure Al_2O_3 spheres. Our data indicate that these particles have existed in the stratosphere at a density of 10^{-2} part. m^{-3} ($\geq$ 5μm) for the past five years. These particles are not extraterrestrial but are generated by solid fuel rocket engines. It may be that these particles can be used as a calibration source in future particle collection programs in the stratosphere. At 4 μm, ∼90% of the collected particles are Al_2O_3. In the submicron range the Al_2O_3 particles are insignificant compared to the sulfate aerosol and for sizes larger than 10μm, extraterrestrial particles are dominant.

Disregarding particles which are largely aluminum, half of the collected particles have elemental abundances which closely match bulk abundances of primitive meteorites or minerals which are common in C1 and C2 carbonaceous chondrite meteorites. These particles have compositions uniquely different from obvious stratospheric, laboratory and aircraft contaminant particles found in the collections (Al particles, skin flakes, TiO_2 paint, Cd plating, etc.). On the basis of elemental abundances we have identified 8 particles from the balloon flights and 115 from the U-2 flights which are almost certainly extraterrestrial. The majority of these particles have Mg, Fe, Si, S, Ca and Ni abundances within a factor of 2 of cosmic abundances. Importantly, we have not seen cosmic-abundance particles which contain detectable amounts of non-cosmically abundant elements (ie. Cu, Cl, Zn, Na, Cd etc.). Because no known terrestrial (or lunar) material can match cosmic abundances for the six most cosmically abundant elements, we feel the cosmic abundance criterion is a very strong diagnostic criterion for identifying undifferentiated extraterrestrial materials. The particles which closely match cosmic abundances we refer to as "chondritic". No genetic association with chondrules or chondritic meteorites is intended.

In addition to the chondritic particles other composition groups have been identified as extraterrestrial by their physical association with chondritic particles. These composition groups have been found as single particles, as particles with chondritic material adhering to their surfaces, as particles imbedded in single chondritic particles and as particles found inside chondritic particles which broke into fragments during collection or were intentionally crushed in the lab. From the observed associations it is clear that all identified extraterrestrial particle groups were at one time in intimate contact with each other.

We have defined three major compositional groups into which nearly all of the collected extraterrestrial particles can be placed. Sixty percent of the particles classify as chondritic, 30% as FSN and 10% as Mg, Fe silicates. The properties of these groups are as follows.

Chondritic - chondritic abundances

Chondritic particles generally have chondritic (cosmic) elemental abundances. Based on morphology and S abundance the chondritic particles fall into two subgroups.

<u>Aggregates</u>: Ninety percent of the chondritic particles are aggregates of 1000 A sized grains (Figure 1). In some particles the component grains are loosely bound and the particle structure is quite porous. Typically the aggregates are compact with little pore space.

The aggregate particles typically have chondritic abundances (factor of 2) for Fe, Mg, Si, S, Ca and Ni (Brownlee et al. 1975A). Mn and Cr are at the limits of detection but can often be detected at concentrations compatible with cosmic abundances. Optically the aggregates are very black suggesting a carbon content of 2% or more. X-ray diffraction of two of these particles has shown definite existence of Fe_3O_4 and FeS. X-ray powder patterns from these two particles are very similar to powder patterns obtained from matrix material from the Murchison (C2) meteorite which had been heated to 450°C (Fuchs et al. 1973). Electron diffraction of two crushed particles implies that most of the constituent grains in the chondritic particles are crystalline.

Ablation spheres: Ten percent of the chondritic particles are spherules which are not porous and do not contain sulfur. The S depletion is probably the result of thermal alteration and the particle shapes imply the particles were molten at one time. In composition and texture these spherules are very similar to fusion crusts of chondritic meteorites. We believe that these particles are secondary micrometeorites produced by ablation of meteoric bodies (Brownlee et al. 1975B).

FSN - An iron-sulfur mineral with a few percent nickel

The FSN particles are roughly similar to meteoritic troilite or pyrrhotite containing a few percent Ni. In many of the particles, sulfur is deficient relative to FeS by factors of 50% or more. The FSN particles may be related to the poorly characterized Fe, S, O, and Fe, S, C phases reported in carbonaceous chondrites or they may be combinations of FeS and FeO.

Unlike the chondritic particles, which have fairly uniform structures, the FSN particles come in a wide variety of forms. The majority of the FSN particles are spheres, but they also have been found as solid irregular masses, aggregates, well defined single crystals (octahedron with cubic truncation), and stacks of platelets (Figure 2). Some of the nonspherical FSN particles show remarkable similarities to magnetite forms found in Cl meteorites as reported by Jebwab (1971). The FSN spheres may be ablation debris, while the irregular shapes are probably not.

Mg, Fe SILICATES - olivine or pyroxene

These particles are iron poor olivine and pyroxenes with clumps of chondritic aggregates adhering to their surfaces (Figure 3). One euhedral crystal has been found but typically they are subhedral to irregular.

The FSN and most of the Mg, Fe silicate particles are believed to be extraterrestrial because they have been found in physical association with (e.g., actually inside) chondritic aggregate particles. Other particle types, although rare, have been found in crushed chondritic aggregates. For example, on one flight a chondritic aggregate particle was collected that broke into ∼100 fragments upon impacting the collectio surface. Most of the fragments were small pieces of chondritic aggregate material but also found in the debris were FSN particles, enstatite, olivine, an opaque high Si mineral (SiC?) and two fragments of a Si, Al, Ca Ti mineral (high temperature condensate?).

Other extraterrestrial particle types probably exist in our collections but have not been identified as extraterrestrial either because they have not been found in physical association with the three major micrometeorite groups or because they do not have distinctive compositio Because almost all of the collected particles are either high Al particl identified micrometeorites, or obvious contaminants, we believe other micrometeorite types probably constitute only a minor component (<10%) o the extraterrestrial particles normally in the stratosphere.

Our collections indicate that the flux of extraterrestrial dust in the stratosphere is 3×10^{-6} particles $m^{-2} s^{-1}$ (diameter $\geq 10\mu m$). In the 2-30 μm size range most of the particles are true micrometeorites and have not melted during atmospheric entry. Although a variety of particle types has been observed it presently appears that they all originate from a common parent body type, possibly typical for interplaneta meteoroids, which consists of an opaque fine-grained matrix material containing minor amounts of inclusions. The matrix is an aggregate of 1000 A sized grains whose cumulative composition is close to cosmic abun dances; it is very black and probably contains >2% finely dispersed carb Imbedded in the fine-grained matrix are occasional micron-sized inclusio primarily Ni bearing iron sulfides, (similar to troilite) and olivines and pyroxenes with compositions clustering towards forsterite and enstatite. The only known materials which are similar to the recovered micrometeorites in elemental abundances, texture, mineralogy and inclusion content are type 1 and the matrix of type 2 carbonaceous chondrite meteorites. If micrometeorites are cometary particles which formed by aggregation of pre-solar interstellar grains, (Cameron 1973) then the similarity of micrometeorites and Cl chondrites may indicate that a significant fracti of interstellar grains formed in environments similar to those which pro duced C. chondrite meteorites.

We are grateful to M. Blanchard, N. Farlow, G. Ferry and H. Shade of NAS Ames for making the U-2 collections and X-ray diffraction analysis possi ble.

References

Brownlee, D.E., Horz, F., Tomandl, D.A., and Hodge, P.W. 1975A. Physical properties of interplanetary grains in IAU Colloquium no. 25, the Study of Comets (in press).

Brownlee, D.E., Blanchard, M.B., Cunningham, G.C., Beauchamp, R.H. and Fruland, R. 1975B, Criteria for identification of ablation debris from primitive meteoric bodies, J. Geophys. Res. (in press).

Cameron, A.G.W., 1973, Interstellar grains in Museums? in IAU symposium 52, Interstellar Dust and Related Topics, D. Reidel.

Fuchs, L.H., Olsen, E. and Jensen, K.J., 1973, Smithsonian Contr. to the Earth Sciences, No. 10.

Jedwab, J. 1971 La magnetite de la meteorite D'Orgueil, ICARUS, 15, p. 319.

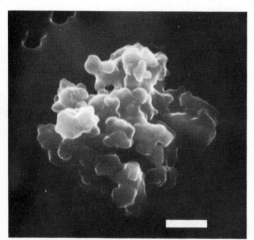

Fig. 1 Typical chondritic aggregate particle. All scale bars = 1 μm.

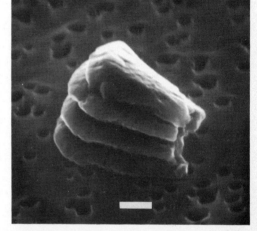

Fig. 2 FSN particle with morphology similar to Cl magnetite (Jedwab, 1971).

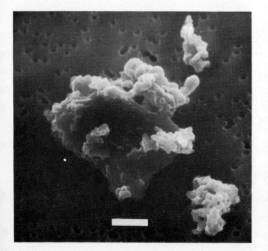

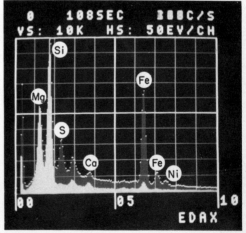

Fig. 3 A 3μm enstatite grain with chondritic material adhering to its surface and lying next to it. The double X-ray spectra (20 KV excitation) are for the enstatite grain (solid bright bars) and for adhering chondritic clump (faint bars topped with dots). The Mg and Si peaks are identical for both spectra. The condritic spectra is an exact match with cosmic abundances. Peak between S and Ca is Pd coating.

2.3.5   MAGELLAN COLLECTIONS OF LARGE COSMIC DUST PARTICLES

D. S. Hallgren and C. L. Hemenway
Dudley Observatory, Albany, N. Y. USA

R. Wlochowicz
Herzberg Institute of Astrophysics
National Research Council of Canada, Ottawa
Ontario/Canada

Abstract

A new balloon collection technique, Magellan, for obtaining large cosmic dust particles $> 50\mu$ is described. The technique utilized a 40 m^2 aperture funnel suspended 300 m below a balloon at an altitude of 25 km. First results indicate that the collected particles are diverse both chemically and physically.

The Magellan collection system evolved from the earlier Sesame program for collecting cosmic dust. Although Sesame provided considerable data[1] concerning particles in the mass range 10^{-9} to 10^{-7} grams, the number of particles collected, being considerably less than anticipated, indicated the need for a system with a much larger area-time product to improve the signal to noise ratio and to extend the measurement to particles larger than 10^{-7} gram (i.e. $> 50\mu$ dia.). The Magellan system utilizes a funnel having a collection area of 40m^2, $\sim$ 2000 times greater than for Sesame, and when flown on an earth orbiting superpressure balloon, can collect a significant number of particles greater than 50μ. The Technique is described in detail in a paper by Wlochowicz et al. The feasibility of such long duration experiments was demonstrated when a superpressure balloon launched in January 1973 from Australia "orbited" the earth several times before landing in the Pacific 210 days later. Unfortunately our collection experiment was not recovered. Since then two successful Magellan collection flights on zero-pressure balloons have been obtained with sampling times indicated in Table 1.

Magellan - 1 May 6, 1974

Sample Number	1	2	3
Collection Time (hours)	-	16	.67
Candidate Particles Collected	-	18	3

Magellan - 2 October 17, 1974

Sample Number	1	2	3
Collection Time (hours)	3	34	3
Candidate Particles Collected	6	126	3

Table 1

The quality of the data from Magellan primarily depends both on the ability of the collection funnel to shed contaminants and on the ability of particles falling into the funnel to reach the collecting surface. The performance was evaluated under several conditions. With the funnel deployed in a hanger, a sample of spherical particles and a sample of irregular particles was dropped into the funnel and in each case 70% of the particles $>50\mu$ was recovered. Humidity was found to affect the experimental collection efficiency; at 33% relative humidity, the smallest particle recovered was 20μ, at 45%, 40μ and at 70% 60μ. Under conditions simulating those at balloon altitude, tests on the funnel fabric indicated that particles greater than 40μ tumbled down the inclined test surface. On the basis of these tests and by considering only particles larger than 50μ for the analysis, it was expected that at altitude contaminants initially in the funnel would be shed while the funnel was open at the bottom and that subsequently, the majority of particles falling on the collection surface would be those falling into the collection funnel.

To test whether the system would shed particles as predicted from the laboratory tests, approximately 15mg of nickel microspheres were dusted into the closed funnel prior to launch of the second flight. An additional contamination test, to determine if the balloon was a source of contamination, involved dusting approximately 15mg of aluminum microspheres on the recovery parachute located just below the main balloon 300m above the experiment. Despite the precautions to avoid and shed contaminants it appears that a significant amount of sand-like material found its way into the funnel and subsequently into the collection boxes. These particles are very similar physically and chemically to soil particles taken from the launch area in Palestine, Texas. Apparently the time constant for the contaminants to clear the funnel was longer than anticipated. Further evidence for a long decay time for contamination comes from the fact that nickel microspheres were found in the first two collection boxes. Three were found in the first box exposed for 4 hours and 72 were found in the box exposed for 34 hours. No microspheres were found in the third box which provides some measure of time for the system to cleanse itself. None of the aluminum test spheres from the parachute were found in the collection boxes.

The possible contaminant material has the appearance of a clear glassy particle either rounded or angular in shape with varying amount of red-brown debris on the surface with silicon being the most abundant element found. Until each questionable particle can be removed from the plexiglas collection slide and examined in detail, all such particles have been, for the present, put aside. The remaining candidate particles must include many real cosmic dust particles and the numbers of candidate particles collected during each exposure are also shown in Table 1.

Figure 1 - 6 illustrates some of the particles considered to be collected cosmic dust particles. The micrographs shown were taken in a scanning electron microscope. Many of the following descriptions include optical observations.

Figures 1 and 2 show examples of particles which have most probably been heated, at least at the surface. Note the fine cracks in the surface of these particles which we believe resulted from the heating and subsequent cooling experienced when the particle entered the atmosphere. Only aluminum and chlorine were detected in the particle in Figure 1 while the particle in Figure 2 contains a significant fraction of iron with lesser amounts of calcium and chlorine plus traces of five other elements.

Several types of spherical particles have been found. Figure 3 shows a

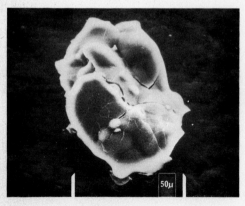

Figure 1
Particle with Heating Cracks

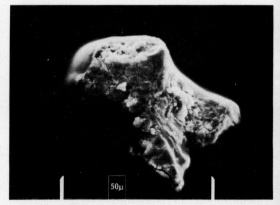

Figure 2
Particle with Heating Cracks

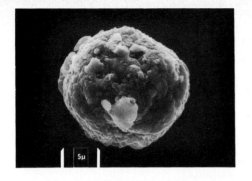

Figure 3
White Spherical Particle

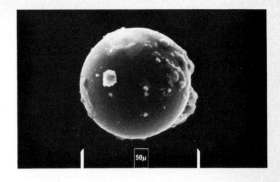

Figure 4
Hollow Glassy Sphere

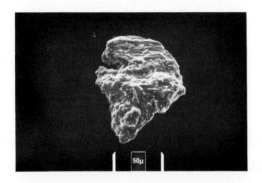

Figure 5
Iron Rich-Particle with Metallic Luster

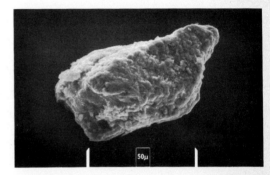

Figure 6
Iron-Rich Particle; Black in Color

particle which appears milky white with small dark inclusions at the surface. The composition of these particles varies widely from particle to particle with Ca, S, Fe, and Ti the major elements. The hollow spherical particle shown in Fig. 4 is water white and contains about equal portions of Na, Si, and Fe. Figures 5 and 6 show particles in which only iron was detected. That in Figure 5 is metallic in appearance whereas Figure 6 shows a black particle with a granular surface. From these few examples which do not represent all of the particle types it is clear that the collected particles are indeed diverse.

Figure 7 shows the S-149 impact crater flux measurement[3] and fluxes computed separately for all of the candidate particles, for the particles which show signs of ablation or heating, and for glassy spheres. From this curve it can be seen that the flux as measured by the Magellan balloon technique is in rough agreement with the flux from Skylab S-149.

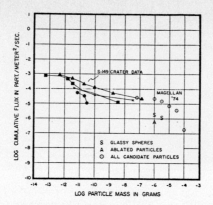

Figure 7
Flux Curve Comparing Magellan Result with S-149 Skylab Data

The candidate particles are greatly different from each other both in morphology and in chemical analysis. The observed morphology is interesting in that there are few particles which show signs of ablation. Micrometeorite theory[4] predicts that 100μ diameter particles or even particles as large as 1000μ can, under special conditions, enter the atmosphere unablated. Blanchard[5] in studies simulating ablation of olivine and magnetite showed that 5 and 20% respectively of the shock fractured fragments failed to show signs of ablation. It is reasonable to find some particles with no signs of melting but to find so few is surprising.

It would be expected that the chemical abundances would bear some resemblence to average abundances determined by other techniques, i.e. the major constituents would be magnesium, silicon and iron. This is not the case. Based only on semi-quantitative analysis aluminum, silicon, calcium and iron are the most abundant elements detected. Nickel has been detected in trace quantities in only two out of 67 candidate particles.

References

1. Hemenway, C. L. et al "New High Altitude Balloon Top Cosmic Dust Collection Technique", Space Research XI, p. 393-395, Berlin, Akademie Verlag, 1971.

2. Wlochowicz, R. et al "Magellan - A Balloon-Borne Collection Technique for Large Cosmic Dust Particles", submitted to Canadian J. of Physics, 1975.

3. Hemenway, C. L. et al "Near Earth Cosmic Dust Results from S-149", AIAA/AGU, Conf. on Scientific Results from Skylab, 1974.

4. Opik, E. J., "Physics of Meteor Flight in the Atmosphere", Interscience, 1958.

5. Blanchard, M. B., et al "Artificial Meteor Studies: Olivine", J. Geophysical Res., V. 79, No. 26, 1974.

2.3.6   SPECIFIC SOURCES OF EXTRATERRESTRIAL PARTICLES

J. Rosinski
National Center for Atmospheric Research
Boulder, Colorado USA 80303

Sampling of magnetic particles in the 5 to 20 μm diameter size range on the earth's surface at different latitudes was used to detect meteor streams composed of particles of masses too small to be detected by other means. Magnetic spheroid concentration peaks appeared simultaneously at different latitudes in the Northern and Southern Hemispheres, proving that the majority of spheroids is of extraterrestrial origin. Extraterrestrial particles consist of two major classes: (1) magnetic particles containing a high concentration of iron oxide in the form of magnetite; and (2) glassy particles containing a low concentration of iron and a large concentration of Si and Al; they contain Ni, Mn, Cr and Ti; the concentration of Ti is approximately one hundred times larger in glassy particles than in magnetic spheroids. Comparison of chemical composition of particles collected from different meteor streams must be very carefully evaluated; chemical composition of surfaces and interiors of magnetic particles is different due to selective fractionation of different chemical compounds between molten and solid phases and due to evaporation during entry of meteoritic particles into the atmosphere. The presence of H_2O, CO_2 and C_nH_{2n+2} in internal cavities of a few spheroids was found to be difficult to explain; those gases were associated with meteoritic particles and consequently they existed in space. The yearly earth flux of extraterrestrial particles can only be determined by the summation of daily fluxes obtained at different latitudes; the daily flux changes from year to year. Fluxes of particles present in January and October were calculated. Particles collected in January seem to originate from sources present between November 6 and January 15; during this period of time S. and N. Taurid, Geminid, Ursid and Quadrantid meteor streams are active. Particles collected in October seem to originate from the unknown meteor stream or from the Giacobini-Zinner meteor stream providing it is preceded by micrometeoritic particles approximately two weeks prior to the visual display.

2.3.7   NEAR-EARTH FRAGMENTATION OF COSMIC DUST

H. Fechtig and C. Hemenway
Max-Planck-Institut für Kernphysik, Heidelberg/F.R.G.

Abstract

Cosmic dust fragmentation in near-earth space is strongly suggested by the high fluxes ($\sim 10^2/m^2s$) of submicron particles observed in the upper atmospheric collection experiments, the lower fluxes ($\sim 10^{-1}/m^2s$) observed in near-earth space by the lower sensitivity detection experiments and the much lower fluxes ($\sim 10^{-4}/m^2s$) observed in the deep-space experiments. Various possible fragmentation mechanisms are discussed.

Evidence for Near-Earth Fragmentation

During particle collection experiments by rockets in the earth's upper atmosphere ($\geq$ 80 km) clusters of particles are found which suggest particle breakup at a higher altitude. Fig. 1 shows a solid-type cosmic dust particle cluster collected with the aid of in-flight shadowing[1]. The particle appears to have broken up at a higher altitude and the breakup energy given to the fragments is small compared to the kinetic energy of the original particle.

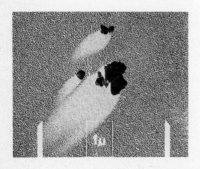
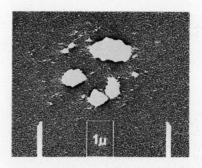

Fig. 1
Collected Particle Cluster

Fig. 2
Collected Hole Cluster

Recovered near-earth satellite exposure of thin films and polished surfaces also provide evidence of particle breakup. Fig. 2 shows a cluster of holes[2] found in a thin film exposed in the Gemini cosmic dust experiment (S-12) at an altitude of 230 km during Gemini 9. The particle appears to be of a similar nature to that shown in Fig. 1 and probably has broken up at a still higher altitude. Similar thin film penetration hole evidence[3,4] has been found in the SKYLAB cosmic dust experiment (S-149) which provided three recovered, month-long exposures at an altitude of 430 km. In addition, evidence of clustering of impact craters has been found in the S-149 experiment. For example, forty 1-4 μ diameter impact craters were found[5] in an area of 4 cm^2 which would be equivalent to a localized particle flux enhancement about 100 times the average near-earth flux[3]. The high sensitivity Prospero micrometeorite detection experiment[6] has found evidence of clusters of submicron particles at altitudes ranging from 547 to 1582 km. At higher altitudes ranging from about 500 to 240.000 km the HEOS coincidence micrometeorite detection experiment has also found evidence of time clustering[7]. Furthermore, the HEOS experiment has shown in near-earth space ($\leq$10 earth radii) that most of the particles (93 %) are parts of clusters and that submicron particles were detected primarily when the HEOS sensor was facing the direction of the earth.

In addition to the type of particle shown in Fig. 1, evidence exists in the collection experiments[8] for the existence of submicron particles which in interstellar space would be called "core-mantle" particles. Fig. 3 shows an example of a particle having a high density core and an extremely low density, fragile mantle. Such particles appear to have been detected at higher altitudes by thin film penetration experiments[2,4] in which the core penetrates the thin film whereas the mantle does not. An example of the resulting "evil-eye" structures from the S-149 experiment is shown in Fig. 4.

Fig. 3   Fig. 4
"Core-Mantle" Particle   "Evil-Eye" Hole Structure

Table 1 shows the fluxes of submicron particles (including "core-mantle" particles) and the fluxes of holes (including "evil-eyes") as a function of altitude. It suggests that the submicron particle influx in the earth's atmosphere is approximately constant with altitude over a remarkably large range of atmospheric pressures and that the particle breakup occurs in the ionosphere. Table 1 also indicates that the near-earth submicron particles flux is more than 100 times the lunar submicron flux[10], thus providing additional evidence of particle fragmentation in near-earth space.

Table 1

Altitude	Experiment	Flux	Pressure	Type of Event
430 km	S-149	$3 \times 10^2 m^{-2} s^{-1}$	3×10^{-8} mb	Submicron Holes
230 km	S-12	$2 \times 10^2 m^{-2} s^{-1}$	7×10^{-6} mb	Submicron Holes
80 km	Pandora Collections+)	$10^2 m^{-2} s^{-1}$	10^{-3} mb	Submicron Particles
26 km	Balloon Detection++)	$10^2 m^{-2} s^{-1}$	20 mb	Submicron Particles

+) The Pandora collections[8] free from special enchancements such as Noctilucent clouds are given here.

++) Submicron particle numerical density data[9] above the Junge layer converted to fluxes through estimated particle falling velocities.

Possible Particle Breakup Mechanisms

<u>Evaporation Breakup:</u> Fig. 5 shows a sketch of a fluffy particle made up of "core-mantle" elements. If the cores were stony and the mantles were icy, such icy mantles would be expected to evaporate and the

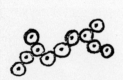

Fig. 5
"Fluffy" Particle

particle might breakup upon entering the earth's atmosphere (or approaching the sun). However, such a breakup mechanism does not explain the relatively high altitude breakup implied by the near-earth observations.

<u>Phase Change Breakup:</u> If the dirty ice of comets were amorphous rather than crystalline, then the particle entry into the earth's ionosphere might trigger the phase change into crystalline ice and the resultant energy release be partially responsible for the near-earth particle fragmentation. Such a process cannot be excluded at present.

<u>Chemical Breakup:</u> If the cores of the schematic fluffy particle shown in Fig. 5 were refractory or stony and the mantles primarily carbon

soaked with hydrogen, chemical breakup as a result of oxidation in the earth's ionosphere would be possible. The carbon would react with the atomic oxygen to make CO or CO_2 and the hydrogen to make H_2O. Both reactions would be expected to release relatively large amounts of thermal energy and the remaining ash of a mantle would be quite fragile as observed in particles such as that shown in Fig. 3. It must be remembered that cosmic dust (no matter what its origin - comets, stellar atmospheres, etc.) is formed in a hydrogen-rich environment and would not be expected to be chemically stable in an oxygen-rich environment such as the ionosphere or the earth's atmosphere. Indeed, submicron cosmic dust particles and "evil-eye" hole samples to be stored in argon in order to be kept without corrosive destruction by atmospheric oxygen. Table 2 shows the number of oxygen atoms that a particle of 1 cm^2 cross section would be expected to encounter during passage through the earth's ionosphere on a $90°$ scattering trajectory as a function of minimum altitude above the earth's surface. The data in Table 2 indicates that passage through the ionosphere would not be expected to seriously oxidize a solid particle of about one gram mass.

Table 2

Min. Altitude	200 km	500 km	800 km
Sunspot Maximum	$1 \times 10^{18} O^+ atoms/cm^2$	$3 \times 10^{16} O^+ atoms/cm^2$	$2 \times 10^{15} O^+ atoms/cm^2$
Sunspot Minimum	$7 \times 10^{17} O^+ atoms/cm^2$	$2 \times 10^{15} O^+ atoms/cm^2$	$4 \times 10^{13} O^+ atoms/cm^2$

However, a fluffy particle such as that shown in Fig. 5 made up of "core-mantle" elements with .1 dimensions with hydrogen soaked, carbon mantles could be broken up by oxidation during transit through the earth's ionosphere. This chemical breakup mechanism would be sensitive to sunspot activity and may be playing a role in larger fluffy bodies such as radio meteors. Furthermore, oxidation of materials exposed in near-earth space has been detected[3].

<u>Collisional Breakup:</u> Collisional breakup would be expected to be less effective for fluffy particles than for solid particles. Particle collisional breakup would not be expected to play a major role in near-earth space, although the ß meteoroids may result in part from such fragmentation between 1 AU and the sun. A fluffy particle would be expected to breakup upon hitting a foil having a thickness approximating the size of the individual elements of the particle or breakup upon encountering the equivalent mass/cm^2 of a spacecraft atmosphere. Such a thin-film effect was discovered by laboratory ex-

periments[12] and found in SKYLAB. This S-149 data clearly shows the existence of fluffy particles in near-earth space[4].

<u>Electrostatic Particle Breakup:</u> Particles entering the earth's ionosphere would be expected to change their charge significantly by increased electron accretion, especially upon entry into the shadow of the earth. Particle breakup by this means requires the particles to be extremely fragile. More work should be done on this possibility.

<u>Crater Particle Formation:</u> This process can be important when a small body hits a large one with high velocity. Small particles are formed during the crater-forming process as noted in laboratory experiments[13]. If the impact occurs in a stony or glassy surface a concoidal fracture pattern may appear around the impact crater and indicates additional fragmentation. The moon is likely providing small particle clusters to near-earth space as a result of particle impacts[7].

Conclusions

Evidence from a variety of experiments indicates a great deal of cosmic dust fragmentation in the ionosphere. Oxidation fragmentation can account for the relatively high fluxes of submicron particles collected in the earth's atmosphere.

References

1. C. Hemenway, Dr. Hallgren, "Time variation of the Altitude Distribution of the Cosmic Dust Layer in the Upper Atmosphere", Space Research X, p. 272-280, 1970.
2. C. Hemenway, D. Hallgren and J. Kerridge, "Results from the Gemini S-10 and S-12 Micrometeorite Experiments", Space Research VIII, p. 521-535, 1968.
3. C. Hemenway, D. Hallgren and D. Tackett, "Near-Earth Cosmic Dust Results from S-149 AIAA/AGU Conference on the Scientific Experiments of SKYLAB, Huntsville, Alabama, Oct. 30-Nov. 1, 1974.
4. D. Hallgren, C. Hemenway and W. Radigan, "Micrometeorite Penetration Effects in Gold Foil", COSPAR, Varna, Bulgaria (in press) 1975.
5. K. Nagel, H. Fechtig, E. Schneider and G. Neukum, "Micrometeorite Impact Craters on Skylab Experiment S-149", this volume, 1976.
6. D. Bedford, "Observations of the Micrometeorite Flux from Prospero", Proc. Roy. Soc. Lon. A. V 343, p. 277-287, 1975.
7. H. Hoffman, H. Fechtig, E. Grün, J. Kissel, "Temporal Fluctuations and Anisotropy of the Micrometeoroidal Flux in the Earth-Moon System Measured by HEOS 2", Planetary Space Science 23, 985 (1975).
8. C. Hemenway, "Collections of Cosmic Dust", Whipple Symposium (in press) 1973.

9. J. Rosen, "The Vertical Distribution of Dust to 30 Kilometer", JGR, V 69, 21, p. 4673-4676, Nov. 1, 1964.

10. E. Schneider, D. Storzer, J. Hartung, H. Fechtig and W. Gentner, "Microcraters on Apollo 15 and 16 Samples and Corresponding Cosmic Dust Fluxes", Proc. Fourth Lunar Science Conf., V 3, p. 3277-3290, 1973.

11. H. Patashnick, G. Rupprecht, D. Schuerman, "Energy Source for Comet Outbursts", Nature, V 250, p. 313-314, 1974.

12. E. Grün, P. Rauser, "Penetration Studies of Iron Dust Particles in Thin Foils", Space Res. IX, p. 147-154, 1969.

13. G. Eichhorn, Dissertation, University of Heidelberg,

3 COMETARY DUST

3.1   DUST IN COMETS AND INTERPLANETARY MATTER

Vladimir Vanýsek
Department of Astronomy and Astrophysics
Charles University
150 00 Praha 5 - Smichov, Czechoslovakia

Summary: Results of many current attemps to estimate the physical and chemical composition of dust particles in comets are reviewed and discussed. It is shown that even the most basic parameters, such as albedo of the cometary dust, are not properly known at the present time. The emission feature in the infrared spectra of comets, which resembles those observed in the interstellar (and circumstellar) clouds and which indicates a relation between the general composition of comets and interstellar matter, is widely ascribed to silicates. Similar features may, however, be also caused by polymerized molecules or hydrocarbons mantles of the dust grains.

1. Introduction

The generally accepted concept that comets are the most efficient source of interplanetary dust, is based on the fact that the mass loss of the dusty material from a moderately bright comet during the perihelion passage at a heliocentric distance $r \leq 1$ AU is about 10^{13} to 10^{15} grams.

The physical properties of the cosmic dust in the cometary environment therefore constitute one of the relevant topics in the study of interplanetary matter.

The determination of the physical and chemical structure of dust particles in the zodiacal light, comets or interstellar space clouds is, however, one of the most difficult tasks in astrophysical research. Information on interplanetary dust based on the study of directly analysed samples is still very scarce and complicated by selection effects. Indirect methods used for the estimation of sizes and refractive indices of cosmic grains are based on the comparison of the observed and computed scattering properties of dust clouds likely to be found in interplanetary (or interstellar) space. Usually, the problems are restricted to the cases of spherical particles and to the use of some simplified assumptions about the size distribution function and the fitting of the computed models (based on the Mie theory) to the observational results.

From this point of view, comets are very suitable as space scattering elements. The cometary head and tail are supposed to be optically thin media where the effects of higher-order scattering are negligible. The phase angle is perfectly defined by the position of the object in its orbit, and the integration along the line of sight involving the phase function can be neglected. The relatively large angular dimensions of the cometary head and tail permit surface photometry of the innermost regions of the coma and the kinematics of the tail provides some information on the dynamical effects of the light pressure on small particles which, of course, involves size and mass of the dust grains.

Light pressure and drag forces as well as the interaction of gas and dust (see for instance Finson and Probstein (1968), Shulman (1969), Sekanina and Miller (1973)), give rise to selection mechanisms separating grains with different sizes and optical properties from each other. But the most promising methods for the study of physical and chemical structure of the cometary dust are 1) direct analysis by space probes, 2) determination of optical properties of small particles including the search of emission or absorption at discrete wavelength intervals.

Only few samples of micrometeoroids which seem to be of cometary origin have until now been directly analysed by space probes (see Grün et al., 1976). The optical parameters of small particles studied by the colorimetry, spectrophotometry and infrared measurements are therefore practically the only sources of information about the grains structure. The limitation imposed upon this problem by comparing the observed properties of cosmic grains with those computed for particles of regular form (namely spheres and cylinders) constitutes, unfortunately, a disadvantage of this method. The Mie theory was successfully used in cases where almost identical particles have narrow-size distribution. In astrophysical objects, however, the variety of size and form of particles is large, and their distribution function, shape and chemical composition cannot be perfectly reproduced either by computed models or laboratory samples or both. Nevertheless, with the help of high-speed computers and idealized scatterers of the Mie theory, we may simulate some of the optical and physical characteristics of scattering aggregates impossible to reproduce in laboratory, but likely to exist in cosmic space. The same holds for the emission or absorption features of solids, particularly in the infrared region.

In the next paragraphs we review the present state of studies concerning the cometary dust by means of colorimetric, polarimetric and infrared observation.

2. Colorimetry

Earlier studies of cometary dust particle characteristics were based on the colorimetry or spectral gradients of the cometary continuum. The classical photometric system UBV is efficient for such a purpose only when the spectrum of the comet is a pure continuum. In fact we know only two such cases: Comets Baade 1954h and P/Schwassmann-Wachmann 1. Both objects were photometrically observed near the opposition with the Sun. The measurement of Comet 1954h made by Walker (1958) in the UBV colour system, indicates a definitely positive excess of about $+0.2^m$ in B-V (relative to the colour of the Sun), i.e. reddening of the scattered light. The same was found for several other comets (see Vanýsek, 1965). Although in other cases the influence of emission C_2-bands in the V-colour is evident, the tendency of "reddening" of the continuum seems to be typical of the bright comet, as well as of a very faint object (Johnson, 1960; Liller, 1970).

It is worth noting that the reddening of light scattered by interplanetary dust is a general phenomenon. Lillie (1972) concluded from the OAO-2 results that zodiacal light is redder than the Sun between 4300-2500 A, resembling a G8 V star. It is an obvious manifestation of the selective scattering of light. In UV below 2500 A the colour of zodiacal light is similar to the colour of B stars, and the albedo of the grains in this spectral region is relatively high. Similar behaviour of the continuum was observed in comets (see Lillie, 1976), but the albedo of the cometary dust is higher than that of the zodiacal cloud. From this point of view comets resemble the interstellar rather than the interplanetary matter. The spectrophotometric results of Comet Arend-Roland 1957 III show that the spectral distribution in the cometary continuum resembles the spectral distribution of G8 V stars and this is due to the selective scattering by small particles. Liller (1970), for instance, concluded that the scattering particles are iron-like conductive particles with diameters of about 0.3 microns. It can be shown, however, that dielectric particles fit the observation quite well provided that more precisely computed scattering properties are used.

The positive colour excess was confirmed recently by Babu and Saxena (1972) and by Vanýsek in an unpublished measurement of Comet Bennett

1970 II. Spectrophotometric results for Comets 1968 I, 1968 V and 1968 VI by Gebel (1970) show, on the other hand, that the reflected or scattered light is "grey" and the continuum energy distributions for these comets follow closely the spectrum of the Sun. Similar conclusions follow also from the measurements made by Johnson et al. (1971) in the continuum of Comet Bennett. These authors, however, used a very wide bandpass; contamination by molecular emissions may therefore significantly distort the assumed form of the continuum.

The discrepancy of the "grey" scattering results with the reddening deduced by other authors is not surprising at all. This may be merely an effect of the reflection of light by a bright central condensation where large particles dominate, in agreement with the colour change along the coma radius found in earlier colorimetric measurements by Vanýsek (1960).

Babu and Saxena (1972) (for Comet 1970 II) found a spectral gradient change with time; this may be caused by the dependence of light scattering on the phase angle or by time-changes in the dominant size or other physical characteristics of the dust particles. The results obtained by Babu (1975) for continuum energy distribution in the head of Comet 1973f indicate that the reddening of the scattered light decreases with phase angle and with heliocentric distance. These measurements were made at very large zenith distances, however, and the uncontrollable influence of anomalous extinction may alter the interpretation of the observed spectral gradients.

Theoretical values of the relative spectral gradient have been determined for different size distributions and different refractive indices for small particles for various models (Rémy-Battiau, 1966). The results show that the spectral gradient remains nearly constant for phase angles $90° \leq \vartheta < 180°$ ($\vartheta = 180°$ is for the backward scattering) and is insensitive to the values of the physical parameters of the particles that have been considered. If $\vartheta = 30°$ the spectral distribution of cometary spectra as well as the intensity of the continuum, are very sensitive to the phase angle. A considerable change in the spectral gradient can therefore be expected if the comet is observed near the inferior conjunction with the Sun where also a strong forward scattering effect may be expected. Similar significant changes can occur even at the phase angle $\vartheta \sim 90°$ if the size distribution of the particles is very narrow.

It seems to be very difficult to obtain reasonable conclusions from the colorimetric observations in the visual spectral range only. The usefulness of the relative spectral gradient Sun-comet (or colour difference Sun-comet) for the determination of the particle size is only limited by our virtual ignorance of the size distribution function. Another source of diffulties is the numerical modelling. Although in the computed models the integration of the intensity of the scattered light over a large grain size interval sweeps out some resonance peaks, and even if the process of the integration actually used modifies the phase only slightly, the differences of computed intensities in two or more wavelengths (which determine the computed colour) may be strongly affected by the accuracy of numerical results.

Nevertheless, precise spectrophotometry of the continuum in cometary spectra along the coma or tail would provide very valuable data. Such data can lead to inferences regarding possible differences between dust particles in the vicinity of the cometary nucleus and in distant regions of the tail.

3. Polarization

The radiation from comets is highly polarized. The polarization of the radiation from any particular object varies considerably with time, phase angle, measured area in the coma or tail and depends on the wavelength bandpass.

The polarization of the cometary light is caused by two mechanisms. One is the polarization of the fluorescence emission in the molecular bands and the other is the scattering of light by small dust particles. The linear polarization in the molecular band is about 8 % and is almost independent of the phase angle, while in scattered light polarization generally depends on the phase angle and may reach 50 %. Results reported by Michalsky (1975) for Comet Kohoutek 1973f were exceptional: the emission was more highly polarized than the continuum.

The presence of nonspherical dust particles in cometary atmospheres may cause circular polarization of the continuum radiation. No positive results have yet been obtained, however, i.e., no circular polarization larger than 0.05 % has so far been detected. Polarimetric measurements appear to be more efficient tools for the study of the physical properties of the dust component in comets than the colorimetric ones.

Just as has been the case with colorimetry, great care must be used when interpreting polarimetric results obtained from wide spectral bandpass photometry. The relative contribution of the continuum total flux in the bandpass used varies with time as well as with distance from the nucleus. The behaviour of the polarization of the large area measured is therefore not necessarily representative of the scattering properties of the cometary dust. The available polarimetric data on comets are still very scarce. Extensive sets of measurements were made for some bright comets (1957 III, 1957 V, 1970 II, 1973f) and a few fainter objects.

The polarization of the scattered light in the coma is, on the average, 15 to 25 % and sometimes increases up to 50-65 % near phase angles of $90°$ (although these extreme values seem to be unique to Comet Ikeya-Seki 1965 VIII). The measurements by Blackwell and Willstrop (1957) and Martel (1960) on Comets 1957 III and 1957 V indicate an increase in the degree of polarization from 5 % near phase angle $\vartheta \sim 145°$ to 30 % for 90°. According to the result obtained by Gehrels (1972) the polarization of Comet 1970 II near $\vartheta \sim 90°$ increases with the wavelength, ranging from 25 % at $\lambda \sim 0.5$ m to 41 % at 0.96 m. Very early studies of the dust characteristics of Comet Arend-Roland (1957 III) based on polarimetric data by Rémy-Battiau (1964) show that the presence of dielectric particles is more likely than that of metallic micrometeorites. A similar conclusion follows from the study of Donn et al. (1967).

Of particular interest is the change in the polarization vector from positive to negative, which means a change of orientation of the electric vector relative to the direction of incident beam (i.e. to the plane defined by Sun-observer-comet). In a study of the polarization on polydisperse cloud models (Vanýsek, 1971) it may be established that near the phase angle $\vartheta = 60°$ the scattering by small particles exhibits a considerable increase in positive polarization with increasing particle conductivity. A high positive polarization (i.e. the electric vector perpendicular to the polarization plane is greater than the one parallel to it) is present for absorbing clouds (even with moderate absorbers) having a maximum between $\vartheta = 60°$ to $90°$, while on the other hand negative polarization near phase angles $150°$-$170°$ is very typical of all cloud models with dielectric particles. This, of course, is not valid for very small particles in the Rayleigh scattering domain. Variations in the polarization, which may help one to distinguish between dielectric and absorbing particles, are very pronounced near small or very large phase angles.

The sharp change of orientation of the polarization plane (orientation of the electric vector) with the phase angle is typical of the behaviour of a polydisperse thin cloud containing particles having a refractive index with a very small imaginary part. This polarization reversal is present in planetary atmospheres (including Earth) and has been observed in the zodiacal light (Weinberg, 1964; Wolstencroft and Rose, 1967; Weinberg and Mann, 1968; Frey, 1975).

From this point of view, the most important results concerning the polarization of cometary light are the ones from multicolour observations made by Weinberg (1974) for the tail of Comet Ikeya-Seki (1965 VIII). Data were obtained at six effective wavelengths and with two different filters centered at the 5577 A emission line of OI.

The measurements made along the tail axis provide information about the change of the degree of polarization with phase angle and neutral point (zero polarization). The phase angle of the neutral point is determined by the size of the particles and their refractive index, their alignment (in case of nonspherical particles) and quality of their surface, or by a combination of all these effects.

The negative polarization found by Weinberg and Beeson (1975) in the tail of Comet 1965 VIII, requires the presence of dielectric grains or that of highly irregularly-shaped particles. The interpretation is more difficult however, when elongated particles dominate the distribution of the cometary dust. Detailed measurements of the polarization made by Martel (1960) (Comets 1957 V and P/Giacobini-Zinner), Osherov (1970) and Clarke (1971) (Comet 1970 II) show that the plane of vibration (or the plane of polarization) sometimes deviates significantly from one of the two possible positions orthogonal to the scattering plane. Such a deviation may be caused by scattering by aligned and elongated particles. It has been shown by Harwit and Vanýsek (1971) that an efficient alignment mechanism might be provided by bombardment with solar wind protons. The extent to which particles become aligned depends also on the gas flow from the nucleus; the plane of polarization near the nucleus is therefore more arbitrarily oriented than in the tail, where the solar wind effect dominates. The polarization measurements of Comet Bennett made by Osherov as well as those by Clarke fitted very well this hypothesis. This means of course, that the models based on the Mie theory using spherical particles are inadequate for the estimation of the dust composition.

4. Infrared Emission of the Cometary Dust

Infrared measurements constitute the decisive method for determining the physical characteristics of cometary grains. The interpretation of such data leeds to estimates of the albedo. The emission and absorption features in the infrared spectrum (outside possible Ballik-Ramsay C_2 emission wavelengths in near infrared) provide some information on the physical and chemical composition of the solid-state component of the cometary atmosphere.

Infrared measurements of the thermal emission from the dust component of the cometary atmosphere have been made for Comets Ikeya-Seki (1965 VIII); Bennett (1970 II); Kohoutek (1973f); Bradfield (1974b) and P/Encke (Becklin and Westphal, 1966; Maas et al., 1970; Kleinmann et al., 1971; Lee, 1972; Westphal, 1972; Rieke and Lee, 1974; Ney, 1974; Gatley et al., 1974; Merrill, 1974; Noguchi et al., 1974; Zeilik and Wright, 1974).

Most of these observations revealed emission features near 10 μm which had been widely ascribed to silicates. Similar features have been observed in infrared spectra of cool stars having circumstellar dust clouds.

By comparing the continuum radiation from the comet in the visual with that in the infrared regions, the optical albedo may be estimated. This can be done by assuming that the infrared emission consists predominantly in the reemission of the absorbed visible solar radiation. The infrared measurements made by Becklin and Westphal (Comet 1965 VIII) and Kleinmann et al. (Comets 1969 VIII and 1970 II) have in this manner been analyzed by O'Dell (1971). He estimated the particles' diameter to be about 0.1 micron and found a value for albedo of $\gamma = 0.3 \pm 0.15$. This method has recently been applied by Ney (1974) to Comets 1973f and 1974b. The albedo of the dust coma found in these objects was low ($\gamma = 0.18 \pm 0.2$) and results for Comet Bradfield 1974b reveal that the "silicate bump" has dissappeared and that the albedo decreases significantly in a few days because the dust in the coma must have changed from small to large particles. These values of albedo γ are, however, not identical with those determined from the ratio of the scattering efficiency Q_s to the extinction efficiency Q_e. If E_s is the measured specific intensity of the scattered light and E_r the radiation of the dust cloud, then for an optically thin case $E_r/E_s = (1-\gamma)/\gamma$ holds. E_s as well as E_r may approximately be defined by the Planckian maxima of the cometary visual continuum (colour temperature 5700 K) and infrared emission of the dust coma (T≥ 300 K, for heliocentric distances r ≤ 1 AU). The scattered sunlight maximum depends only slightly on the selectivity of the scattering process, but the absolute value of E_s is a function of the phase angle. The submicron particles with a Mie parameter $x < 10$ and moderate absorption are the most efficient scatterers. For instance, a particle with the diameter of about one micron and the refractive index 1.7 - 0.05i at $\lambda \sim 0.5$ m has the ratio Q_s/Q_e 1:2; the albedo is then about 0.5. But the radiance of the particle at the phase angle $\vartheta = 90°$ is only 10^{-1} of its radiance at $\vartheta = 30°$ and about 10^{-5} of its radiance in the forward direction ($\vartheta = 0°$).

For a dust cloud containing such particles one can find a high value for E_r/E_s (low albedo) if the phase angle is somewhere between $\vartheta = 30°$ to 150°. Because of the prevalence of strong forward scattering, the value of γ increases up to 1 at $\vartheta = 0°$. In the case of small reflecting (and slightly absorbing) particles a similar effect exists for backscattering angles. This phase effect could even be significant for clouds with a large variety of grain compositions and size distribution. The quantity is therefore some kind of phase albedo which is defined as $\gamma(\vartheta) = \psi(\vartheta) Q_s/Q_e$ and in real cases $\psi(\vartheta)$ is an unknown function. It is difficult to estimate the value of the true albedo but for phase angles $60 < \vartheta < 120°$, is considerably smaller than the ratio Q_s/Q_e; $\gamma \leq 0.2$ obtained by Ney (1975) must be regarded as the lower limit of the grain albedo in the cometary atmosphere.

5. Grain Composition and Structure

From the preceding paragraphs it is obvious that knowledge of the physical structure as well as the chemical composition of dust particles in the cometary atmosphere is still fragmentary. A comparison of the available photometric and polarimetric data with results from the computed models of scattering media can lead only to uncertain conclusions regarding the absorptivity and approximate grain sizes. It seems that the submicron particles are more likely composed of low conductivity than of metallic-like material.

At small heliocentric distances even the less volatile grains vaporize and in the spectra of the Sun-grazing Comet 1965 VII, taken at a heliocentric distance r = 0.14 AU, many atomic emission lines, particularly of neutral Na, K, Fe, Ni, Cu have been observed (Preston, 1967; Spinrad, 1968). Relative abundances indicating a very low K/Na and a high Cu/Fe ratio have been found. Data on abundances are unfortunately, not quite representative of the light elements and cannot be used for a compilation of the "true" composition of dust grains. Dust particles are obviously carriers of sodium and are responsible for the appearance of the Na emission in the cometary tail far from the nucleus. Free neutral Na atoms have certainly short lifetimes until ionization and are unable to reach the observed distances from the nucleus. But the relative abudance of Na in the solids may be almost the same as in the cosmic mixture.

The emission peak near $\lambda \sim 10$ m and another one close to 20μ m are, at the present time, the most interesting features in infrared spectra of comets and circumstellar dust clouds. They were attributed to metallic-silicate grains such as $MgSiO_3$ or similar compounds. On the other hand, there is no convincing evidence as to the existence of emission features near $\lambda = 1 \mu m$; such emission might be expected from ferro-silicate material. Although silicate grains constitute an acceptable model for cometary dust, it cannot be ruled out that a considerable fraction of the submillimeter particles in comet atmospheres are polymerized molecules. Vanýsek and Wickramasinghe (1975) recently proposed polymerized formaldehyde for such a possible constituent. Formaldehyde polymers are expected to condense on silicate grains at temperatures ≤ 40 K. (Wickramasinghe, 1974, 1975). The polymerization reaction is exothermic with an exothermicity ~ 15 K cal mole^{-1} and is expected to occur spontaneously under interstellar conditions. The resulting polymer chains could be of variable length, and stabilized by the addition of monovalent atoms or ions. These chains are, in general, helically wound into crystal structures and are therefore endowed with considerable mechanical strength. The melting temperature depends on the degree of polymerization as well as on the nature of the end-groups in the chains; it is typically within the range of 450 - 500 K, or even somewhat higher. These particles grow as long whiskers and possess the optical properties required for interstellar and circumstellar grains.

Assuming that the polymerization hypothesis is correct, it must be concluded that a considerable amount of formaldehyde may be present in comets in the solid-state. It could also be a substantial reservoir of OH or CN radicals which serve to terminate the polymer chains, and which may be released when depolymerization occurs at high temperatures.

The optical properties of crystalline polymer $(CH_2O)_n$ (known as poly-oxymethylene = POM) in the visible spectral range correspond closely with those of dielectric particles having a refractive index $n \simeq 1.5$; these are thus consistent with cometary data.

More important, however, is the 10 μm feature in infrared spectra. The absorption spectra of POM films show absorption bands in the range 8-12 μm. The strong optical activity in the 8-12 μm wavelength band and the variation of these bands with temperature could be important in explaining the behaviour of the 10 m-emission in cometary dust. The two principal bands at 9.2 μm and 10.7 μm are due to vibrational modes of bonds C - O - C in the polymer chains.

Formaldehyde in a solid-state polymer form may occur in large quantities even though its direct detection in comets appears most difficult. In the thermally radiating cometary dust the strongest emission bands of H_2CO are expected in region 8-12 μm, where, in fact, the most pronounced peak of infrared excess emission is observed. Unfortunately, an ambiguity arises because a similar feature is expected for thermal silicate emission. Likewise, the 3.4 μm band of Polyoxymethylene (which may be nearer 3.1 μm for $H(CH_2O)_nOH$) falls in the region of ice grain features. One possibility of distinguishing the emission of silicates from that of formaldehyde polymers is infrared measurement in the waveband ~18-20 μm. There should be a stronger peak near 20 μm for silicate dust than for POM polymers.

Although no direct evidence for the presence of H_2CO in gaseous or polymer form in comets exists as yet, its presence should be considered probable in the cometary models. The stability of formaldehyde polymers, and particularly the high cosmic abudance of H, C and O compared with the abudance of Si, Mg and Fe suggests that formaldehyde polymer grains may be the major constituents not only of interstellar dust, but also of the outer regions of the circumstellar dense clouds, protostellar clouds, and also cometary matter. These grains originated from "starting" nuclei containing silicate or heavy elements.

Another constituent responsible for the emission feature at 10 μm waveband may be hydrocarbon molecules. The presence of CH bands in the visual cometary spectra provides evidence that saturated molecules as CH_4, C_2H_4, C_3H_4 ... C_4H_{10} can be expected in comets too. The possibility that the "silicate bump" observed in carbon stars is caused by hydrocarbons was recently discussed by Tarafdar and Wickramasinghe (1975). Most of these compounds have strong broad absorption (or emission) bands centred mainly at 11 μm. An infrared spectrum arising from mixture of hydrocarbon type C_nH_{n+2} will give a broad band centred at 9-11 μm. The source of a 10 μm feature in infrared spectra of cosmic dust clouds could either be the gas phase or it could be hydrocarbon mantles on solid particles, or both forms. Thus the "silicate bump" in the infrared spectra of comets is by no means conclusive for the presence of silicate-like particles in these objects.

It its worth noting that polymers are common in carbonaceous chondrites, and practically all carbon in such chondritic material is bound in the form of aromatic polymers with -OH and -COOH groups. Chondrites may also be typical ingredients of cometary meteoroids. Besides, in high-resolution spectra of bright meteors, bands of C_2 or CN are also observed (Ceplecha, 1971). Spectra photographed with the image orthicon technique show faint band structures in early parts of meteor trajectories (see Millman, 1976). Therefore, the presence of a high percentage of light elements H; C; N and O in meteoroids is highly probable.

The bulk densities of meteoritic particles, derived from meteor trajectories are very low, mostly below 1.5 g cm^{-3} with the lowest value 0.01 g cm^{-3}. These densities are considerably lower than would be appropriate for silicate material. Since the bulk density depends on the internal strength and porosity of the meteoric matter, a fairly "soft" binding of silicates and metallic grains with some kind of polymers cannot be ruled out.

In the central part of the cometary head, the presence of larger particles, probably having a rather complicated structure, must be expected. In an attempt to explain the discrepancy between the computed and observed life-time of the assumed parent molecules of observed radicals Delsemme and Miller (1970) developed a model. This model consists of clathrates of CH_4 in icy grains of diameter 0.1 to 1 mm in the halo of dusty material in the inner coma. The contribution of ice-like particles to the light scattering is significant in large heliocentric distances but becomes almost negligible at $r \leq 1$ AU.

The two-component or multicomponent characteristics of the solid-state compounds in cometary atmosphere are also suggested by the infrared measurements. Ney (1974) observed in Comets Kohoutek (1973f) and Bradfield (1974b) at least two different types of dusty material: One is characterized by "10 μm band" and somewhat higher albedo, the other with lower albedo and without emission features. The "10 μm" component may be ascribed to particles with higher albedo and having sizes smaller than $\sim 2 \mu$m, the other may be ascribed to low albedo micrometeoroids with diameters of about $\geq 20 \mu$m. The brightness of the high-albedo cloud in visual and near infrared range is very high even if it represents a small fraction of all the solid state compounds produced by the nucleus.

The indirect methods applied to the determination of the cosmic dust characteristics provide only rough qualitative information on this problem. The dust components in comets have to be regarded as a mixture of particles with different sizes and compositions. Some particles may be relatively unstable. The light elements are more abundant in cometary solids than in the zodiacal light particles. The chemical composition and the physical structure of comets seems to be very similar to that of the circumstellar environment and the composition of grains may be identical. Cometary dust contributing to interplanetary light must evidently be depleted from light elements almost immediately after the release from the parent body. Only a direct analysis of dust samples collected "in situ" by space probes moving slowly along with the comet would be an irreplaceable method for a decisive analysis of the cometary grain composition.

References

Babu, G.S.D.: 1975, in Study of Comets (IAU Coll. No. 25) to be published.
Babu, G.S.D. and Saxena, P.P.: 1972, Bull.Astr.Inst.Czech. 23, 346.
Becklin, E.E. and Westphal, J.A.: 1966, Astrophys.J. 145, 445.
Blackwell, D.E. and Willstrop, R.V.: 1957, MN RAS 117, 590.
Ceplecha, Z.: 1971, Bull.Astr.Inst.Czech. 22, 219.
Clarke, D.: 1971, Astron.Astrophys. 14, 90.
Donn, B., Powell, R.S., Rémy-Battiau, L.: 1967, Nature 213, 379.
Delsemme, A.H. and Miller, D.: 1970, Space Sci. 18, 717.
Finson, M.L. and Probstein, R.F.: 1968, Astrophys.J. 154, 327, 353.
Frey, A.: 1975, Astr. and Ap., in press.
Gatley, I., Becklin, E.E., Neugebauer, G. and Werner, M.W.: 1974, Icarus 23, 561.
Gebel, W.L.: 1970, Astrophys.J. 161, 765.
Gehrels, T.: 1972, in Comets, Proc. of Tucson Comet Conference ed. Kuiper, G.P. and Roemer, E., p. 152.
Grün, E.: 1976, this volume.

Harwit, M. and Vanýsek, V.: 1971, Bull.Astr.Inst.Czech. 22, 18.
Johnson, H.M.: 1960, Publ.Astron.Soc. Pacific 72, 10.
Johnson, T.V., Lebovsky, L.A. and McCord, T.B.: 1971, Publ.Astron. Soc. Pacific 83, 93.
Kleinmann, D.E., Lee, T., Low, F.J. and O'Dell, C.R.: 1971, Astrophys.J. 165, 633.
Lillie, F.Ch.: 1972, in The Scient. Results from OAO-2, ed. Code, A.D., NASA SP-310, Washington D.C., p. 95.

Lillie, F.Ch.: 1976, this volume.

Liller, W.: 1970, Astrophys.J. <u>132</u>, 867.

Lee, T.A.: 1972, in <u>Comets</u>, Proc. of Tucson Comet Conference 1970, ed. Kuiper, G.P. and Roemer, E., p. 20.

Maas, R.W., Ney, E.P. and Woolf, N.J.: 1970, Astrophys.J. <u>160</u>, L 101.

Martel, M.T.: 1960, Ann.Astrophys. <u>23</u>, 480 and 498.

Merrill, M.K.: 1974, Icarus <u>23</u>, 566.

Michalsky, J.: 1975, in <u>Study of Comets</u>, Proc. IAU Coll. No. 25, to be published.

Millman, P.: 1976, this volume.

Ney, E.P.: 1974, Icarus <u>23</u>, 551.

Ney, E.P.: 1975, in <u>Study of Comets</u>, Proc. IAU Coll. No. 25, to be published.

Noguchi, K., Sato, S., Maihara, T., Okuda, H. and Uyama, K.: 1974, Icarus <u>23</u>, 545.

Osherov, R.S.: 1970, Komety i meteory No. 19, 17 (in russian).

O'Dell, C.R.: 1971, Astrophys.J. <u>166</u>, 675.

Preston, G.W.: 1967, Astrophys.J. <u>147</u>, 718.

Rémy-Battiau, L.: 1964, Bull.Acad.R.Belg., Cl.Sci., Sér. 5, <u>50</u>, 74.

Rémy-Battiau, L.: 1966, Bull.Acad.R.Belg., Sér. 5, <u>52</u>, 1280.

Rieke, G.H. and Lee, T.A.: 1974, Nature <u>248</u>, 737.

Sekanina, Z. and Miller, F.D.: 1973, Science <u>179</u>, 565.

Shulman, L.M.: 1969, Astrometry Astrophys. Kiev <u>4</u>, 101 (in russian).

Spinrad, H. and Miner, E.D.: 1968, Astrophys.J. <u>153</u>, 355.

Tarafdar, S.P. and Wickramasinghe, N.C.: 1975, Astrophys. and Space Sci. <u>35</u>, L 41.

Vanýsek, V.: 1960, Bull.Astr.Inst.Czech. <u>11</u>, 215.

Vanýsek, V.: 1965, Acta Univ.Carol. Prague; sec. Maths. and Phys. <u>1</u>, 23.

Vanýsek, V.: 1971, Acta Univ.Carol. Prague; sec. Maths. and Phys. <u>13</u>, 85.

Vanýsek, V. and Wickramasinghe, N.C.: 1975, Astrophys. and Space Sci. <u>33</u>, L 19.

Walker, M.W.: 1958, Publ.Astron.Soc. Pacific <u>70</u>, 191.

Weinberg, J.L.: 1974, in <u>Study of Comets</u>, Proc. IAU Coll. No. 25 (abstract).

Weinberg, J.L.: 1964, Ann. d'Ap. <u>27</u>, 718.

Weinberg, J.L. and Beeson, D.E.: 1975, in <u>Study of Comets</u>, Proc. IAU Coll. No. 25, to be published.

Weinberg, J.L. and Mann, H.M.: 1968, Astrophys.J. <u>152</u>, 665.

Westphal, J.A.: 1972, in <u>Comets</u>, Proc. of Tucson Comet Confer. 1970, ed. Kuiper, G.P. and Roemer, E., p. 56.

Wickramasinghe, N.C.: 1974, Nature <u>252</u>, 465.

Wickramasinghe, N.C.: 1975, Monthly Not. RAS **170**, 11.
Wolstencroft, R.D. and Rose, L.J.: 1967, Astrophys.J. **147**, 271.
Zeilik, M. and Wright, E.L.: 1974, Icarus **23**, 577.

3.2   THE PRODUCTION RATE OF DUST BY COMETS

by
A. H. Delsemme
Department of Physics and Astronomy
The University of Toledo
Toledo, Ohio 43606

ABSTRACT: The present set of short-period comets produces only 250 kg/sec of dust whereas the flux of long-period and "new" comets produces an average of 20 tons/sec.

1. Production of Gas and Dust by Comets

The meteoritic dust cloud present in the solar system is self-destructive. Rather reliable estimates of the dissipating mechanisms (Whipple,1967) suggest that a source of some 10 tons per second would be required for maintaining an approximate steady state. Whipple (1955) has proposed, that cometary dust be the major source maintaining the meteoritic complex. Whipple (1967) has revised his assessment more recently, concluding that Comet Encke could have been, over the past several thousand years, the major support for maintaining the quasi-equilibrium of the zodiacal cloud.

However, recent evidence has changed several of the parameters he used, all in the sense of diminishing the cometary contribution. In particular, 1) recent evaluations of large albedos for the nuclei of three comets (Delsemme and Rud,1973), despite the fact that they may be upper limits (Sekanina,1975), suggest that the efficiency of the solar radiation for vaporizing snows was overestimated by a factor of two or more; 2) water snows, or snows of solid hydrates, seem to control the vaporization of the short period comets, and of many, but not all long-period comets (Delsemme and Swings,1952; Delsemme,1965; Delsemme and Miller,1970,1971; Marsden et al.,1973). Solid hydrates (clathrates) and water ice have a larger heat of vaporization than that used by Whipple (1967). Last but not least, the radiation losses of the nucleus back to space cannot be neglected any more for distances larger than 1.5 AU, because water snow sublimates there at approximately 200°K. As cometary dust is dragged away by vaporizing gases, the gas production by comets must be estimated first.

2. Production Rate of Comet Halley

In order to reassess the whole situation, the water vaporized by Comet Halley has been integrated along its trajectory, by using the model proposed by Delsemme (1965) and developed by Delsemme and Miller (1970,1971). The albedo of the cometary nucleus for the visible light and for the radiative losses to space (near 15µ) have not to be specified, but have been assumed to be the same. The hydrogen production rate of Comet Bennett is assumed entirely produced by water dissociation and normalized for Comet Halley, in proportion to the absolute brightnesses of the two comets. (H_{10} = 4.5 for Bennett, 4.6 for Halley). No evaluation of the size or of the albedo of any cometary nucleus is needed.

The integration, extended to the whole orbit of Comet Halley, yields an _average_ production rate of 63 kg/sec of water; (22 tons/sec at perihelion, 7.6 tons/sec at 1 AU, 800 kg/sec at 2 AU, but less than 1 gram/sec at Jupiter's distance).

3. Short-Period Comets

In order to extend the results to all other comets, a criterion (admittedly crude) was developed to pick up those comets whose production rate is not negligible. In the range where the vaporization is very large, a parabolic approximation of the ellipse can be used. Then, the exposure times per perihelion passage are obviously proportional to $q^{3/2}$, q being the perihelion distance, and the vaporization rate per unit area is proportional to q^{-2}, at least up to 1.5 AU. If the brightness at 1 AU is in proportion to the vaporizing rate, then the gas production rate G per century is

$$G = G_o \; 10^{-0.4\Delta H} \; q^{0.5} \; p^{-1}$$

where G_o is the gas production rate of Comet Halley, ΔH the difference between the absolute magnitudes of the comet considered and Comet Halley, q the perihelion distance and p the period, both in Comet Halley units. This index G, normalized for G_o = 100 for Comet Halley, has been computed for all 97 short-period comets (Marsden,1975). The major producers of gas and dust of the last two centuries are listed per century in Table I. To avoid overestimates of 18th and 19th century brightnesses, Comet Halley's 1835 passage has been again normalized to 100.

Swift-Tuttle is listed in the 1870-1970 period because the averaging procedure ignores where it is on its orbit (it comes back in 1982). As a matter of fact, more than half of the production rates of the short-period comets comes from four comets with rather long periods, namely Swift-Tuttle (120 years), Halley (76 years). Pons-Brooks (71 years) and Olbers (69 years).

The set of short-period comets produces about 0.25 ton/sec. The observed fluctuations from one century to the other are understood, because large comets brighter than Halley's are not likely to be captured more often than once in two centuries (see Delsemme, 1973). At any rate, the production of the short-period comets does not contribute substantially to the production of gas and dust.

TABLE I

Gas Production Rates of Periodic Comets per Century

(p/Halley = 100)

1870-1970		1770-1870	
Swift-Tuttle	120	Swift-Tuttle	120
Halley	100	Halley	100
Olbers	30	Pons-Brooks(1)	70
Pons-Brooks	24	Olbers	47
Encke	16	Encke	44
Faye (2)	7	Faye (1)(2)	40
Gale	5	Biela	31
Schaumasse	4	Brorsen	12
Pons-Winnecke	4	Tuttle	10
Herschel-Rigollet	3	Pons-Winnecke	10
All Others	15	All Others	15
Total	328	Total	499

(1) the gas production of these comets was integrated from their discovery date, since it is not certain that they were in the inner solar system before.

(2) special integration because q>1.2.

4. Long-Period Comets

If we use the same formula (obviously without the p^{-1} term) to compare long-period comets to one passage of Comet Halley, the summation of all long-period comets for the last one hundred years (1870-1970) yields an average production rate of 4.9 tons/sec.

TABLE II

Gas Production Rates of Long-Period Comets per Passage

(One passage of Comet Halley = 100)

Six Recent Comets		Six Very Bright Comets	
Mrkos (1957V)	316	Comet 1577	66,000
Ikeya-Seki (1965VIII)	915	Comet 1729	16,500
Bennett (1970II)	148	Comet 1744	7,100
Kohoutek (1973XII)	135	Comet 1747	5,600
Arend-Roland (1957V)	85	Comet 1402	5,400
Seki-Lines (1962III)	68	Comet 1811 I	5,100

Table II shows the gas production of a few recent comets. If all comets at least as bright as magnitude zero are excluded, the average of any particular century remains about constant at 5.0 tons/sec. However (see Table II) the rare bright comets like 1811 I ($H_{10}=0$), 1747 ($H_{10}=-0.5$), 1729 ($H_{10}=-3.0$) or 1577 ($H_{10}=-1.8$) must be included with their average frequency. This frequency is given by Vsekhsviatskii's (1964) "restored distribution" of the brightnesses, more exactly, by the average slope of the log N versus brightness diagram established from his data. Moreover, the Great Comet of 1729 must be excluded because perihelion was at 4.05 AU and the fourth power law used in H_{10} is probably not applicable; (its H_o probably is -0.5). The result depends now on the cutoff limit for the brightest comets. We accept $H^o_{max}=-2$, (almost observed in 1577). Vsekhsviatskii's curve predicts one comet like Comet 1577 in four centuries.

The new production rate yields 15.6 tons/sec for all comets at least as bright as magnitude zero, therefore 20.8 tons/sec for all comets. Faint comets play an insignificant role, therefore incompleteness does not introduce any corrections. No assessment of the production rate of the gases other than water vapor can be realistically tried at this stage, although there is little doubt that "new" comets (in Oort's sense) and anomalous comets like Humason, Morehouse, or Comet 1729 must contain materials much more volatile than water.

5. Dust Production Rate

Finson and Probstein's (1968) approach shows that the mass ratio dust/gas was of the order of 1.67 for Comet Arend-Roland near perihelion. For Comet Bennett, using Finson and Probstein's method, Sekanina and Miller (1973) find a ratio of 0.5. Since most long-period comets are very dusty, it is therefore reasonable to accept an average ratio of 1.0.

We conclude that comets provide an average source of dust of 20 tons/ sec, although the absence of bright comets has diminished it to 5 tons/sec for the last century. As the characteristic time for the establishment of the steady state is around 10^4 years, the rate to be used for the average source should clearly be rather 20 tons/ sec than 5 tons/ sec.

However, a large fraction of this dust may be lost at once on hyperbolic orbits. If comets are indeed the only source of dust, arguments must be found to demonstrate that approximatively half of this dust remains within the solar system.

I have reviewed the possible arguments on the final day of this colloquium. They are given hereafter under the title ; "Can Comets be the only source of Interplanetary Dust ? "

NSF Grant GP 39259 is gratefully acknowledged.

REFERENCES

Delsemme, A. H. (1965) Colloq. Internat. Astrophys. Univ. Liege 37, 77., also Mem. Soc. R. Sci. Liege 12, 77 (1966).

Delsemme, A. H. (1973) Astron. Astrophys. 29, 377.

Delsemme, A. H., and Miller, D. C. (1970) Planet. Space Sci. 18, 717.

Delsemme, A. H., and Miller, D. C. (1971) Planet. Space Sci. 19, 1229.

Delsemme, A. H., and Rud, D. A. (1973) Astron. Astrophys. 28, 1.

Delsemme, A. H., and Swings, P. (1952) Ann. Astrophys. 15, 1.

Finson, M. L., and Probstein, R. F. (1968) Astrophys. J. 154, 327 and 353.

Marsden, B. G. (1975) Catalogue of Cometary Orbits, 2d edition, I.A.U. Central Bureau for Astronomical Telegrams, Cambridge, Mass.

Marsden, B. G., Sekanina, Z., Yeomans, D. K. (1973). Astronom. J. 78, 211.

Sekanina, Z. (1975) IAU Colloquium No. 25, "The Study of Comets," G.S.F.C., Greenbelt, Maryland (in press), NASA-SP, Washington, D.C.

Sekanina, Z., and Miller, F. D. (1973) Science 179, 565.

Vsekhsvyatskii, S. K. (1964) Physical Characteristics of Comets, English translation, NASA and NSF, Washington, D.C.

Whipple, F. (1955) Astrophys. J. 121, 750.

Whipple, F. (1967) "The Zodiacal Light and the Interplanetary Medium," p. 409, edit. Weinberg, J. L., NASA-SP 150, Washington, D.C.

3.3 CAN SHORT PERIOD COMETS MAINTAIN THE ZODIACAL CLOUD ?

S. Röser

Max-Planck-Institut für Astronomie

Heidelberg-Königstuhl, FRG

The combined effects of the Poynting-Robertson drag, collisions and sputtering are destroying the interplanetary dust cloud. The mass-losses estimated by different authors reveal great discrepancies. The estimations range from 1 t sec^{-1} (purely Poynting-Robertson loss) over some 10 t sec^{-1} given by Whipple (1967) to a value of 100 t sec^{-1} which is propagated by Bandermann (1967). The latter derives this high value from the assumption that the main mass loss is due to self-collisions.

As dust sources for the Zodiacal Cloud asteroids and comets have been discussed repeatedly. This paper deals with comets and discusses the effect of radiation pressure on the trajectories of the ejected particles. We calculate the total dust production of a comet along its orbit and the percentage of dust that does not leave the solar system immediately.

The new orbital elements of the particles after their separation from the comet are calculated under the assumption that the particles have no additional velocity component at ejection. This assumption is supported by investigations of Finson and Probstein (1968), who derived ejection velocities of only about 0.2 km sec^{-1} for Comet Arend-Roland. Then the formulae for the new orbital elements a' and e' read as follows

$$a' = \frac{a\mu}{1 - 2 \cdot (1-\mu) \cdot (\frac{a}{r})}$$

$$e'^2 = 1 - \frac{1-e^2}{\mu^2} + \frac{2(1-\mu)}{\mu^2} \cdot \frac{p}{r}$$

where a, e, r are the semi-major axis, eccentricity and radius vector of the comet resp., $p = a(1-e^2)$ and $\mu = 1 - \frac{\text{rad. pressure}}{\text{gravity}}$

We get the particle diameter using the formula

$$1 - \mu = \frac{C}{\rho d}$$

where ρ is the mass-density, d the diameter of the particle and C = 1.19 x 10^{-4} g cm^{-2}. The size-distribution function of the ejected particles was taken from Finson and Probstein (1968), who derived this distribution from measurements of Comet Arend-Roland. The distribution has its maximum at $\rho d = 3 \times 10^{-4}$ g cm^{-2}, a sharp cut-off to smaller values of ρd and follows a power law (exponents between -3 and -5) for the decay to larger values.

We have considered two cases for the law of decay of the dust production with increasing distance from the Sun: dust production is proportional (I) to the brightness of the comet (r^{-4}) or (II) to the flux density of the solar radiation (r^{-2}). Using these laws the emission rates have been integrated along the whole orbit of the comet.

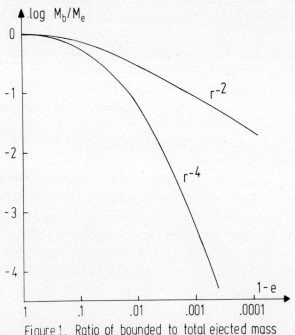

Figure 1. Ratio of bounded to total ejected mass as a function of the comet's eccentricity

In Figure 1 we have plotted the percentage of mass with e' < 1 against the original eccentricities for both cases. The shape of the curves is only valid under the assumptions made above about ejection velocity and size distribution. The ratio of the mass bound to the Solar System decreases rapidly with increasing eccentricity of the comet, though the relatively strong weight of aphelion production prevents a steeper decay. This ratio has to be taken into account if one considers the contribution of dust from long-period comets. These always produce the bound material during their passage through the aphelion parts of their orbits, that means at great distances from the Sun, and we do not know what kind of emission law can be applied there. Further, it is still an open question, whether this material can cross the barriers of the great planets and penetrate into the inner Solar System.

To investigate the mass supply by short period comets, absolute mass production rates are needed. At present we know the absolute mass production only of a few long period comets. Finson and Probstein (1968) derived the mass production of Comet Arend-Roland and found a rate of 75 t sec^{-1} at the perihelion. For Comet Bennett Sekanina and Miller (1973) gave a value of some 20 t sec^{-1}. These two values agree sufficiently well using the formula

$$-2.5 \log E_q = H_{10} + 10 \log q$$

where E_q stands for the relative emission rate at perihelion, H_{10} for the comet's absolute magnitude and q for its perihelion distance. The formula is equivalent to the assumption that the comet's emissivity at perihelion is proportional to its brightness there ($10^{-0.4 \times H_{10}} q^{-4}$). Using this formula we calculated the perihelion mass production of the 55 short period comets of Wachmann's list (1965). So we linked the mass production of the short period comets to the production rate of two long period resp. new comets. Integrating now the emission rates along the orbits yields mean production rates for each comet.

Name	total	elliptic	total	elliptic
Halley	58.4	14.6	154.0	77.0
Pons-Brooks	23.7	8.1	62.1	37.3
Encke	22.6	16.5	56.7	48.2
Biela	12.0	10.1	28.7	26.7
Brorsen	11.8	9.3	29.2	26.3
Olbers	7.3	3.5	19.0	13.3
All others	12.5	10.7	27.3	24.4
Sum	148.3	72.8	377.0	253.2
Assumed law for emission rate	r^{-4}		r^{-2}	

Table 1. The best dust producers of the short period comets for different laws of the emission rate. Masses are given in kg sec^{-1}.

In Table 1 we list the best dust producers among the short period comets. Comet Halley turns out to be the best dust source of all, but because of its high eccentricity most of the dust is lost from the Solar System, whereas Encke produces a relatively higher amount of dust having elliptic orbits. All short period comets together produce not more than 72 resp. 253 kg $\sec^{-1}$ of dust remaining in the Solar System. This present rate is by far too low to give an essential contribution to the Zodiacal Cloud, and we conclude that short period comets are not an important source for interplanetary dust.

References

Bandermann, L.W.; Physical Properties and Dynamics of Interplanetary Dust, University of Maryland, Technical Report No 771 (1967)

Finson, M.L. and Probstein, R.F.; A Theory of Dust Comets, ApJ <u>154</u>, 327 (1968)

Sekanina, Z. and Miller, F.D.; Comet Bennett (1970 II), Science <u>179</u>, 565 (1973)

Wachmann, A.; in "Landold - Börnstein", Zahlenwerte und Funktionen, New Series, Group VI, Vol. <u>1</u>, 176 (1965)

Whipple, F.; in "The Zodiacal Light and the Interplanetary Medium", ed. by Weinberg, J.L., NASA SP <u>150</u>, 409 (1967)

3.4 OPTICAL PROPERTIES OF COMETARY DUST

S. Hayakawa, T. Matsumoto and T. Ono
Department of Physics, Nagoya University, Nagoya, Japan

Optical features of comets suggest that the dust in the coma is composed of a micron size core and a mantle formed by an aggregate of submicron grains.

1. Introduction

Photometric observations of Comets Kohoutek (1973f) and Bradfield (1974b) over wide wavelength ranges (Ney, 1974a,b; Noguchi et al, 1974; Iijima et al, 1975) and polarization measurements of Kohoutek in the near infrared (Noguchi et al., 1974) and in the optical range (Tanabe, 1974) show the following features.

(1) The spectrum longward of 2.2 µm is Planckian plus an enhancement at about 10 µm. The Planckian part gives a rather high temperature, suggesting higher absorption in the optical range than in the infrared range.

(2) The spectrum shortward of 2 µm is similar to the solar spectrum, and the scattering coefficient depends only weakly on the phase angle in the range $55^\circ - 93^\circ$.

(3) The degrees of polarization at phase angles of about 90° are 12 - 25% with the electric vector nearly perpendicular to the ecliptic plane and apparently show a slow increase with wavelength.

These features may be accounted for in terms of micron size grains with a mantle of aggregate structure. If submicron grains are responsible (O'Dell, 1971), an unrealistic wavelength dependence of the refractive index would have to be assumed. The present analysis imposes a rather stringent condition on the optical properties of cometary materials in the coma; for example, the crystalline formaldehyde polymers recently proposed by Vanýsek and Wickramasinghe (1975) would have to have a large absorption efficiency in the optical range, although they look transparent. In the present work we attempt to derive quantities which are free from the column density of coma materials and can be directly compared with optical parameters. Quantitative comparison is made with the Mie theory, whereas only qualitative properties are discussed for effects of the surface roughness, leaving details for a separate paper.

2. Thermal Emission

The infrared emission longward of 2.2 μm is considered to be the thermal emission from cometary dust heated by the solar radiation. The grain temperature T at the heliocentric distance r in AU is given by

$$<Q_A>_s F/r^2 = 4G <Q_A>_g \sigma T^4, \qquad (1)$$

where F is the solar flux at 1 AU, σ the Stefan-Boltzmann constant, G the geometrical factor representing a deviation from the uniform sphere. The absorption efficiencies Q_A are averaged over the solar spectrum and the emission spectrum on the left and right hand sides, respectively. The grain temperatures are obtained at several heliocentric distances by fitting the observed emission spectrum to the Planckian distribution (Ney, 1974a,b).

From the observed values of the grain temperature we obtain

$$R \equiv G<Q_A>_g/<Q_A>_s = F/4\sigma T^4 r^2, \qquad (2)$$

as shown in Figure 1. The value of R decreases from 0.6 to 0.3 as the grain temperature increases from 350 K to 950 K.

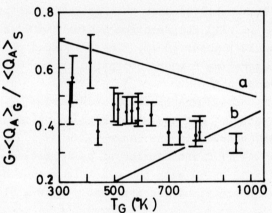

Fig.1. Ratio of the absorption efficiencies obtained from Equation (2). The solid line: (a) Mie theory for kx>1, a = 1.6 μm, but the absolute value is reduced by a factor 3; (b) Mie theory for kx<<1, k = 1.4, a = 0.16 μm.

Since the emission spectrum can be approximately simulated by the Planckian distribution over the above temperature range, the wavelength dependence of Q_A is very weak in the infrared range concerned.

The Mie theory applied to the spherical grain of radius a predicts that the absorption efficiency is nearly proportional to kx for kx < 1, where the index of refraction is m = n - ik and x = 2πa/λ, λ being the wavelength, passes a maximum at about kx = 1, and decreases slowly as x increases. The wavelength independence results if the grain size is so large that kx > 1. Large grains are also favoured to account for the wavelength independence of the scattering efficiency. Small grains of constant k show a temperature dependence opposite to that observed, as shown in Figure 1, whereas $k \propto \lambda^{1/2}$ simulating small metallic grains gives a too low R because of a high infrared reflectance.

If $kx \gtrsim 1$ is taken for granted for $\lambda = 2 - 8$ μm beyond which emission bands affect the spectrum, the grain size is given as

$$a = x\lambda/2\pi \gtrsim \lambda/2\pi k \sim 1.3/k \text{ μm}. \qquad (3)$$

Namely, cometary dust is of micron size or larger. This is in contrast to the conclusion reached by O'Dell (1971) who has argued for submicron grains based mainly on the effect of radiation pressure.

A difficulty arises if $kx \gtrsim 1$ is assumed. The value of $<Q_A>_g/<Q_A>_s$ greater than unity is obtained from the Mie theory, whereas the value of R observed is appreciably smaller than unity. The discrepancy could be accounted for by the geometrical factor G. The value of G can be smaller than unity if the area facing the earth is smaller than that facing the sun. This would require the alignment of nonspherical grains. However, the direction of alignment is expected to be opposite, since solar wind protons tend to align the long axis parallel to the solar wind direction. The small value of R may result from a deviation from the Mie theory due to the surface roughness.

The surface roughness is characterized by two parameters, the rms vertical deviation from the average surface, σ, and the mean correlation distance of roughness, α (Beckmann, 1963). If σ is not too small compared with λ and α, the reflectance is smaller than that expected from the Fresnel reflection because of multiple reflection in pores (Houchens and Hering, 1967). This increases the absorption efficiency in the optical range but by a factor smaller than two, since Q_A is already as large as unity. If the rough surface is formed by a mantle as an aggregate of submicron grains with a small index of refraction, the mantle is highly transparent for infrared radiation, and the absorption efficiency is therefore essentially equal to that of the core. The core radius as small as or smaller than 2/3 of the mantle radius could account for a small absorption efficiency in the infrared range in comparison with that in the optical range. The aggregate structure of the mantle results in a small bulk density, which is in favour of the dominant effect of radiation pressure (O'Dell, 1971).

3. Scattering

The scattering efficiencies at $\lambda = 1.2$ μm and at phase angles θ, $Q_S(\lambda, \theta)$, are shown relative to the absorption efficiency $Q_A(\lambda)/4\pi$ in Figure 2. The results at other wavelengths are similar. The phase angle dependence can be fitted to the Mie theory only for $x \gtrsim 10$ as indicated by theoretical curves in Figure 2, whereas the spectrum of scattered light is similar to the solar spectrum if $kx > 3$.

The theoretical curve for $x = 10$ is close to that of the Fresnel reflection. If the surface roughness is taken into account, the observed behaviour could be reproduced for smaller values of x.

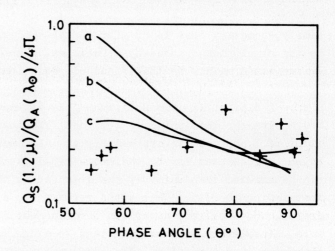

Fig.2. Phase angle dependence of the scattering efficiency. Crosses represent observed results for $\lambda = 1.2$ μm based on Ney (1974b). Solid curves represent results of the Mie theory with $m = 1.7 - 1.0i$. a: $x = 2.5$, b: $x = 5.0$, c: $x = 10$.

The absolute intensity of scattered light requires a rather large value of the refractive index; since the real part cannot be too large, the imaginary part has to be as large as or larger than 0.6 for the smooth surface. The aggregate structure would increase the reflectance and, consequently a little smaller value of k may be permissible.

4. Polarization

If the grain is of micron size, its optical properties are approximately accounted for in terms of the Fresnel reflection. Then the polarization at the phase angle 90° may be as large as 60% (Matsumoto, 1973). The degree of polarization is reduced if diffuse scattering is taken into account. For $\sigma \ll \lambda$, the wavelength dependence of polarization is appreciable, whereas only a weak dependence has been observed in the range 0.5 - 1.7 μm. The observed polarization is reproduced for $\sigma \sim 0.8$ μm, if reference is made to an experiment (Torrance et al, 1966).

Finally, we remark that the grain model proposed here is favourable to account for zodiacal light, in particular the spectrum similar to the solar spectrum in the optical and near infrared ranges (Hayakawa et al.; 1970, Nishimura, 1973; Matsumoto, 1973). It may, however, be almost needless to mention that this does not necessarily rule out the existence of submicron grains since their contribution to brightness is of minor importance.

References

Beckmann, P., 1963, The Scattering of Electromagnetic Waves from Rough Surfaces (Macmillan, New York).
Hayakawa, S., Matsumoto, T. and Nishimura, T.,1970, Space Research, X, 248.
Houchens, A.F. and Hering, R.G., 1967, Astronaut. Aeronaut., 29, 65.
Iijima, T., Matsumoto, T., Oishi, M., Okuda, H. and Ono, T., 1975, Publ. Astro. Soc. Japan, in press.
Matsumoto, T., 1973, Publ. Astr. Soc. Japan, 25, 469.
Ney, E.P., 1974a, Astrophys. J. Letters, 189, L141.
Ney, E.P., 1974b, Icarus, 23, 551.
Nishimura, T., 1973, Publ. Astr. Soc. Japan, 25, 375.
Noguchi, K., Sato, S., Maihara, T., Okuda, H. and Uyama, K., 1974, Icarus, 23, 545.
O'Dell, C.R., 1971, Astrophys. J., 166, 675.
Tanabe, H., 1974, Tokyo Astronomical Bulletin, Second Series, No. 235.
Torrance, K.E., Sparrow, E.M. and Birkebak, R.C., 1966, J. Opt. Soc. Am., 56, 916.
Vanýsek, V. and Wickramasinghe, N.C., 1975, Astrophys. Space Sci., 33, L19.

3.5 The Dust Coma of Comets

K.W. Michel and T. Nishimura
Max-Planck-Institut für Extraterrestrische Physik,
8046 Garching/München

Abstract: Analytical expressions for the dust entrainment in the limits of low and high gas production from cometary nuclei are derived. The resulting volume density is used to evaluate the dust production of Comets Encke and Kohoutek from IR-measurements. It is predicted, that the dust in solar direction was diffuse for Comet Kohoutek during approach to the Sun, but sharp for Comet Encke and nearly independent of heliocentric distance. The dust production rate of Encke ist not sufficient to sustain the Zodiacal Cloud of micron sized particles. Particle impact probabilities and hazards are estimated for a space probe in a fly-by mission to Encke.

Introduction: Whereas the visible coma of a comet is largely determined by the fluorescence of radicals, i.e. the photolysis fragments of parent molecules, the infrared brightness results from light scattering and thermal emission of dust grains, entrained by gases vaporizing from the surface of the comet's nucleus. The production rate of dust is of significance for an understanding of the surface of the nucleus, and thus simple expressions are desirable which relate the dust production rate to the infrared brightness, obtained in a number of recent measurements (Ney, 1974; Rieke and Lee, 1974; Barbieri et al., 1974; Noguchi et al., 1974). The dust in the coma is also at the same time a hazard for fly-by mission as well as a scientific objective of particle impact detectors on such a space probe. These problems require the column density of dust particles in the vicinity of the nucleus, whereas the density in the far tail has been studied extensively by Finson and Probstein (1968).

The grain density of given size distribution is obtained from their trajectories, which can for the present purposes with sufficient accuracy be taken as parabolic, the parameters being determined by the ejection velocity v_a from the inner coma and the solar radiation pressure $\pi a^2 Sq/c R_s^2$ (a = particle radius, S = solar constant, q = relative light-pressure cross-section integrated over the solar spectrum, R_s = heliocentric distance). The grain ejection velocity depends on the entrainment by gases of number density n, molecular weight m, radial velocity v and mass production rate per solid angle $\dot{M}_g = mnv/r^2$, where r is the distance from the comet's nucleus. v is given by the initial enthalpy of the gas $H = \gamma kT/(\gamma -1)$ and by the degree of conversion of H into radial energy of motion $1/2\ mv^2$. H again depends on the transport processes near the comet's surface.

The Comet's Surface:

The nucleus of old comets like P/Encke or of young ones after perihelion seems to be covered by a crust of grains, too heavy to be entrained by vaporizing gases (mostly H_2O and CO_2). For P/Encke the thickness of this dust layer can be estimated from the heliocentric distance, where the vaporization becomes steady, viz. where $\dot{M}_g \sim R_s^{-2}$, which is $R_s \leq 1$ AU (Mumma, 1975). If this comet of aphelion temperature $T_{aph} = (S/4\delta)^{1/4} R_s^{-1/2}$ approaches the sun, the surface of the spinning nucleus, reradiating energy only at the light absorbing surfaces, rises according to $T_s = (S/2\delta)^{1/4} R_s^{-1/2}$, and a temperature wave travels into the interior on the basis of radiative diffusivity $\varkappa_r = 16 \delta T_{eff}^3 a_e/D C_d$, where $D \approx 3$ g/cm^3 the specific weight of dust, $C_d = 0.8 \times 10^7$ erg/g x degr. its specific heat and $T_{eff} = \frac{1}{2}(T_s + T_{aph})$. For P/Encke at heliocentric distances $2 > R_s > 1$ the increase of surface temperature may well be described by $T_s - T_{aph} = 14.6 \times 10^6$ t, such that $t_1 = 10^7$ sec at 1 AU, where the thermal wave has reached the icy core at distance $\Delta r = \sqrt{4 \varkappa_r t_1}$ from the surface. The effective particle size a_e in the dry dust layer is obtained by matching the steady heat flux density $16 \delta a T_s^4/\Delta r$ to the power necessary to explain the vaporization of the observed mass production rate $\dot{M}_g = 3.6 \times 10^3/R_s^2$ at a radius of the nucleus $r_N = 1$ km (Mumma, 1975). The result is $a_e = 10^{-2}$ cm and $\Delta r = 10$ cm (Michel and Nishimura, 1975).

From the ensuing balance of energy fluxes follows that practically all of the solar energy absorbed at albedo 0.1 by the comet's nucleus is reradiated into space by the nearly black surface of temperature $T_s = 330/\sqrt{R_s}$ °K. Only a small fraction of the incident power (1 %) penetrates by radiative diffusion into the interior, where it serves for vaporization. The vaporizing gases are slowly heated up when diffusing through the dry dust layer and escape finally at surface temperature T_s into the coma. The escape velocity from the crevices of the surface occurs under free molecular flow conditions and is given by $\dot{r}_s = \sqrt{kT_s/2\pi m} = 0.36$ km/sec at Encke's perihelion $R_s = 0.34$ AU where $T_s = 560°K$. The upper limit of grain sizes, which can be swept out by this rarefied flow against gravitational attraction is given by (Hübner, 1970):

$$a_{max} = \frac{9}{16\pi} \frac{\dot{M}_g \dot{r}_s}{GD\varrho\, r_N^3}$$

where G is the gravitational constant and $\varrho = 1$ g/cm^3 the mean density of the entire nucleus. At Encke's perihelion $a_{max} = 1$ cm. The smallest particle size is probably $a_{min} = 3 \times 10^{-6}$ cm, corresponding to the smallest interstellar grains (Pecker, 1974).

<u>The Gas Velocity:</u>
Molecules escaping from the nucleus at velocity $\dot{r}$ into the coma collide with other molecules and are partly backscattered toward the surface where they pick up further energy, which is added to the flow. So, at a distance from the surface which is of the order of a mean free path, local sound velocity $\hat{v}_\ell = \sqrt{\frac{2\gamma}{\gamma+1} \frac{kT_s}{m}}$ is established (Michel and Nishimura, 1975). In the subsequent continuum expansion, thermal energy is converted into directed flow energy with an efficiency which is dictated by the dependence of the terminal Mach number on Knudsen number $M_t = 2/\sqrt{Kn}$. The residual temperature after complete expansion is given by $T_t/T_s = [2(\gamma-1)/Kn+1]^{-1}$, which implies for Encke at $R_s = 1$ AU with $Kn = mv_1 r_N/2\delta\dot{M}_g < 10^{-2}$, that practically all of the available enthalpy is converted into radial flow energy of terminal velocity $v = \sqrt{\frac{2\gamma}{\gamma-1} \frac{kT_s}{m}} \approx 1.1$ km/sec at $R_s = 1$ AU. From the continuity equation for steady, spherically symmetric and isentropic expansion follows readily that 90 % of the terminal gas velocity is reached within 6.2 r_N from the nucleus for a ratio of specific heats $\gamma = 1.33$ (H_2O and CO_2). Further change in the gas velocity due to heat addition by both thermal conduction from grains and by energetic photolysis products can be shown to be trifle in the vicinity of the nucleus, where also

dust entrainment and acceleration by molecular drag takes place. Since grains require longer paths for drag acceleration than the gases themselves, the gas velocity can be considered constant.

Grain Trajectories:
Grains are subject to molecular drag and light pressure. The former acts strongly in the immediate vicinity of the nucleus. The force on dust particles of velocity v_a is under steady conditions

$$\tfrac{4}{3}\pi a^3 D v_a \frac{dv_a}{dr} = \pi a^2 (v - v_a)^2 mn$$

giving upon integration with v = const. and for $\dot{M}_g/aDvr_N < 0,2$ or $v_a \ll v$

$$v_a^2 = \tfrac{3}{2}\frac{\dot{M}_g v}{D a r_N}\left(1 - \frac{r_N}{r}\right).$$

This gives for $r = \infty$ a terminal dust speed in excellent agreement with the more detailed numerical calculations of Finson and Probstein (1968) even for the case of high dust-to-gas ratios (N.B.: the effect of high dust content on the gas velocity has been neglected here). Thus, if dust particles of radius $a = (0.2 - 1) \times 10^{-4}$ cm have been ejected from Comet Kohoutek at $R_s = 3.8$ AU at velocity $v_a = 0.5$ km/sec (Grün et al., 1976) and nuclear radius $r_N = 5$ km (Rieke and Lee, 1974), it follows readily that the gas production rate must have been $\dot{M}_g = (0,5 - 2,5) \times 10^6$ g/sec sr, i.e. of the same order as the estimated dust production rate. By comparison, the gas production rate at a preperihelion distance of $R_s = 0.64$ AU has been estimated to be $\dot{M}_g = 2.7 \times 10^6$ g/sec x sr (Barbieri et al., 1974) and the gas molecules are probably mostly H_2O and CO_2 (corresponding to a formation temperature of the comet of $250 - 400°$ K in the equilibrium approximation). The high gas loss at large heliocentric distance where $T_s = 144°$K can only be explained by evaporation of very volatile compounds of low heats of vaporization. Hence, after its formation at $T > 250°$K, the nucleus was coated probably at $T < 100°$K by a hoar-frost of CH_4 or CO.

Since the terminal grain velocity is established effectively at $r_N/r = 20$, where the grains also decouple from the gas, the solar radiation pressure results in strictly parabolic trajectories, the envelope of which is a paraboloid, characterized by $X_o - X = Y^2/4X_o$ where the apex distance from the nucleus on the connecting line toward the Sun

$$X_o = \frac{2}{3} \frac{a D c R_s^2}{S} \frac{v_a^2}{q}.$$

For all grain sizes $a > 0.4\,\mu$ the radiation pressure efficiency is $q = 1$, for $0.05 < a < 0.2\,\mu$ one finds $q \approx 15 \times 10^8\, a^2$ (compare Lamy, 1974). Inserting the appropriate expression for v_a in the approximation $v_a < v$, one obtains for P/Encke at $R_s < 1$ AU, i.e. where $\dot{M}_g \sim R_s^{-2}$, that grains of all sizes $a > 0.4\,\mu$ produce the same stand-off distance $X_o = 800$ km and the same enveloping paraboloid independent of heliocentric distance. So Encke should feature a sharp dust front, and a fly-by mission for dust impact studies should pass within 800 km in front of the nucleus. For Kohoutek one calculates with the post-perihelion data of Barbieri et al., (1974) at $R_s = 0.64$ AU and $\dot{M}_g = 2.5 \times 10^5$ g sec^{-1} sr^{-1} a sharp dust front at $X_o = 4500$ km.

This refocussing of initially divergent grain trajectories by radiation pressure takes place only if $v_a \sim 1/\sqrt{a}$, which is not the case if $\dot{M}_g/aDvr_N > 1$. Then all grains, obeying this unequality, are accelerated uniformly to the terminal gas velocity v. In this case, the particle size does not cancel anymore in the expression for X_o and we expect a diffuse dust front, as for Comet Kohoutek during approach to the Sun, where at $R_s = 0.64$ $\dot{M}_g = 2,7 \times 10^6$ g/sec x sr with $a < 0.3\,\mu$.

The Surface Density of Grains:

The exact density profile, given by the parabolic grain trajectoties yields an untractable expression. For the present purpose, however, the mean surface density averaged over a slab at distance X from the nucleus yields for particles in the size range between a and $a + da$ in the sunward direction from the nucleus

$$\frac{d\bar{N}}{da} da = \frac{dQ_a}{da} da \frac{\pi}{v_a X_o} \frac{\ln(\sqrt{\frac{X_o}{X}} + \sqrt{\frac{X_o}{X} - 1})}{\sqrt{1 - \frac{X}{X_o}}}$$

where dQ_a/da is the number production rate in this size range. Polarization measurements (Noguchi et al., 1974 made these on the inner dust coma, where alignment of non-spherical particles by solar wind or radiation pressure is negligible) suggest a size distribution $dN/da \sim a^{-4}$, so that the infrared thermal emission for Comet Encke at 3.5 and 4.8 µ (Ney, 1974) can be evaluated with optical constants and grain temperatures from Lamy (1974). The result is that the mass ratio of dust-to-gas production rate is $\dot{M}_d/\dot{M}_g = 0.1$ for this Comet (Michel and Nishimura, 1975). With

$$\frac{d\bar{G}_a}{da} = \frac{3\sqrt{a_{min}}}{8\pi D} a^{-4.5} \dot{M}_d$$

and the expression for v_a, one obtains then the mean surface density of dust grains on the sunward side

$$\bar{N} = \int_{a_m} \frac{d\bar{N}}{da} da = \frac{1}{8\pi} \frac{\dot{M}_d}{\dot{M}_g} \sqrt{\frac{2\dot{M}_g r_N a_{min}}{3 D v}} \cdot \frac{\ln\left(\sqrt{\frac{x_o}{x}} + \sqrt{\frac{x_o}{x}-1}\right)}{x_o\sqrt{1-\frac{x}{x_c}}} \cdot \frac{1}{a_m^3} = 0.76 \cdot 10^{-10} \frac{1}{a_m^3}$$

for Encke at perihelion at a distance of 400 km from the nucleus. Hence, one may expect about 10^5 impacts per cm^2 for particles in the range $0.1 < a < 1\mu$, detectable by an impact detector, but only 10^{-4} cm^{-2} for grains with radius $a_m > 0.1$ mm, which can cause serious damage.

References:

Barbieri, C., Cosmovici, C.B., Michel, K.W., Nishimura, T., 1974, Icarus 23, 568

Finson, M.L., and Probstein, R.F., 1968, Astroph. J. 154, 327 and 353

Grün, E., Kissel, J., Hoffmann, H.-J., 1976, this volume

Hübner, W.F., 1970, Astron. and Astroph. 5, 286

Lamy, Ph.L., 1974, Astron. and Astroph. 35, 197

Michel, K.W. and Nishimura, T., 1975, to be published

Mumma, M.J., 1975, in "Ballistic Intercept Missions to Comet Encke, NASA TMX-72542, March 1975

Ney, E.P., 1974, Icarus 23, 551

Noguchi, K., Sato, S., Maihara, T., Okuda, H. and Uyama, K., 1974, Infrared Photometric and Polarimetric Observations of Comet Kohoutek 1973f, Icarus 23, 545

Pecker, J.C., 1974, Astron. and Astrophys. 35, 7

Rieke, G.H. and Lee, T.A., 1974, Nature 248, 737

DUST EMISSION FROM COMET KOHOUTEK (1973f) AT LARGE DISTANCES FROM THE SUN

E. Grün and J. Kissel

Max-Planck-Institut für Kernphysik, Heidelberg, FRG

and

H.-J. Hoffmann

Physikalisches Institut, Rheinische Friedrich-Wilhelms-Universität Bonn, Bonn, FRG

Abstract

During a period of 60 days around the time when the HEOS 2 satellite penetrated the orbital plane of Comet Kohoutek (1973f) the micrometeoroid experiment on board registered an excessive particle flux. These particles are identified with those originated in Comet Kohoutek. Orbit calculations show that their emission occurred outside 3.8 AU from the sun. The ratio of the force of radiation pressure to that of gravity of these particles was determined to $\beta = 1 \pm 0.1$, their mass has been measured from the satellite data (10^{-13} g to 10^{-11} g). The velocity and the rate of dust particles emitted from the comet is studied on the basis of the theory of dust comets formulated by Finson and Probstein. An emission rate of appr. 1.2×10^{18} particles per second in the size range corresponding to $0.9 \lesssim \beta < 1.1$ and an emission velocity of appr. 0.5 km/sec match best the observed data.

From optical observations of the shape and occurrence of the anti-tail of Comet Kohoutek (1973f) Sekanina (1974) concluded that appreciable dust production took place as far as 4 AU from the sun. These grains found in the anti-tail are in the order of 0.1 to 1 mm in size. The presence of particles with diameters in the order of 1 μm has been shown in the coma and tail of Comet Kohoutek by infrared observations (Ney, 1974 and Noguchi et al., 1974). This paper discusses some consequences of in situ measurements of dust particles from Comet Kohoutek reported by Hoffmann et al. (1975b) which give evidence for emission of micron sized particles at heliocentric distances of approx. 4 AU from the sun.

During a period of 60 days around the time when the HEOS 2 satellite penetrated the orbital plane of Comet Kohoutek in June 1974, an excessive flux of particles in the velocity range of 15 to 20 km/sec and mass range of 10^{-13} to 10^{-11} g was registered. The average rate of these particles measured during preceding years was 1 to 2 particles in the same time interval compared with 7 particles in 1974. The chance for such an accidental clustering is less than 0.01;

therefore, an additional source for these particles is required. Nazarova and Rybakov (1975) also reported an enhancement of the dust particle flux detected by the Luna-22 space probe which occurred in the same period as observed by HEOS 2. The measured properties of the particles detected by HEOS 2 are:
- ecliptic longitude of detection by HEOS 2: $235°$ - $295°$
- apparent radiant: cone of $60°$ half angle around earth apex
- mass range: 10^{-13} g - 10^{-11} g
- speed range: 15 km/sec - 20 km/sec.

The moon and the upper atmosphere of the earth as source for the particles (Hoffmann et al., 1975a) are ruled out because the sensor was looking away from these objects most of the time when it recorded the impacts. Particles from meteor showers which are observed during the months of May through August have apparent radiants which do not fit the radiants of the registered particles. Additionally in 1973 during a similar period no such flux increase was detected; therefore, shower meteors cannot supply these particles. As the nodes of all comets (apart from Comet Kohoutek) observed in 1973 and 1974 do not fit the required interval of ecliptic longitudes or have perihelion distances outside 1 AU, they cannot supply these particles either. Only particles from Comet Kohoutek meet all requirements set by the measurements. A simple orbit consideration already shows that dust particles released from the comet at a distance of approx. 4 AU come very close to the earth at the time when it passes the line of nodes if their ß (ratio of radiation pressure over gravity) is close to 1. These particles would have a speed relative to the earth of approx. 20 km/sec and an apparent radiant $40°$ away from the earth apex, which is well within the field of view of the sensor.

Orbit calculations for dust particles, taking into account an initial release velocity v_i, show that particles with $0.9 \leqslant ß < 1.1$ and release distances outside approx. 3.8 AU can reach the satellite at its passage through the comet's orbit. From the magnitude of the time interval during which the particles were recorded, Hoffmann et al. (1975b) concluded that the initial velocity was in the order of several 100 m/sec. Only micron sized particles can be accelerated to such a high speed by the outstreaming gas from the comet at large heliocentric distances (Delsemme and Miller, 1971). Therefore, an emission of these particles from the comet inside millimeter sized "dirty-ice" grains and subsequent evaporation of the large grains, as suggested by Sekanina (1974 and 1975a), must be ruled out for the particles detected by HEOS 2. In the following it is assumed that the particles have not changed in size during their flight path.

A theory developed by Finson and Probstein (1968a and b) was applied in order to compare the number of particles observed with the dust emission rate, the initial speed distribution and the size distribution of dust particles released from the comet. Finson and Probstein assume that dust particles are continuously expelled from a cometary nucleus by evaporating gases, and that the subsequent motion of the particles is controlled only by solar gravity and solar radiation pressure. Instead of the diameter d of dust-particles it is convenient to use their radiation pressure ratio as a measure of their size:

$$ß \sim 1/d$$

The number of particles emitted in the time t to t + dt with sizes in the range ß to ß + dß is given by

$$\dot{N}_d (t) \, dt \, h(ß) \, dß$$

where $\dot{N}_d (t)$ is the dust emission rate in particles per second and h (ß) is the corresponding size distribution function. After emission this group of particles forms a spherical shell which expands with the speed v_i which is a function of the particles' size and time

$$v_i = v_i (ß, t).$$

If one denotes the time from emission t_e to the time of observation t_o with τ:

$$\tau = t_o - t_e$$

the shell has a radius of $v_i \tau$. The number of particles seen by the satellite experiment while penetrating this shell is given

$$\frac{\dot{N}_d \, h(ß) \, dß \, d\tau}{4 \pi (v_i \tau)^2} \quad \frac{dA}{dA_\perp} \quad p(\delta)$$

where dA is the differential surface area of the shell, $dA_\perp$ is the projection of dA on a plane perpendicular to the flight direction of the satellite and p (δ) is the sensitivity of the micrometeoroid detector for particles coming from an apparent direction with respect to the sensor axis, indicated by the angle δ (Hoffmann et al., 1975a).

Integrating numerically over all such spheres of different τ and ß values the impact rate on the micrometeoroid detector is found. Since only a small range of sizes $0.9 \leqslant ß \leqslant 1.1$ can reach the detector, the size distribution is taken to be constant

$$h(ß) = \text{const. for } 0.9 \leqslant ß \leqslant 1.1$$

Fig. 1 shows the calculated impact rate as a function of observation time t_o for a set of different initial conditions. The dust emission rate $\dot{N}_d$ is kept constant over the time of consideration, whereas the emission speed v_i is varied from 100 m/sec to 700 m/sec. At 100 m/sec the expected impact rate should peak during a narrow time interval around the earth's passage through the line of nodes. At higher emission speed the enhancement of the impact rate becomes broader and more flat during the time of observation. For comparison the actual observation dates of the particles are indicated at the bottom of the figure. An emission speed of 500 $\pm$ 200 m/sec best fits the observed data.

In Fig. 2 the effect of dust emission rate and initial velocity varying in time is examined. The variation of both functions with the heliocentric distance r, which is equivalent to a time variation, does not change the rate profile very much. This is due to the fact that more than 50 % of the observable particles are released within a narrow distance interval of 3.8 AU $\leq$ r $<$ 4.4 AU.

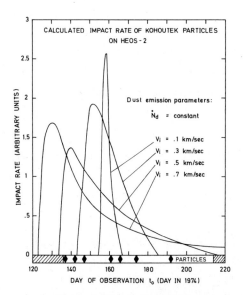
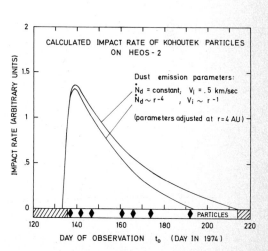

Fig. 1: Calculated impact rate for constant emission rate $\dot{N}_d$ and constant emission speed v_i compared with measured particles.

Fig. 2: Effect of varying dust emission rate $\dot{N}_d$ and varying emission speed v_i on the impact rate.

The absolute dust emission rate $\dot{N}_d$ is obtained by integrating the calculated impact rate over the observation time t_o and comparing it with the number of particles observed. A dust emission rate of $\dot{N}_d = 1.2 \times 10^{18}$ particles per second is found to account for the particles detected by HEOS 2. Taking a mass of 10^{-13} to 10^{-11} g per particle, a dust emission of 0.1 to 10 tons per second is estimated. This mass was released by the comet in the form of particles in the size range corresponding to $0.9 \leq \beta \leq 1.1$.

Due to the large uncertainty of the mass determination no information can be gained on the particles' material composition or density by a comparison of the observed values for β and mass (Sekanina, 1975b). However, ice, as a material of the registered particles, must be excluded because of the long flight time between emission from the comet and detection by the satellite.

References:

Delsemme, A.H., and Miller, D.C. (1971), "Physico-chemical phenomena in Comets. III. The continuum of Comet Burnham (1960 II)", Planet. Space Sci. 19, 1229.

Finson, M.L., and Probstein, R.F. (1968a), "A theory of dust comets. I. Model and equations", Astrophys. J. 154, 327.

Finson, M.L., and Probstein, R.F. (1968b), "A theory of dust comets. II. Results for Comet Arend-Roland", Astrophys. J. 154, 353.

Hoffmann, H.-J., Fechtig, H., Grün, E., and Kissel, J. (1975a), "Temporal fluctuations and anisotropy of the micrometeoroid flux in the earth-moon system measured by HEOS 2", Planet. Space Sci. 23, 985.

Hoffmann, H.-J., Fechtig, H., Grün, E., and Kissel, J. (1975b), "Particles from Comet Kohoutek detected by the micrometeoroid experiment on HEOS 2". in "The Study of Comets", IAU Colloquium No. 25, in press.

Nazarova, T., and Rybakov, A. (1975), "The meteoric matter investigation on Mars-7 and Luna-22 spaceprobes", presented at the XVIIIth COSPAR Meeting, Varna, Bulgaria.

Ney, E.P. (1974), "Multiband photometry of Comets Kohoutek, Bennet, Bradfield and Encke", Icarus 23, 551.

Noguchi, K., Sato, S., Maihara, T., Okuda, H., and Uyama, K. (1974), "Infrared photometric and polarimetric observations of Comet Kohoutek 1973 f", Icarus 23, 545.

Sekanina, Z. (1974), "On the nature of the anti-tail of Comet Kohoutek (1973f). I. A working model", Icarus 23, 502.

Sekanina, Z. (1975a), "A Study of the icy tails of the distant comets", Icarus 25, 218.

Sekanina, Z. (1975b), private communication.

3.7

PREDICTED FAVORABLE VISIBILITY CONDITIONS FOR ANOMALOUS TAILS OF COMETS

Zdenek Sekanina
Center for Astrophysics
Harvard College Observatory and Smithsonian Astrophysical Observatory
Cambridge, Massachusetts 02138, U.S.A.

It was shown elsewhere (Sekanina, 1974) that the observability from the earth of an anomalous tail (antitail) of a comet can be rather straightforwardly predicted from the dynamical and geometric conditions. The physical presence or absence of the antitail at a precalculated time is then a measure of the comet's production rate, at the relevant emission times, of relatively heavy dust particles (mostly of submillimeter size) that constitute such an antitail. Because the large grains are emitted from the nucleus at very low velocities (typically meters or tens of meters per second), an antitail is essentially a two-dimensional formation in the orbit plane of the comet and can be recognized best when projected edge-on, i.e., when the earth crosses the nodal line of the comet's orbit. In general, however, this condition is not essential for the recognition of antitails (cf., e.g., Comet Kohoutek 1973 XII).

Since the emission rate of heavy dust particles is a potentially significant parameter for a physical classification of comets, we made use of the visibility conditions to list the comets that should have displayed a sunward tail around the time of the earth's passages through the orbit plane. This type of the antitail observability will be termed the nodal appearance. A computer program executing the conditions for a nodal appearance was applied to the Catalogue of Cometary Orbits (Marsden, 1975), starting with the comets of 1737. However, we excluded all comets that were at the critical times located near the antisolar point in the sky (elongations exceeding 135°), where the definition of the sunward direction becomes meaningless. We also excluded all cases at heliocentric distances larger than 2 AU in order not to confuse the antitails with the icy tails (Sekanina, 1973, 1975) that are observed far from the sun and point fairly frequently in the general direction of the sun.

The statistics of the nodal appearances of antitails of comets, whose conditions were satisfied within, or not more than 5 days outside, the period of observation, are listed in Table I, separately for nearly-parabolic comets (revolution periods more than 200 years) and for short-period comets. The calculations were done for dust particles with a ratio $1-\mu$ of radiation pressure to solar gravity of 0.01 (known to be common in observed antitails) and for two different starting emission times. Whereas the choice of $1-\mu$ is not crucial, Table I shows that the time of onset of dust production affects the statistics substantially. The comets with a sunward tail reported to have been detected near the predicted time are listed in Table II, where columns 2 to 4 give, respectively, the perihelion distance, the reciprocal value of the original semimajor axis (for P/Encke the revolution period), derived from Marsden (1975) and from Everhart and Raghavan (1970), and the absolute magnitude (Vsekhsvyatsky, 1958). We remark that with the exception of 1937 IV the comets have perihelia well inside the earth's orbit, and that apart from the controversial case of P/Encke (see details below) the comets' revolution periods are longer than 7000 years and their absolute magnitudes brighter than 8.

Table I points consistently to a conclusion that only about 20 to 30% of the nearly-parabolic comets that should have displayed an antitail at the node were actually observed to do so. Indeed, if we count only the comets with nearly-ideal observing conditions, the figure is 22% for the onset of emission at 4 AU and increases to 30%, if the condition is relaxed to 2 AU. If we count all comets that were observed near the node, the fraction of positive observations is lower, as can be expected, but not very substantially: we find 19% for 4 AU emissions and 23% for 2 AU emissions.

The results are dramatically different for short-period comets. Although there were numerous opportunities for observing a nodal appearance of an antitail, we do not yet have a single clearly positive observation. The only promising case so far is that of P/Encke in 1964, for which Roemer (Roemer and Lloyd, 1966) secured a pair of plates only 2.5 days after the earth's nodal passage; the comet was 88 days after perihelion. A close inspection of the plates by Dr. Roemer and the writer revealed two extensions emanating from the weak, nearly stellar image of the comet in the opposite directions, one of them pointing right toward the sun. Although this sunward tail does not, in the writer's opinion, resemble the gas jets, frequently observed in P/Encke before perihelion, there is still no more than a 50% chance that it is a true antitail.

Table I. Statistics of predicted nodal appearances of antitails of comets in the past ($1-\mu = 0.01$)

Comets	Nearly-parabolic			Short-period	
Assumed sun-comet distance at ejection (before/at perihelion)	4 AU	2 AU	2 AU	Perihelion
(A) Number of comets (apparitions) whose predicted nodal appearances of antitail lie within the observed arc of orbit	69	45	21 (32)	6 (7)
(B) Number of comets (apparitions) under (A) that were observed near the node	48	30	16 (23)	6 (7)
(C) Number of comets (apparitions) under (B) with significant predicted characteristic length of antitail	38	28	15 (20)	2 (2)
(D) Number of comets (apparitions) under (C) with elongations exceeding 30° from the sun at the node	30	20	15 (19)	2 (2)
(E) Number of comets (apparitions) under (D) that were free from severe moonlight interference at the node	18	10	7 (10)	2 (2)
(F) Number of comets (apparitions) under (B) whose antitail was actually observed	9	7	1?(1)?	1?(1)?
(G) Number of comets (apparitions) under (E) whose antitail was actually observed	4	3	1?(1)?	1?(1)?

Table II. Comets with antitails observed at node

Comet	q (AU)	$(1/a)_{orig}$ (AU)$^{-1}$	H_{10} (mag)	Date of node crossing	Antitail seen	Conditions at node and remarks
1823	0.23	.	4.2	1824/1/24	1/22-25, 27, 31	Very favorable conditions
1844 II	0.86	+0.001007	4.9	1844/10/25	11/3, 8	Close to sun; moonlight interfering
1844 III	0.25	+0.002592	4.9	1845/1/18	1/11, 16, 25, 27	Moonlight interfering
1895 IV	0.19	-0.000168	5.2	1896/2/9	2/15, 19-21	Close to sun
1937 IV	1.73	+0.000063	6.0	1937/7/31	7/30, 8/1	Favorable; but early emissions only
1954 VIII	0.68	+0.000051	7.0	1954/7/25	7/30, 8/1, 3, 6-7	Node crossed 3 days before discovery
1957 III	0.32	+0.000009	5.4	1957/4/25	4/22-30	Very favorable conditions
1961 V	0.04	+0.002211	7.5	1961/7/21	7/25-26, 8/1	Node crossed 2 days before discovery; close to sun; moonlight interfering
1964 IV	0.34	(3.30 yr)	13-15	1964/8/27	8/30	P/Encke; nature of tail not clear
1969 IX	0.47	+0.000507	5.8	1970/1/2	12/26-28, 30-31, 1/2	Antitail short; early emissions only

The general absence of antitails among the short-period comets appears to be incompatible with the existence of meteor streams known to be associated with many of these comets. Unfortunately, at their observed returns, the parent comets of the three spectacular-storm producing meteor streams — P/Biela, P/Giacobini-Zinner and P/Tempel-Tuttle — were never placed favorably enough for a nodal appearance of an antitail. And, of all the other comets known to be related to meteor streams, only two had such very favorable apparitions: P/Encke in 1878, 1888 and 1964, and P/Pons-Winnecke in 1909, although P/Pons-Winnecke is not apparently associated with a permanent stream (Cook, 1973). The other comets with favorable conditions were P/Tempel 1 in 1867, P/Finlay in 1919, P/Kopff in 1945, P/Grigg-Skjellerup in 1947, and P/Schaumasse in 1952 and 1960. Streams that could be associated with P/Finlay or P/Grigg-Skjellerup have never been reported; the other comets have perihelia well beyond 1 AU.

With one doubtful and two negative results in the three nearly-ideal returns, P/Encke presents probably the most solid evidence to date against the positive correlation between the antitails and the meteor streams. In order to obtain more data, positive or negative, on the occurrence of the antitails, we investigated their visibility conditions in the future returns of the short-period comets. Among 166 returns of 60 comets with perihelia within 2 AU between 1976 and 1999 (orbital elements courtesy of Dr. Marsden), the following instances — most of them outside nodal areas — are considered as most significant: P/d'Arrest in 1976/77, P/Encke in 1977 and 1987, P/Schwassmann-Wachmann 3 in 1979, P/Honda-Mrkos-Pajdušáková in 1980, P/Grigg-Skjellerup in 1982 and 1987, P/Crommelin in 1984, P/Pons-Winnecke in 1989/90, and P/Giacobini-Zinner in 1999.

This work was supported by grants NGR 09-015-159 and NSG 7082 from the National Aeronautics and Space Administration.

References

Cook, A. F. (1973), in "Evolutionary and Physical Properties of Meteoroids", Hemenway, C. L., Millman, P. M., and Cook, A. F., Eds., NASA SP-319, Washington, D.C., p. 183.
Everhart, E., and Raghavan, N. (1970), Astron. J. 75, 258.
Marsden, B. G. (1975), Catalogue of Cometary Orbits. (2nd edition.) IAU Central Bureau for Astronomical Telegrams, Smithsonian Astrophys. Obs.
Roemer, E., and Lloyd, R. E. (1966), Astron. J. 71, 443.
Sekanina, Z. (1973), Astrophys. Lett. 14, 175.
Sekanina, Z. (1974), Sky and Tel. 47, 374.
Sekanina, Z. (1975), Icarus 25, 218.
Vsekhsvyatsky, S. K. (1958), Physical Characteristics of Comets. (In Russian.) Moscow, p. 51. (English transl.: Jerusalem, 1964.) Follow-ups: Soviet Astron. 6, 849 (1963); 10, 1034 (1967); and 15, 310 (1971).

3.8 STUDY OF THE ANTI-TAIL OF COMET KOHOUTEK FROM AN OBSERVATION ON 17 JANUARY 1974

Ph. Lamy
Laboratoire d'Astronomie Spatiale, Marseille

S. Koutchmy
Institut d'Astrophysique, Paris

Abstract. As part of our program of observation of Comet Kohoutek at Pic-du-Midi Observatory, we obtained, on January 17.8 UT, 1974 a photograph in polarized light showing dramatically the (dust) anti-tail extending for almost $1°$ from the Comet's head (reported in Sky and Telescope, June 1974); indeed the comet is visible in polarized light further away than in total light as noticed by Weinberg and Beeson (IAU Colloquium No. 25, 1974) for Comet Ikeya-Seki. A photometric and polarimetric study was performed (Bücher, A., Robley, R., and Koutchmy, S., 1975, Astron. Astrophys. 39, 289) showing that the anti-tail is strongly polarized (up to 50 %). These large degrees of polarization are of the same order of magnitude as those reported for the tail of Comet Ikeya-Seki by Matjagin Sabitov and Kharitonov (1967, Astron. Zh. 44, 1075) and by Weinberg and Beeson (op. cit.). As discussed by these latter authors, particle alignment is precluded as a significant contributor to polarization in the tail of comets. Polarization by large spheres as obtained from the Fresnel reflection coefficients applies only in the case of perfect surface, a circumstance very unlikely in interplanetary space; the scattering is in fact controlled by the surface microstructures (Van de Hulst, private communication). Therefore we hypothesized that submicronic grains should play an important role in the anti-tail. The classical method of Finson and Probstein (1968, Astrophys. J. 154, 327, 353) was used to draw the sky plane view of the syndynes for the day of observation. Since the ratio ß of the radiation pressure force to the gravitational attraction is proportional to the third power of the grains' radius s for $s \lesssim 0.1$ μ, submicronic grains with typical radii of 0.02 μ may indeed be present in the anti-tail and provide a straight-forward explanation of the observed polarization. This size is of the same order of magnitude as that inferred for interstellar grains which may well be embedded in the comet's nucleus as well as meteorites for which there exists good evidence. Our conclusion does not rule out the presence of millimeter-size grains as proposed by Sekanina and Gary and O'Dell in their preliminary investigations (1974, Icarus 23, 502, 519) which did not take into account the polarimetric result; such grains may well coexist with the submicronic ones. Finally, the line of maximum intensity is close to a synchrone corresponding to a time of emission 100 days before perihelion passage. This supports the synchronic formation of anomalous tails and possibly of tails as proposed by Vsekhsvyatsky (1932, Astron. Zh. 9, 166).

3.9   Condensation Processes in High Temperature Clouds

Bertram Donn*
Department of Astronomy
Cornell University
*On Leave, Astrochemistry Branch, Goddard Space Flight Center
Greenbelt, Md. USA

Neither thermodynamic calculations nor standard nucleation theory describe condensation processes in cosmic clouds, but rather are a first approximation. Condensation is an irreversible kinetic, not equilibrium, process. The nominal molecules of condensed minerals (e.g. Al_2O_3, $CaTiO_3$, $MgAl_2O_4$) occur only in the bulk lattice not in cosmic gases. Reactions among gas molecules must be an intrinsic part of condensation. Molecules yielding a particular condensate are a small fraction, less than 10^{-3}, of condensible and reactive (e.g. H, H_2, H_2O) molecules. No theoretical or experimental results are available for condensation under such conditions. There is no basis for expecting nearly pure, known minerals to form in cosmic clouds. A generalization of nucleation theory following Hirschfelder (JCP, 61, 2690, 1974) can be made, in principle. No thermodynamic data is known for most of the possible compositions of nucleating clusters and no calculations are possible. Grain compositions should fall between thermodynamic and kinetic calculations. An experimental study appears feasible and is planned. The following conclusions are drawn from the theoretical analysis. (1) An interpretation of interstellar grains may not be made using only terrestrial minerals. (2) Condensation temperatures and compositions of interstellar grains or primordial solar system condensates may not be deduced from equilibrium calculations.

MARINER MISSION TO ENCKE 1980

C. M. Yeates, K. T. Nock, R. L. Newburn

California Institute of Technology
Jet Propulsion Laboratory
Pasadena, California

I. INTRODUCTION

    A. Why Study Comets?

The planetary program has always been conducted with the hope that the results would reveal great insight into the early period of solar system history and perhaps into the actual formation processes themselves. However, little knowledge has been gained of this very early stage for several reasons: The intense surface bombardment of all larger bodies, particularly in the inner solar system during that period; the subsequent differentiation of large bodies; and atmospheric effects and continued bombardment of the surface. The most promising approach to acquisition of knowledge pertaining to the early state of the solar system, its origin and evolution, therefore seems to be in the study of small bodies, e.g., comets and asteroids.

    B. Do we need a space probe?

Ground based comet observations have yielded relatively little information regarding their composition and behavior, and this section will point out the need for a close-up look at comets.

There is little hope of verifying the existence of a nucleus without close-up viewing with ~100 meter resolution cameras, in fact no one has seen a cometary nucleus as anything more than a point of light its approximate size implied from apparent brightness. Also, the nucleus size, shape and albedo (all important for thermal balance calculations) cannot be determined without a space probe.

Even though within the last two years 3 probable parent species (H_2O, HCN, and CH_3CN) have been detected by ground based radio observations, the source of most observed radicals and ions is not known. Furthermore, other "parent" species must be present as well as other molecular species which arise as a result of molecular interactions and photo-processes. The whole question of molecular abundances must remain open without in situ study by a mass spectrometer.

A cometary flyby mission should enable the determination of the gas to dust ratio (necessary for thermal balance calculations) as well as the composition of solids. The composition will be compared to cosmic abundances which may provide an idea of the region of comet formation, whether from the outer regions of the solar nebulae or from interstellar material. If the nuclear bodies represent the primitive material of the solar nebula, it represents the oldest and least modified of any material we are ever likely to study.

In spite of efforts to understand the formation of ion tails and energization processes therein, the whole question of solar wind interaction remains open. Several interaction modes are possible: ionospheric, atmospheric, unipolar induction, and possibly a vestigial cometary field. These questions will remain unanswered until fields and particles measurements are made which will verify the existence (or nonexistence) and location of bow shock and contact surface caused by the comet as it acts as an obstacle to the solar wind flow. Quite possibly one could also observe changes in the comet tail structure and behavior in response to traveling interplanetary phenomena such as blast waves and other discontinuities in the solar wind.

The mission described in this paper is based upon the use of the spare Mariner 10 Spacecraft with as little science and hardware changes as possible (1,2).

C. Why Encke?

Three general criteria largely determine the quality of a cometary flyby: the activity of the comet, the relative velocity of the encounter, and the approach distance. All long period comets are excluded from consideration since their positions are never well enough known in advance to maneuver a spacecraft sufficiently near to them. Encke's perihelion distance is less than any other short period comet, and its intrinsic activity is quite high so that its activity as determined by its visual brightness is larger than any other suitable comet except P/Halley.

Flyby velocity relative to P/Encke at perihelion can be as low as ~7 km sec^{-1} while relative velocities of other short period comets with perihelia near 1 AU are typically ~15 km sec^{-1}. The ballistic intercept of P/Halley has a relative velocity of 55 km sec^{-1}.

Because P/Encke is brighter at encounter than other available comets, it can be found earlier so that terminal guidance techniques can be used, which can result in an approach distance of a few hundred km.

II. EXAMPLE EXPERIMENTS AND HOW GOOD ARE THEY FOR P/ENCKE?

   A. Volatile Analysis

   One of the critical areas mentioned was that of volatile analysis. It is important to measure number densities down to say 10^{-4} that of H_2O which for P/Encke at ~0.4 AU means that number densities as small as 10^2 cm^{-3} must be measurable. This can be accomplished with a double-focusing, magnetic sector mass spectrometer having a unit duty cycle detector (measuring all masses simultaneously).

   To obtain unit duty cycle performance, one can use a microchannel array at the focal plane of the spectrometer and the electrons produced are accelerated to a phosphor coated surface, the photons produced here can be directed to a photodiode array by fiber optics. The overall gain is ~10^6 with mass range of 1-100 amu being read out simultaneously. The ionizing efficiency may be improved over present mass spectrometers by using higher ionizing beam currents. One ion per 10^2 neutrals cm^{-3} instead of one per 10^3 cm^{-3} is probably realizable. The spectrometer can be operated with a retarding potential to create a barrier against molecular fragments and products from spacecraft contamination.

   The P/Encke H_2O source strength was measured at 0.71 AU (assuming that all of the atomic hydrogen results from dissociation of H_2O). Scaling this measurement to 0.4 AU and using typical values of the lifetime and initial velocity of H_2O we obtain the H_2O density as a function of nuclear distance.

   Table 1 gives the number density of H_2O and of other species with various lifetimes having mixing ratios of 10^{-2} to 10^{-4} that of H_2O at three distances from the nucleus. Also given are the integration times required to give a SNR of three for the spectrometer described above. Clearly, the desired sensitivity with short integration times is easily achieved with such an instrument.

B. Non-Volatile Analysis

Chemical Analysis of the non-volatile components is one of the most important objectives and one of the most difficult to achieve. Typical dust detectors use a target of gold or tungsten which upon impact volatilizes and partially ionizes the incident particle, after which the ions are analyzed by some form of mass spectrometry for example time of flight. This type of instrument would have a mass range of 10^{-10} to 10^{-15} grams and hopefully a mass resolution ($\Delta m/m$) of 60 from perhaps 12 to 70 amu.

The dust model assumes a distribution derived from observations of Arend-Roland (3) and Bennett (4). The model was scaled to Encke by assuming a total mass of dust equal to the total mass of volatiles, which is probably near the upper limit of solids. Table 2 shows the fluence and density for various size particles. The peak of the distribution is a density of 25×10^{-6} cm^{-3} at a 200 km miss distance so that a 100 cm^2 target area would result in counting rates of 10^3 per second so that saturation would not be a problem. Particles of mass 10^{-10} grams (4×10^{-4} cm diameter) and smaller can be detected with this analyzer but for larger particles an additional particle detector is required, e.g., optical, microphone, cell penetration or capacitor discharge detectors. An instrument designed with a mass range of 10^{-3} to 10^{-6} grams and a detector area of 10^4 cm^2 would result (for a trajectory 200 km from the nucleus) in count rates of 10^{-1} to 10^2 sec^{-1}. These dust and particle detectors would complement each other and leave a gap in particle size of only one order of magnitude between 10^{-3} and 10^{-4} cm.

C. Imaging

The primary scientific goal of the imaging system is to furnish high resolution pictures of the nucleus. The cameras used in this study consist of dual cameras one of which is the Mariner 10 150 cm f/8.43 system and one is the Mariner 9 50 cm f/2.35 system. For design purposes we assume Encke has a nuclear diameter of 3 km and $pr^2 = 0.24$ where p is the geometric albedo an r is the nuclear radius. Taking into consideration the image smear due to spacecraft attitude drift and the relative velocity of the comet and spacecraft, and assuming a lunar phase function we calculate the nucleus resolution as a function of distance (Figure 1). Also shown is the time to encounter, the exposure time, and image size. The camera system takes a picture every 42 seconds with alternate cameras (a picture every 84 seconds per camera),

Table I

Mass Spectrometer Operation in P/ENCKE at 0.4 AU
(Molecular density and integration time to give SNR = 3)

Number Density Lifetime	Distance		
	200 km	500 km	1000 km
H_2O 7.272×10^4 s	6.8×10^6 cm^{-3} 1.3×10^{-4} s	1.1×10^6 cm^{-3} 8.2×10^{-4} s	2.6×10^5 cm^{-3} 3.5×10^{-3} s
10^{-2} H_2O $\tau \geq 5 \times 10^5$ s	9.3×10^4 cm^{-3} 9.7×10^{-3} s	1.5×10^4 cm^{-3} 6.0×10^{-2} s	3.7×10^3 cm^{-3} 2.4×10^{-1} s
10^{-3} H_2O $\tau \geq 5 \times 10^5$ s	9.3×10^3 cm^{-3} 9.7×10^{-2} s	1.5×10^3 cm^{-3} 6.1×10^{-1} s	3.7×10^2 cm^{-3} 2.6 s
10^{-4} H_2O $\tau \geq 5 \times 10^5$ s	9.3×10^2 cm^{-3} 9.9×10^{-1} s	1.5×10^2 cm^{-3} 6.8 s	3.7×10^1 cm^{-3} 37 s
10^{-2} H_2O $\tau = 1.476 \times 10^4$ s	6.5×10^4 cm^{-3} 1.4×10^{-2} s	9.4×10^3 cm^{-3} 9.6×10^{-2} s	2.0×10^3 cm^{-3} 4.5×10^{-1} s
10^{-2} H_2O $\tau = 2.52 \times 10^3$ s	4.8×10^4 cm^{-3} 1.9×10^{-2} s	4.4×10^3 cm^{-3} 2.1×10^{-1} s	4.4×10^2 cm^{-3} 2.1 s

Table II

Dust Concentration and Fluence, Nominal Model

Diameter, cm	Velocity, ms^{-1}	Concentration, m^{-3} at 10 km*	Fluence, m^{-2} at 10 km†
0.925×10^{-4}	277	3.62×10^2	1.14×10^7
0.975×10^{-4}	272	8.46×10^2	2.66×10^7
1.125×10^{-4}	248	6.80×10^3	2.14×10^8
1.5×10^{-4}	212	1.01×10^4	3.17×10^8
2.175×10^{-4}	177	6.83×10^3	2.15×10^8
3.3×10^{-4}	141	2.94×10^3	9.24×10^7
5×10^{-4}	114	9.84×10^2	3.09×10^7
9.5×10^{-4}	82	5.92×10^2	1.86×10^7
1.975×10^{-3}	56	1.22×10^2	3.84×10^6
3.325×10^{-3}	43	1.98×10^1	6.22×10^5
6×10^{-3}	32	5.00	1.57×10^5
1×10^{-2}	24	5.17×10^{-1}	1.62×10^4
2×10^{-2}	16	9.75×10^{-2}	3.06×10^3
5×10^{-2}	8.6	5.10×10^{-3}	1.60×10^2
1.2×10^{-1}	4.6	2.61×10^{-4}	8.21
3.34×10^{-1}	1.9	1.31×10^{-5}	0.41

*Concentration $\propto R^{-2}$
†Fluence $\propto R^{-1}$

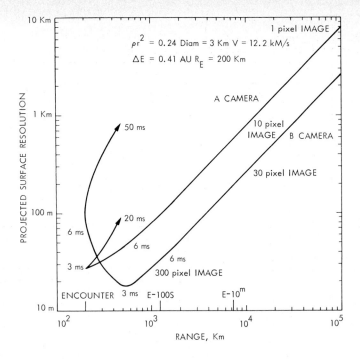

Figure 1. Nucleus Resolution

so we can estimate the total number of pictures obtainable as a function of resolution. For example we obtain 4 pictures with better than 100 meter resolution and about 70 pictures with better than 1 km resolution.

D. Ultraviolet Spectrometer

The ultraviolet spectrometer can obtain two dimensional maps of the coma and corona from which we can obtain production rates, and observe the coma structure and tail symmetry relative to the sun comet vector. It can also hopefully resolve mass ambiguities in the mass spectrometer measurements, e.g., CH_4, NH_2 and O all at mass 16; NH_3 and OH at mass 17; and CO and N_2 at mass 28.

By slight modifications to an existing instrument designed for the MJS Mission (5) an instrument suitable for comet missions can be obtained. An example calculations shows that for the NO bands 1909-2260 Å with a production rate Q of 10^{-3} that of H_2O a SNR of 3 is obtained with an integration time of ~4 minutes.

E. Magnetic Fields and Plasma Experiments

The fluxgate Magnetometers and electrostatic analyzers already on the Mariner 10 Spacecraft are capable of detecting a bow shock and contact surface if they exist. They would measure changes in the magnetic field strength and direction as well as fluctuations up to 10 Hz, and the energy, flow direction, and density of the electrons and ions. Also energization processes in the tail could perhaps be identified.

F. Other experiments have been considered some of which are perhaps desirable and others which are probably not useful for a comet mission. Those which are not useful include infrared spectrometry for lack of instrument sensitivity, mass determination by use of radio tracking, and radio measurements of electron densities. Those instruments which should be considered further are an infrared radiometer, a plasma wave detector, and a plasma ion spectrometer.

III. MISSION DESIGN

Due to the unique relationship between the Earth and Encke orbit geometries during the 1980 apparition, there is an opportunity for a short flight time-low flyby velocity mission. This situation will not occur again for 33 years. Due to spacecraft thermal control constraints, the encounter was chosen to occur at a solar distance of 0.4 AU.

A. Mission Profile

A single spacecraft is launched in early September 1980 by a Titan launch vehicle. The spacecraft encounters Encke in late November within a few hundred kilometers of the nucleus. The encounter sequence begins with periodic imaging a few days before closest approach. Analysis of the gaseous and solid material in coma continues as the spacecraft passes to the anti-sun side of the nucleus and enters the tail. Imaging ceases shortly after closest approach.

B. Trajectory and Targeting

Figure 2 shows a typical trajectory as seen from the north ecliptic pole. The spacecraft orbit is nearly in the plane of Encke's orbit which is inclined about 12° to the ecliptic plane. The comet approaches the slower moving spacecraft from the outside, i.e., at small phase angles, overtakes it and continues inward towards perihelion. The spacecraft aimpoint was selected to be along the anti-sun line as close as possible to the nucleus such that the 3 σ error eclipse just touches the nucleus (Figure 3).

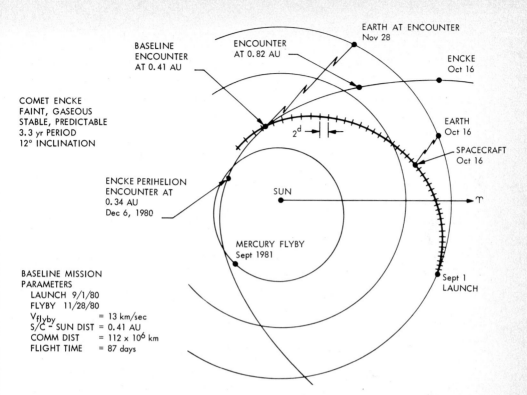

Figure 2. Mariner Encke 1980 - Trajectory Profiles

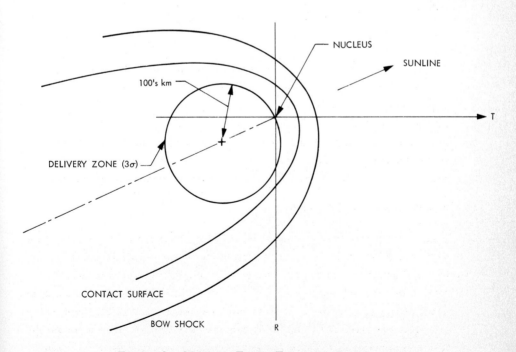

Figure 3. Mariner Encke Targeting Requirements

Analysis shows that nine days before encounter, the onboard TV cameras will acquire the comet and begin the optical navigation phase. The final spacecraft maneuver will occur between $E-6^d$ and $E-4^d$.

C. Spacecraft

Acceptable solutions were found for all the problems caused by the constraint of retaining the spare Mariner 10 Spacecraft. The two major problems were thermal control of the spacecraft (no more than 6 sun intensity allowed) and approach navigation to within a few hundred kilometers of the nucleus. This spacecraft appears to be capable of supporting a wide variety of science experiments. Figure 4 shows the Mariner 10 Spacecraft.

IV. CONCLUSIONS

The mission presented here represents a reconnaissance type mission which is necessary before more elaborate comet missions are attempted such as a rendezvous or flyby of Halley. The mission design itself is based on limited observations and theory. In spite of these constraints, the available spacecraft and science instruments discussed here will provide a wealth of new information which will result in answers to many of the questions discussed in Section I.

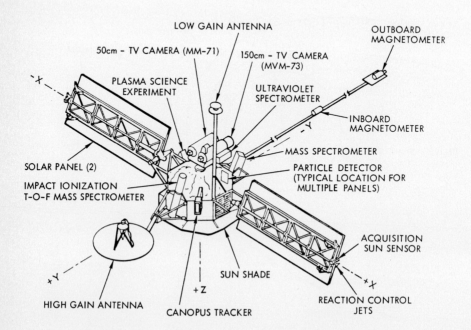

Figure 4. Mariner Encke 1980 Baseline Configuration

REFERENCES

1. Nock, K. T. and Yeates, C. M., "Early Mariner Comet Flyby," AIAA Paper No. 75-198, presented at the AIAA 13th Aerospace Sciences Meeting, Pasadena, California, January 1975.

2. Wilson, J. N., Study Leader, Mariner Encke 1980: Study Report, June 30, 1975 (JPL Internal Document, to be published)

3. Finson, M. L. and Probstein, R. F., "A Theory of Dust Comets. II. Results for Comet Arend-Roland," Ap. J. $\underline{154}$, 353, 1968.

4. Sekinina, Z., and Miller, F. D., "Comet Bennett 1970 II," Science $\underline{179}$, 565, 1973.

5. Broadfoot, A. L., et al., "Ultraviolet Observations of Venus from Mariner 10: Preliminary Results," Science $\underline{183}$, No. 4131, 1974.

6. The work described in this paper was supported by NASA Contract NAS 7-100.

4 METEORS AND THEIR RELATION TO INTERPLANETARY DUST

4.1 METEORS AND INTERPLANETARY DUST

Peter M. Millman
Herzberg Institute of Astrophysics
National Research Council of Canada
Ottawa, Canada K1A 0R6

ABSTRACT

The contribution of meteor observations to our knowledge of meteoroids and interplanetary dust is reviewed under four headings - flux, mass distribution, physical structure and chemical composition. For lower limits of particle mass ranging from 1 g to 10^{-5} g the mean cumulative flux into the earth's atmosphere varies from 2×10^{-15} to 6×10^{-9} particles m^{-2} s^{-1} $(2\pi ster)^{-1}$, and the mean size distribution of these particles is given by $\log N = C - 1.3 \log M$, where N is the cumulative number of particles counted down to a lower mass limit M, and C is a constant. The physical structure of meteoroids in the above range is essentially fragile, with generally low mean bulk densities that tend to increase with decrease in mass. A minor fraction, about 10 or 15 per cent, with orbits lying inside that of Jupiter, have densities several times the average densities, approaching those of the carbonaceous chondrites. The mean chemical composition of meteoroids seems to be similar to the bronzite chondrites for the elements heavier than number 10, but with the probable addition of extra quantities of the light volatiles H, C and O.

INTRODUCTION

In this brief review I propose to summarize our current knowledge of interplanetary dust, acquired from the observational data of meteors. Direct ground-based observations of meteors have been made primarily by four techniques - classical visual recording; conventional photography utilizing the photographic plate or film; use of radar equipment, introduced since 1945; and, more recently, the electronic image-intensification systems where data is recorded on standard video-tape. In all of these techniques the actual observational data consists of a recording of a portion of the kinetic energy of the meteoroid, the solid particle which impacts on the upper atmosphere at high velocity. The kinetic energy appears as light, heat or ionization, and one of the major problems in meteor physics has been to relate this observed energy to the pre-impact mass of the meteoroid.

The observational parameters of meteors relevant to the present review may be grouped under four headings, as related to the complex of interplanetary particles at the earth's mean distance from the sun:-

        (a) flux
        (b) mass distribution
        (c) physical structure
        (d) chemical composition.

A statistical study of meteor orbits can give some information about variations in the above parameters as we move closer to the sun or proceed outward through the asteroid belt, but the problems of observational selection make this extrapolation difficult.

It is important to take note of the particle mass ranges over which meteor observations have given us reasonably reliable data. These are indicated in a general way only in Figure 1, since it is impossible to set exact limits of mass for the various types of data. At the lower end of the mass range lack of sensitivity in the recording equipment produces a cut-off, while at the upper end the observational events are too infrequent to give good statistics.

FLUX

It has become conventional to describe the flux of interplanetary particles by giving the total number that pass through a square metre of space in one direction (that is from one hemisphere = 2π steradians) per second, counted down to a given lower mass limit. Meteor observations have shown very clearly that for particle masses greater than about 10^{-5} g there are marked variations in the observed flux from day to day, after diurnal and seasonal effects resulting from the earth's motions have been removed. These variations are due chiefly to the superposition, on a relatively steady background flux, of particular groups of particles moving along orbits that correspond closely to periodic comet orbits. These groups of particles produce the well-known annual meteor showers, and in the mass range near 1 g the background flux may be increased by factors of 10 to 100, or even to 1000 in extreme cases, near the centre of a major meteor stream.

Most of the early meteor counting was performed visually and unfortunately it is difficult to reduce such records to standard flux figures since the visual recording is very incomplete for the faint meteors yet these, being more numerous than the bright meteors, are more influential in determining rates. Counting down to specified mass limits can be carried out more reliably when instrumental data are used. Good examples

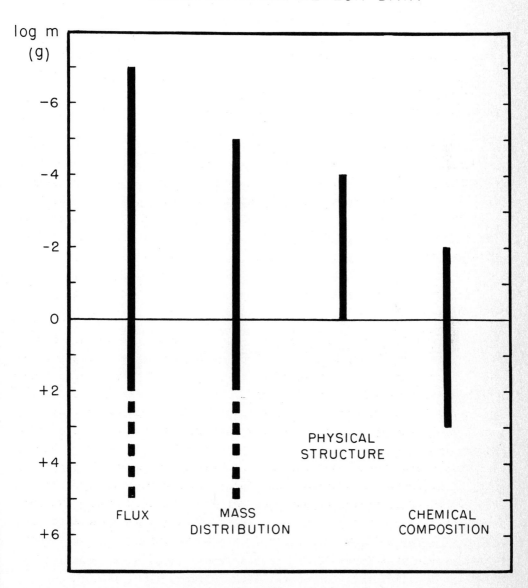

FIGURE 1

of programs employing the three basic instrumental techniques referred to earlier are found in papers by Hawkins and Upton (1958), Hawkins (1963), Nilsson and Southworth (1968), Naumann and Clifton (1973) and Clifton (1973). It is reassuring that these completely independent methods of instrumental meteor observing give background fluxes in reasonable agreement. When care is taken to select visual counting that is statistically complete (Millman, 1970a), this also gives a flux close to the instrumental values. Making use of various compilations of meteor data we find that, in round figures, the mean background flux is 2×10^{-15} m^{-2} s^{-1} $(2\pi \text{ ster})^{-1}$ counted down to a mass of 1 g, and 6×10^{-9} m^{-2} $s^{-1}(2\pi \text{ ster})^{-1}$ counted down to a mass of 10^{-5} g.

Counting down to a lower mass limit of 1 g, the 12 major meteor showers have a total annual flux that is equal to 20 per cent of the total annual background flux. Examples of shower fluxes at the centres of some of the major meteor streams are listed in Table 1. These fluxes are based on material from Millman and McIntosh (1964), Millman (1967) and McIntosh and Millman (1970). As noted in earlier review papers (Whipple, 1967; Millman, 1975) most of the mass of the meteoritic complex is in the form of particles with masses near 10^{-5} g, and here the major showers merge into the background (Millman, 1970b). Southworth and Sekanina (1973) estimate the space density near the earth of all particles smaller than meteorites as 4×10^{-22} g cm^{-3}.

MASS DISTRIBUTION

The mass distribution of interplanetary particles may be defined by any one of a number of indices (Millman, 1973). Here I will use the integrated mass index "S", which is simply the negative slope of the mean line drawn through the plot of the $\log_{10}$ of the standard cumulative fluxes against the $\log_{10}$ of the lower mass limit to which the flux refers. As previously pointed out (Millman, 1973) year-round meteor programs by all four observational techniques show good agreement in mass distribution for background particles in the mass range from 10^{-5} to 1 g. The integrated mass index "S" increases from a value of 1.15 at 10^{-5} g to 1.35 at 1 g (McIntosh and Simek, 1969; Elford, 1967; Kaiser, 1961; Clifton, 1973; Hawkins and Upton, 1958; McCrosky and Posen, 1961; Kresáková, 1966; Millman, 1970a). For heavier particles there is some evidence from the photographic fireball networks that the mass index drops again (McCrosky, 1968) but so far the number statistics are low and there are uncertainties in the meteor theory that is available for meteoroids near a mass of 10^5 g (McCrosky and Ceplecha, 1970).

TABLE 1

Peak fluxes of meteor showers, counted down to particles of mass one gram

shower	date	flux in units of background = 2×10^{-15} $m^{-2} s^{-1} (2\pi ster)^{-1}$
Quadrantids	January 3	15
Lyrids	April 22	1
Perseids	August 12	1.5
Orionids	October 21	0.5
Leonids (1833)	November 13	? 500 - 1000
Leonids (1966)	November 17	200
Geminids	December 14	25

TABLE 2

Physical structure of meteor showers

shower (associated comet)	shower centre	no. meteors	class	no. meteors	ρ g cm^{-3}	χ^*	B
$\mathrm{S}\iota$ Aquarids	Aug. 5	13	A	5	0.30	0.17	0.8
Geminids	Dec. 14	77	B	20	1.06	0.20	0.5
δ Aquarids	Jul. 30	22	B	9	0.27	0.54	1.6
Quadrantids	Jan. 3	17	B	10	0.20	0.41	0.8
Taurids (P/Encke)	Nov. 8	105	C_1	23	0.28	0.06	0.01
α Capricornids (1954 III Honda-Mrkos-Pajdusakova)	Jul. 30	21	C_1	13	0.14	0.38	1.0
Orionids (P/Halley)	Oct. 21	49	C_2	6	0.25	0.40	0.4
Perseids (1862 III Swift-Tuttle)	Aug. 12	45	C_2	10	0.29	0.22	0.03
Giacobinids (P/Giacobini-Zinner)	Oct. 9	2	above C_1	2	<0.01	1.9	4.0

* Values of χ are reduced to standard initial mass $m_\infty = 0.8$ g.

Mass indices for the meteor showers are more difficult to determine since on many programs it is not easy to make a clear separation between the shower meteors and the background meteors. Also, there is no guarantee that any one meteor shower will exhibit the same mass indices at all positions in the stream complex. In general, meteor showers have lower mass indices than the background meteors (Millman, 1972a), and Dohnanyi (1970) has shown that this is to be expected if collisional erosion is the dominant factor in the fragmentation of interplanetary particles, and if the background meteors are primarily former shower meteors that have dispersed into the general meteoritic complex.

PHYSICAL STRUCTURE

Since at most only one or two per cent of the meteoroids which produce visible meteors seem to be associated with the type of interplanetary material that reaches the earth as meteorites and is available for study in the laboratory, the physical structure of a normal meteor-associated meteoroid (a cometary meteoroid) becomes of considerable interest. Included under the term physical structure I am thinking of parameters such as the bulk density, the grain density, strength of the meteoroid in regard to the breaking off of major portions and to the fragmentation into dust particles, the speed of sublimation or melting and the shape of an average particle, both large and small.

Quantitative analysis of meteor photographs has made it possible to estimate the original mass of a meteoroid on entry into our atmosphere. This can be done by using the integrated luminosity (photometric mass) or by a study of the deceleration of the meteoroid (dynamic mass). The differences between these two give us a function of the bulk density of the meteoroid, since its surface area enters directly into the calculation of dynamic mass but does not for photometric mass. Verniani (1967, 1969) took 220 precisely reduced background meteors, photographed with Super-Schmidt cameras on a Smithsonian Observatory program (Jacchia et al., 1967) and found a mean bulk density of 0.28 g cm^{-3}. However, there was good evidence of a bimodal density distribution among these data and removing 31 high-density examples, mean density 1.38 g cm^{-3}, the remaining 189 had a mean density of only 0.21 g cm^{-3}. It is noteworthy that effectively all the high-density group were meteoroids with orbits whose aphelia lay within 5.4 a.u. distance from the sun, or inside the orbit of Jupiter, while some 60 per cent of the low-density meteoroids had aphelion distances greater than 5.4 a.u. It may be that, owing to uncertainties that still exist in meteor theory, Verniani's bulk densities are all somewhat too low, but the important result is the differential between the two groups of background meteoroids.

A much more direct body of evidence is found in a study of the beginning heights of Super-Schmidt meteors. Ceplecha (1967, 1968) has used the observational data for 2529 Super-Schmidt meteors (Jacchia and Whipple, 1961; McCrosky and Posen, 1961). The primary influence on the beginning height of the visible path of a meteor is the velocity with which the meteoroid enters the atmosphere "V_∞". Ceplecha made a statistical study of beginning heights versus velocity for 1848 background meteors and found good evidence for a bimodal height distribution, after the effect of velocity had been allowed for. At any given velocity the separation in height between the two peaks was roughly 10 km. Meteors with low initial heights were classified as A, those with high initial heights as C_1 (low velocity) and C_2 (high velocity), and those with a height distribution in between the primary peaks as B. On examining various possibilities for these beginning-height differences Ceplecha concluded that they resulted primarily from differences in what I have termed physical structure. Ceplecha also looked briefly at the classification of the shower meteors in these data, and recently Cook (1973a) has made a more detailed statistical study of the beginning heights of shower meteors. He finds that when sufficient observational data are available, individual showers can be classified as one of A, B, C_1, C_2, $A+C_1$, and above C_1, and suggests that as we progress from A to above C_1 we sample material progressively more porous and fragile, with lower bulk densities and from layers that had been progressively nearer to the surface of the parent comet nucleus. To illustrate the differences among some of the major showers I have given Cook's classification for nine showers in the first part of Table 2. Except for the ι Aquarids these are the same as those given by Ceplecha (1967). It is probably not a coincidence that the first four showers listed lack an associated comet. It has been suggested that Class B showers come from comets so small that they have lost all their outer icy layers and the nucleus that remains, if it still exists, is like a very small asteroid, difficult to detect. It should be added that showers of Class C_2 definitely favour orbits with greater aphelion distances than those of classes A, B or $A+C_1$. In the last-named type of shower we are evidently sampling material from two different levels in the parent comet, perhaps a result of some major fragmentation in the past history of the complex.

In the second part of Table 2 I have listed mean shower data from the precisely reduced Super-Schmidt meteors. The bulk densities "ρ" are those given by Verniani (1967). The progressive-fragmentation index "χ" is a logarithmic quantity that depends on the amount by which the increase in deceleration along a meteoroid's path exceeds the theoretical, single-body, increase in deceleration. This difference from single-body theory is generally attributed to progressive fragmentation of the meteoroid. Optical evidence of this effect is seen in photographs taken with an occulting shutter as blending "B" where, towards the end of the meteor path, the segments produced by the shutter are extended in length or blend completely due to the light of the trailing dust particles. The progressive-fragmentation index "χ" and the qualitative figure for blending "B" show reasonably good correlation and are listed in Table 2 as given by Jacchia et al. (1967). Of particular note are: the high density of the Geminid meteoroids, though these seem to crumble as readily as the Perseids and the S ι Aquarids; the resistance to crumbling of the lower-density Taurid meteoroids; and the exceptional character on all counts of the Giacobinid meteoroids. It should be remarked here that, although the actual figures given for the Giacobinids in Table 2 depend on only 2 Super-Schmidt meteors, there is good support for the exceptional character of these meteoroids from other photographic evidence (Jacchia et al., 1950).

Verniani (1973) has summarized the information on densities from 5759 meteors recorded on the Harvard Radio Meteor Project. These data give mean bulk densities of 0.8 g cm^{-3} for meteoroids of mass 10^{-4} to 10^{-2} g, with a higher density portion among the meteoroids that have small aphelion distances from the sun. The mass range as published by Verniani has been adjusted in the light of new information on ionizing efficiency (Cook et al., 1973a; Früchtenicht and Becker, 1973). These results, when compared with those for the brighter photographic meteors, indicate a trend towards higher densities for smaller masses, and this trend is confirmed by a study of microcratering on lunar rocks and high-altitude particle collecting (Millman, 1975). A model of fragile, porous structures in gram sizes breaking up into smaller, more solid particles at 10^{-5} gram sizes, fits in with the observed density trend.

The meteoroids associated with fireballs, such as those recorded by the Prairie Network in the mass range 10^2 to 10^7, do not seem to fit any simple pattern of physical structure (McCrosky et al., 1971). Theoretical bulk densities for these objects range from 0.1 to 1.5 g cm^{-3} with a mean value near 0.5 g cm^{-3} (McCrosky and Ceplecha, 1970), and on this basis the only recovered meteorite from the Prairie Network, Lost City, would have been considered of low density. Actually, Lost City had a bulk density of 3.7 g cm^{-3}. If initially these fireball meteoroids entered the atmosphere with appreciably flattened shapes, and moved flat side forward, the theoretical densities, which assume a nearly rounded shape, could be too small by factors of 5 to 8. Some fireball data, but not all, can be explained by this assumption. Do flattened shapes extend to smaller masses? We have little information on this at present until we reach the mass range of 10^{-3} to 10^{-15} g. Here, microcratering on lunar rocks indicates that the particles are not either platelets or needles (Hörz et al., 1975).

CHEMICAL COMPOSITION

Chemical abundances are normally given either by relative numbers of atoms of the various elements or by the relative weights of the elements. In discussing solids it is usual to employ the latter method, and only abundances by weight will be used here. Since the interplanetary dust is part of the solar system it is logical to discuss the composition of this dust in relation to that of the entire system. One of the most recent compilations of the mean chemical composition of the solar system is given in a review paper by Cameron (1973). The relative abundances of the 25 commonest elements from this paper have been plotted in Figure 2 on a logarithmic scale, normalized to 6.0 for Si. H and He account for over 98 per cent of the total mass in the solar system while the next eight elements, down to S, make up the greater part of the remainder.

Meteorites are the only samples of interplanetary material for which we have detailed chemical information. Of the total of recorded meteorite falls 84 per cent are of the type known as chondrites. The mean composition of the commonest type of chondrites, the olivine-hypersthene chondrites, is plotted in Figure 2 as given by Mason (1962, 1965). Mason has pointed out that if you add 30 per cent more Fe, Ni and Co (also plotted in Figure 2) to the abundances of the hypersthene chondrites you have almost exactly the average composition of the other large group of chondrites, the olivine-bronzite chondrites. The relative elemental abundances, estimated for both the solar system and for meteorites, are not independent since the latter have been used, along with other data, to determine the former. For the common elements the chief difference between solar-system and meteorite abundances is the depletion in meteorites of the light volatiles H, O, C N and S, and of the inert

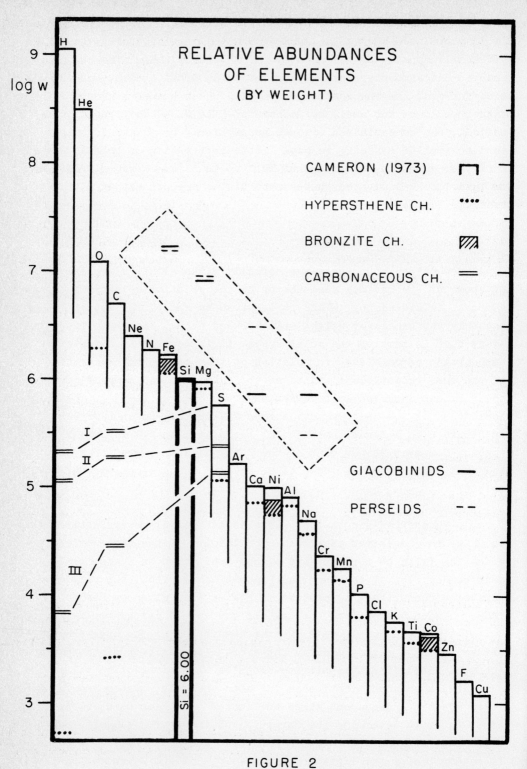

FIGURE 2

gases He, Ne and Ar. In this connection one small group of meteorites, the carbonaceous chondrites, that account for only 4 per cent of the falls, is of interest. These meteorites have similar relative abundances to the olivine-bronzite chondrites, except for higher abundances of H_2O, C and S. Mean abundances for H, C and S relative to Si in the three classes of carbonaceous chondrites, I, II and III, as given by Mason (1971), are shown in Figure 2.

As noted earlier, most of the meteoroids we observe as meteors are probably cometary fragments and are so fragile that they do not reach the earth in pieces of any size. We have no a-priori reason to assume that their elemental composition will match that of the meteorites. Early meteor spectroscopy (Millman, 1963) gave qualitative information on the chemical composition of the cometary meteoroids and now a total of some 18 or 19 elements have been identified in meteor spectra, including 16 of the first 22 plotted in Figure 2. Absent are He, Ne, Ar and S, P, Cl. Unfortunately it is difficult to derive quantitative abundance data from a photometric study of meteor spectra since the atomic-line and molecular-band radiation are produced by collisional excitation, and the required collisional cross sections are largely unknown for the physical conditions under which the light of meteors is produced. Thanks to work by Savage and Boitnott (1973) it has been possible to find relative abundances for the elements Fe, Mg, Ca and Na in a few Giacobinid and Perseid meteors (Millman, 1972a, 1972b). Since these cannot be normalized to Si they are plotted in Figure 2 on a log scale with an arbitrary zero above the other abundance plots. Their significance is in the trend of abundance for these four elements, which corresponds closely with that for the olivine-bronzite chondrites and the carbonaceous chondrites.

No quantitative measures of the light volatiles are available as yet for meteor spectra, but H and O appear strongly. Data published by Ceplecha (1971) and Millman and Clifton (1975) show evidence for CN and possibly CH in both a bright fireball and in Geminid meteors, and the presence of C provides a good explanation for the peculiar spectrum of a meteor observed with the image-orthicon equipment (Cook et al., 1973b). Thus, there is certainly a strong suggestion that carbon is present in many of the meteoroids that produce visible meteors.

It is significant that, after the β Taurid meteor shower, associated with Comet Encke, Goldberg and Aikin (1973) found an enhancement of the ions of Na, Mg, Si, K, Ca, Cr, Fe and Ni at a height of 114 km, in abundances that agreed closely with chondritic meteorites. The residue in a microcrater found in a plate exposed on Skylab IV also showed abundances of Fe, Si, Mg, Ca, Ni, Cr and Mn similar to carbonaceous chondrites (Brownlee et al., 1974).

We can summarize the current knowledge of the average chemical composition of meteoroids in a broad mass range near one gram as apparently quite similar to the carbonaceous and bronzite chondrites for elements heavier than number 10, and with the light volatiles H, C, O probably present in appreciable quantities.

A few particles (less than one per cent) that behave like iron meteoroids or asteroidal fragments have been detected (Halliday, 1960; Griffin, 1975; Cook et al., 1963). Iron and sodium are two of the commonest elements in meteor spectra, but in a systematic program of meteor photography Harvey (1973) has found a few examples of iron free and sodium deficient meteoroids. However, these are exceptions to a rule of basic uniformity in most meteor spectra, the general character of which is governed primarily by velocity in the atmosphere and original mass.

DISCUSSION

The fluxes listed in this paper lead to a total of roughly 30 metric tons of interplanetary dust swept up by the earth each day integrated from the smallest particles up to masses of 10^3 g (Millman, 1975), and this order of magnitude is confirmed by the meteoritic residue found in lunar soils (Anders et al., 1973). For larger masses there are still too many uncertainties in both observation and theory to make a good quantitative estimate possible. Most of the material observed in meteor streams moves in orbits similar to, or identical with, periodic comets but a minor fraction may be associated with unusual earth-crossing asteroid orbits (Cook, 1973b; Southworth and Sekanina, 1973). There is no marked difference in physical parameters between the stream meteoroids and the background meteoroids and the observed mass distributions are consistent with a gradual dispersion from the streams into the background. The entire complex seems to consist primarily of rather fragile, low-density particles but with a significant high-density component among those particles in the mass range 10^{-4} to 10 g with orbits lying inside that of Jupiter. Since various erosional and fragmentation processes are active in the meteoritic complex it can only be maintained if there is a continuous source for this material. The disintegration of comets perturbed into short-period orbits by the major planets seems adequate to satisfy this requirement (Whipple, 1967). What information we have concerning the chemical composition of the meteoroids is consistent with this hypothesis.

REFERENCES

Anders, E., Ganapathy, R., Krähenbühl, U. and Morgan, J.W. 1973 Moon 8, 3.
Brownlee, D.E., Tomandl, D.A., Hodge, P.W. and Hörz, F. 1974 Nature 252, 667.
Cameron, A.G.W. 1973 Space Sci. Rev. 15, 121.
Ceplecha, Z. 1967 Smithsonian Contrib. Astrophys. 11, 35.
----- 1968 Smithsonian Astrophys. Obser. Spec. Rep. No. 279.
----- 1971 Bull. Astron. Inst. Czech. 22, 219.
Clifton, K.S. 1973 J. Geophys. Res. 78, 6511.
Cook, A.F. 1973a Smithsonian Contrib. Astrophys. No. 14.
----- 1973b Evolutionary and Physical Properties of Meteoroids (eds. C.L. Hemenway, P.M. Millman and A.F. Cook) NASA SP-319, U.S. Government Printing Office, Washington, D.C., p. 183.
Cook, A.F., Jacchia, L.G. and McCrosky, R.E. 1963 Smithsonian Contrib. Astrophys. 7, 209.
Cook, A.F., Forti, G., McCrosky, R.E., Posen, A., Southworth, R.B. and Williams, J.T. 1973a Evolutionary and Physical Properties of Meteoroids (eds. C.L. Hemenway, P.M. Millman and A.F. Cook) NASA SP-319, U.S. Government Printing Office, Washington, D.C., p. 23.
Cook, A.F., Hemenway, C.L. Millman, P.M. and Swider, A. 1973b Evolutionary and Physical Properties of Meteoroids (eds. C.L. Hemenway, P.M. Millman and A.F. Cook) NASA SP-319, U.S. Government Printing Office, Washington, D.C., p. 153.
Dohnanyi, J.S. 1970 J. Geophys. Res. 75, 3468.
Elford, W.G. 1967 Smithsonian Contrib. Astrophys. 11, 121.
Früchtenicht, J.F. and Becker, D.G. 1973 Evolutionary and Physical Properties of Meteoroids (eds. C.L. Hemenway, P.M. Millman and A.F. Cook) NASA SP-319, U.S. Government Printing Office, Washington, D.C., p. 53.
Goldberg, R.A. and Aikin, A.C. 1973 Science 180, 294.
Griffin, A.A. 1975 J. Roy. Astron. Soc. Canada 69, 126.
Halliday, I. 1960 Astrophys. J. 132, 482.
Harvey, G.A. 1973 Evolutionary and Physical Properties of Meteoroids (eds. C.L. Hemenway, P.M. Millman and A.F. Cook) NASA SP-319, U.S. Government Printing Office, Washington, D.C., p. 131.
Hawkins, G.S. 1963 Smithsonian Contrib. Astrophys. 7, 53.
Hawkins, G.S. and Upton, E.K.L. 1958 Astrophys. J. 128, 727.
Hörz, F., Brownlee, D.E., Fechtig, H., Hartung, J.B., Morrison, D.A., Neukum, G., Schneider, E., Vedder, J.F. and Gault, D.E. 1975 Planet. Space Sci. 23, 151.
Jacchia, L.G., Kopal, Z. and Millman, P.M. 1950 Astrophys. J. 111, 104.
Jacchia, L., Verniani, F. and Briggs, R.E. 1967 Smithsonian Contrib. Astrophys. 10, 1.
Jacchia, L.G. and Whipple, F.L. 1961 Smithsonian Contrib. Astrophys. 4, 97.
Kaiser, T.R. 1961 Ann. Géophys. 17, 60.
Kresáková, M. 1966 Contrib. Astron. Obser. Skalnaté Pleso 3, 75.
Mason, B. 1962 Meteorites, John Wiley and Sons, New York, N.Y., p. 151.
----- 1965 Am. Museum Novitates, No. 2223.
----- 1971 Meteoritics 6, 59.

McCrosky, R.E.   1968   Smithsonian Astrophys. Obser. Spec. Rep., No. 280.
McCrosky, R.E. and Ceplecha, Z.   1970   Bull. Astron. Inst. Czech. $\underline{21}$, 271.
McCrosky, R.E. and Posen, A.   1961   Smithsonian Contrib. Astrophys. $\underline{4}$, 15.
McCrosky, R.E., Posen, A., Schwartz, G. and Shao, C.-Y.   1971   J. Geophys. Res. $\underline{76}$, 4090.
McIntosh, B.A. and Millman, P.M.   1970   Meteoritics $\underline{5}$, 1.
McIntosh, B.A. and Šimek, M.   1969   Canadian J. Phys. $\underline{47}$, 7.
Millman, P.M.   1963   Smithsonian Contrib. Astrophys. $\underline{7}$, 119.
----- 1967   The Zodiacal Light and the Interplanetary Medium (ed. J.L. Weinberg) NASA SP-150, U.S. Government Printing Office, Washington, D.C., p. 339.
----- 1970a   J. Roy. Astron. Soc. Canada $\underline{64}$, 187.
----- 1970b   Space Res. $\underline{10}$, 260.
----- 1972a   Nobel Symposium No. 21, From Plasma to Planet (ed. A. Elvius) Almqvist and Wiksell, Stockholm, p. 157.
----- 1972b   J. Roy. Astron. Soc. Canada $\underline{66}$, 201.
----- 1973   Moon $\underline{8}$, 228.
----- 1975   The Dusty Universe (eds. G.B. Field and A.G.W. Cameron) Neale Watson Academic Pubs. Inc., New York, N.Y., p. 195.
Millman, P.M. and Clifton, K.S.   1975   Canadian J. Phys. $\underline{53}$, 1939.
Millman, P.M. and McIntosh, B.A.   1964   Canadian J. Phys. $\underline{42}$, 1730.
Naumann, R.J. and Clifton, K.S.   1973   Evolutionary and Physical Properties of Meteoroids (eds. C.L. Hemenway, P.M. Millman and A.F. Cook) NASA SP-319, U.S. Government Printing Office, Washington, D.C., p. 45.
Nilsson, C.S. and Southworth, R.B.   1968   Physics and Dynamics of Meteors (eds. L. Kresák and P.M. Millman) D. Reidel Pub. Co., Dordrecht-Holland, p. 280.
Savage, H.F. and Boitnott, C.A.   1973   Evolutionary and Physical Properties of Meteoroids (eds. C.L. Hemenway, P.M. Millman and A.F. Cook) NASA SP-319, U.S. Government Printing Office, Washington, D.C., p. 83.
Southworth, R.B. and Sekanina, Z.   1973   NASA Contract Rep., NASA CR-2316, Nat. Tech. Information Service, Springfield, Va., U.S.A.
Verniani, F.   1967   Smithsonian Contrib. Astrophys. $\underline{10}$, 181.
----- 1969   Space Sci. Rev. $\underline{10}$, 230.
----- 1973   J. Geophys. Res. $\underline{78}$, 8429.
Whipple, F.L.   1967   The Zodiacal Light and the Interplanetary Medium (ed. J.L. Weinberg) NASA SP-150, U.S. Government Printing Office, Washington, D.C., p. 409.

METEOROID DENSITIES

B.A. Lindblad

Lund Observatory
S-222 24 Lund, Sweden

Introduction

The composition and density of meteoroids has been the subject of a number of investigations in recent years. Analysis of photographic meteor data have been presented by Jacchia et al (1967), Verniani (1964, 1967, 1969), Benyuch (1968, 1973, 1974), McCrosky and Ceplecha (1969) and others. The density of radio meteors have been studied by Verniani and Hawkins (1965) and Verniani (1966). In the present paper the dependence of meteoroid density on the type of the orbit is discussed.

Luminosity- and drag equations

It is generally assumed that the luminous flux of a meteor is at any instant proportional to the kinetic energy of the ablated meteor atoms. The luminosity equation may then be written

$$I = -\frac{\tau}{2} v^2 \frac{dM}{dt} \qquad (1)$$

where I is the luminous flux, τ the luminous efficiency coefficient and M and v are mass and velocity of the body. Present knowledge about the luminous efficiency coefficient has been summarized by Verniani (1965, 1967). If the photographic observations give intensity and velocity as a function of time, a photometric mass can be determined from

$$M = \frac{1}{\tau} \cdot \frac{2}{\bar{v}^2} \int_{t_E}^{t} I \, dt \qquad (2)$$

where t_E is the time at which the meteor luminosity ends (assumed zero mass), and $\bar{v}$ is a value of v at some instant between t_E and t.

The drag equation may be written in the form

$$\frac{dv}{dt} = -\frac{C_D}{2} \cdot A_o \cdot \rho_m^{-2/3} \cdot v^2 \cdot \rho_a \cdot M^{-1/3} \qquad (3)$$

where C_D is the drag coefficient, A_o the shape factor, ρ_m the meteoroid density, ρ_a the atmospheric density and M the dynamic mass of the meteoroid.

The drag coefficient, shape factor and meteoroid density are physical properties of the meteoroid body. It is convenient to introduce a characteristic constant C defined as

$$C = \frac{C_D}{2} \cdot A_o \cdot \rho_m^{-2/3} \qquad (4)$$

The drag equation may then be written

$$-\frac{M^{1/3}}{v^2} \cdot \frac{dv}{dt} = C \cdot \rho_a \qquad (5)$$

For a given point on a photographic meteor trail it is in principle possible to determine v and $\frac{dv}{dt}$. For the mass M we may substitute the photometric mass from eq. (2). The atmospheric density ρ_a may be taken from the US Standard Atmosphere (1962). The characteristic constant C may then be computed for any given point on the meteor trail. Finally, if the drag coefficient and shape factor are known, the meteoroid density may be computed from eq. (4).

Mean meteoroid density

Jacchia et al (1967) have analyzed 413 precisely reduced atmospheric trajectories of meteors photographed by the Harvard Super-Schmidt cameras. For determining a mean meteoroid density 350 meteors with several measured decelerations each were selected. A weighted logarithmic average over each trail and over the 350 objects gave log C = 0.603. Jacchia et al assumed C_D = 2.2, A_o = 1.5 and thus from eq. (4) obtained for the mean bulk density of the precisely reduced photographic meteors $\bar{\rho}_m$ = 0.26 g cm^{-3}. Stream meteors and sporadic meteors, studied separately, yielded the same density value. Verniani (1967, 1969) has summarized the information on meteoroid densities from the Super-Schmidt meteors. The analysis suggests that the meteoroids are of a low-density, porous structure. This is in agreement with the generally accepted model of a comet (Whipple 1951) which predicts low-density meteoroids.

Individual meteoroid densities and dimensions

Jacchia et al (1967) give for each meteoroid a quantity $\Delta \log \rho_{corr}$, which is related to the meteoroid density ρ_m by the equation

$$\log \rho_m = -1.5 \, \Delta \log \rho_{corr} + \log \bar{\rho}_m \qquad (6)$$

where $\bar{\rho}_m$ is the mean meteoroid density of the sample. Using the weighting procedures of Jacchia et al (1967) we computed individual densities for all meteors for which reliable values of deceleration and fragmentation were available. It is obvious that individual values of ρ_m per se have little significance and that only group averages merit any degree of confidence. Mean values of log ρ_m for 11 meteor streams of short period are plotted in fig. 1 versus the mean inverse semi-major axis of the stream orbit. Datum points for the Virginids, Northern Iota Aquarids and Draconids are based on two meteors each and are of low significance. Fig. 1 shows a progressive increase in the mean density of a stream meteoroid as the size of the orbit decreases.

In a second study all short period stream and sporadic meteors were combined and grouped in intervals of 0.05 $(a.u.)^{-1}$ in inverse semi-major axis. The weighted mean log density for each group is plotted in fig. 2. Again we see a progressive change in meteoroid density with $1/a$.

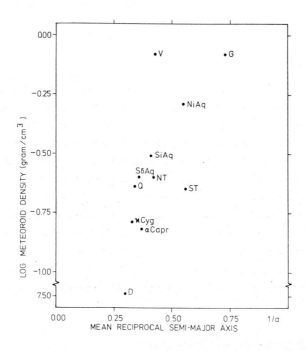

Fig. 1. Mean log meteoroid density g/cm^3) vs mean inverse semi-major axis $(a.u.^{-1})$ for 11 streams of short period.

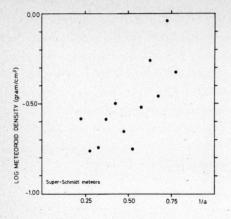
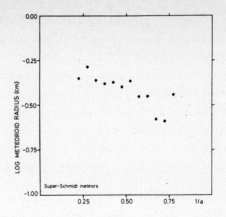

Fig. 2. Mean log meteoroid density (g/cm^3) vs inverse semi-major axis (a.u.$^{-1}$). All meteors with orbits of short period.

Fig. 3. Mean log meteoroid radius (cm) vs inverse semi-major axis (a.u.$^{-1}$). All meteors with orbits of short period.

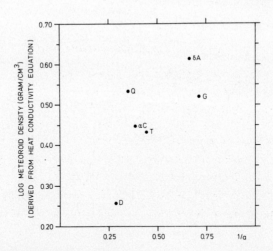

Fig. 4. Log meteoroid density (g/cm^3) as derived from the equation of heat conductivity vs inverse semi-major axis (a.u.$^{-1}$) for six short period streams.

Mean log meteoroid mass for the selected Super-Schmidt meteors was plotted against $1/a$. There was no appreciable variation in the photometric mass over the range of $1/a$ values studied by us. Knowing meteoroid mass and density it is in principle possible to compute an effective particle "radius" for the meteoroid. Mean particle "radius" is plotted against $1/a$ in fig. 3. The mean log meteoroid radius varies by a factor of about two over the range of $1/a$ studied by us.

Meteoroid densities determined from the equation of heat conductivity

The equation of heat conductivity may be written in the form

$$T_B = \frac{\Omega}{2} v_\infty^{5/2} \rho_a (\lambda \rho_m c\, b \cos z_R)^{-1/2} \qquad (7)$$

where T_B is the surface temperature of the meteoroid, Ω = heat transfer coefficient, v_∞ = no atmosphere velocity, ρ_a = air density, λ = heat conductivity of the meteoroid, ρ_m = density of the meteoroid, c = specific heat of the meteoroid, b = air density gradient and z_B = zenith distance of the radiant. The suffix B refers to the beginning point of the luminous meteor trail. The equation of heat conductivity has been discussed by Levin (1961), Ceplecha and Padavet (1961) and Benyuch (1968). The equation may be written as

$$\log \frac{2T_B}{\Omega} + \frac{1}{2} \log (\lambda \rho_m c\, b) = \log \rho_a^B + \frac{5}{2} \log v_\infty - \frac{1}{2} \log \cos z_R \qquad (8)$$

The right hand member, which includes only direct measureable quantities, is denoted k_B. The k_B parameter has been extensively used in statistical studies by Ceplecha (1958, 1966, 1967).

The meteoroid density may be determined from eq. (8) provided the parameters T_B, k_B, Ω, λ, c and b are known. This approach to density determination has been used notably by Benyuch (1968, 1973, 1974). In fig. 4 the mean density is plotted against $1/a$ for 6 short period streams listed by Benyuch (1974). A primary feature of fig. 4 is a dependence of density on $1/a$ of a similar nature to that found in fig. 1.

The use of eq. (8) for determination of density has been commented on by Kruchinenko (1969), Kruchinenko and Tryashin (1970) and Kolomiets (1971). The "mineralogical densities" computed by Benyuch are of an order of a magnitude higher than the "bulk" densities derived from the drag and luminosity equations. The reason for this discrepancy is not fully understood.

Discussion

In order to get a sample of meteoroids with a similar evolutionary history we studied only meteoroids for which $a < 5.20$ a.u. These objects may be considered related to, and possibly remnants of, short period comets of the Jupiter family. The dispersion in the orbital elements of short period meteor streams has been studied by Lindblad (1972, 1974). The dispersion increases with $1/a$, i.e. very short period orbits generally represent more dispersed and therefore presumably older meteor streams.

Fig. 1, 2 and 4 show that meteoroid density increases with decreasing size of the orbit. The progressive increase in density with $1/a$ may be ascribed to differences between the conditions under which the meteoroids were formed. Meteoroids in very short period orbits may have originated in short period comets, and the now observable members of stream may consist of inner, and more dense comet core material. The composition and mass may be the same for all short period meteoroids, but the difference in pressure to which the different layers of the parent comet have been subjected is responsible for the spread in particle density.

References

Benyuch, V.V., Vestnik Kiev, Gos. Univ. Ser. Astron., 10, pp. 51-58, 1968.
Benyuch, V.V., Astron. Vestnik, 7, pp. 21-29, 1973.
Benyuch, V.V., Astron. Vestnik, 8, pp. 96-101, 1974.
Ceplecha, Zd., B.A.C., 9, pp. 154-159, 1958.
Ceplecha, Zd., B.A.C., 17, pp. 96-98, 1966.
Ceplecha, Zd., Smithson. Contr. Astrophys., 11, pp. 35-59, 1967.
Ceplecha, Zd. and Padavet, V., B.A.C., 12, pp. 191-195, 1961.
Jacchia, L., Verniani, F. and Briggs, R., Smithson. Contr. Astrophys., 10, No. 1, 1967.
Kolomiets, A., Astron. Circ., No. 620, 1971.
Kruchinenko, V., Astron. Circ., No. 510, 1969.
Kruchinenko, V. and Tryashin, S., Astron. Circ. No. 566, 1970.
Levin, B., Physikalische Teorie der Meteore und die meteoritische Substanz in Sonnensystem, Akademie-Verlag, Berlin, 1961.
Lindblad, B.A., Meteor and Asteroid Streams, in From Plasma to Planet, Proc Nobel Symposium No. 21, 359, Almqvist & Wiksell, Uppsala, 1972.
Lindblad, B.A., The System of Short Period Meteor Streams, in Asteroids, Comets and Meteoric Matter, Proc. IAU Coll. 22, Bucharest, 1974.
McCrosky, R. and Ceplecha, Zd., Smithson. Astrophys. Obs. Spec. Rep., 305, 1969.
USA Standard Atmosphere, US Gov. Printing House, Washington, 1962.
Verniani, F., Nuovo Cimento, 33, pp. 1173-1184, 1964.
Verniani, F., Smithson. Contr. Astrophys., 8, pp. 141-172, 1965.
Verniani, F., J. Geophys. Res., 71, pp. 2749-2761, 1966.
Verniani, F., Smithson. Contr. Astrophys., 10, pp. 181-195, 1967.
Verniani, F., Space Science Reviews, 10, pp. 230-261, 1969.
Verniani, F. and Hawkins, G., Harvard Radio Meteor Proj. Res. Rep. No. 12, 1965.
Whipple, F., Astroph. Journ., 113, pp. 464-474, 1951.

4.3

POSSIBLE EVIDENCE OF METEOROID FRAGMENTATION IN INTERPLANETARY SPACE
FROM GROUPING OF PARTICLES IN METEOR STREAMS

Vladimír Porubčan
Astronomical Institute
Slovak Academy of Sciences
Bratislava, Czechoslovakia

The clustering of micrometeoroid impact events on satellite-borne detectors (see e.g. Hoffmann et al., 1975) poses a question whether similar phenomena can be identified as well among larger particles observable as meteors. In fact, there is a strong conviction maintained by many observers that meteors within the streams are observed to be clustered in pairs or larger groups more frequently than one could expect from chance distributions. The rate of the dispersive effects indicates that the lifetime of any group of meteor particles is very limited. If real, the groups must be due to recent fragmentation of larger meteoroids.

Analyses of the visual observations from this point of view have been applied to several streams. The Andromedids were examined by Kleiber (1888) who concluded that about 14 % of them appeared in pairs. However, later analysis of his data with different sampling intervals revealed contradictory results (Kresák and Vozárová, 1953). The Leonids were studied by Millman (1936) who obtained negative results for this stream. Simultaneous observations of the Perseids on a long base-line in 1950 and 1951 indicated that the shower was composed of separate meteor clouds of different size (Savrukhin, 1951). Results slightly in favour of a non-random grouping of the Perseids were obtained by Millman (1936), as well as by Subbotin and Agazdanova (see Astapovich, 1958). On the other hand, a thorough analysis of the Perseids 1952 by Kresák and Vozárová (1953) showed that the distribution of meteors within this stream was essentially at random.

Analyses based on radio measurements are much more conclusive and nine studies, obtained at different stations with different instrumental equipment, have been concerned with this problem. A common result of all these analyses, except for those by McCrosky (1957) and Wylie and Castillo (1956), was the absence of grouping of meteors over a random level (Bowden and Davies, 1957; Briggs, 1956; Shain and Kerr, 1955; Poole, 1965; Porubčan 1968).

As for the Poisson distributions, in all cases of sampling intervals between 30 and 0.1 seconds the probability of the distribution being at random (resulting from the chi-square test) was $p > 0.20$. For the distributions of the time intervals between successive echoes, the probability was everywhere greater than 0.10 except for the two distributions obtained by Poole (1965), one of $p = 0.03$ and the other of $p < 0.01$. This was probably caused by the timing accuracy being inconsistent with the high frequency of meteors. Corrections on this effect may be quite considerable in similar cases (Porubčan, 1968).

Of the two investigations claiming the reality of grouping, the first, reported by Wylie and Castillo (1956; see Bowden and Davies, 1957), is lacking accurate quantitative data. The authors have found a significant excess of 30 second sampling intervals containing five or more echoes, over the number expected from the Poisson distribution. Bowden and Davies (1957) believe that the equipment of Wylie and Castillo was not capable of distinguishing satisfactorily between fluctuating echoes and close groups, and this has led to an excess of spurious groups in their results.

The second positive evidence has been put forward by McCrosky (1957), who had analysed the maxima of the Perseids and Geminids 1948 and a sporadic night from May 1949. The number of meteors in one-second and half-minute intervals was compared with that predicted by the Poisson distribution. A very low value of $p = 0.0013$ was obtained for the bulk of data at one-second sampling intervals, which suggested that there was a significant excess of observed pairs and triplets. However, if McCrosky's data are treated for each night separately, in 30 minute periods, the median values of the probabilities for the Perseids, Geminids and the sporadic night are found to be 0.47, 0.38 and 0.41, respectively in agreement with a random distribution.

All the preceding analyses, except for the early work of Kleiber on the Andromedids, refer to the observations of permanent meteor showers, i.e. to stream structures at their middle and late evolutionary stages. For all these streams of considerable dispersion (high age), the result seems to be definitely negative. Therefore it appeared desirable to apply a similar analysis on a shower of recent origin, the Leonids 1969, where the conditions of the dispersion process at the earliest evolutionary phase may be different.

Excellent data on the 1969 Leonid display were obtained by the Springhill Meteor Observatory high-power radar around the shower maximum on November 17 (8:30 - 9:55 UT, 14160 echoes). According to the low-power radar data by McIntosh (1971), this maximum had a higher proportion of short-duration echoes than that of 1966. Three methods of analysis were used (Porubčan, 1974). First, the frequency distributions of time intervals between successive echoes were compared with that expected for randomly appearing echoes, represented by an exponential law. In the second method the data were distributed into 0.1 and 1 second intervals, and the number of intervals containing n echoes (n = 0,1,2,...) was compared with that resulting from the Poisson distribution. The third method was based on the distribution of the slant--range differences between pairs of successive echoes.

To eliminate the steep progressive changes of meteor rates, the data were divided into five-minute sets and, moreover, those around the maximum were combined into sets of approximately equal one-minute rates. As the antenna of the Springhill meteor radar is omnidirectional, the observed region was confined to a narrower zone by a slant-range limitation of the echoes ($R \leq 200$ km).

The deviations of the observed distributions from both the exponential distribution of time intervals and the Poisson distribution of meteor numbers are entirely consistent with random fluctuations both for the ascending and descending branch of the shower activity. However, there is an excess in a short period around the maximum (8:55 - - 9:15 UT), both for the total data and for a sample of $R \leq 200$ km. Median values of the probability of a random distribution being as irregular as observed are less than 0.01 for the bulk of data, irrespective of the methods used. For the sample of $R \leq 200$ km, the Poisson distribution yields $p = 0.01$, and the time distribution $p = 0.27$. The latter, relatively high value is apparently due to mutual blending of the echoes in the densest part of the record; according to a rough estimate the loss amounted to 20 % at the time of the maximum. As the Poisson distributions for one-second sampling intervals gave negative results, the dimensions of the non-random groups are small; according to the distributions $\Delta t / \Delta R$ they may be up to 40 km in diameter.

An analysis of the Poisson distributions for 0.1 second sampling intervals around the shower maximum shows that at least 10 % of the population is associated in close groups. The thickness of the layer where clustering occurs is comparable with the diameter of the Earth.

This finding may be indicative of a fragmentation process continuing after the release of meteoroids from their parent comet, in the central region of the stream which is most densely populated. A confirmation of this effect in observations of similar showers would be of considerable interest. Unfortunately, occasions of meeting the dense core of a meteor stream of recent origin are exceedingly rare.

The author is greatly indebted to Dr. B.A. McIntosh, Herzberg Center for Astrophysics, NRC of Canada, for the unique radar records of the 1969 Leonid display used in the present study.

REFERENCES

Astapovich, I.S. (1958). Meteornye yavleniya v atmosfere Zemli, Moskva, 360.
Bowden, K.R.R. and Davies, J.G. (1957). J. Atmos. Terr. Phys. 11, 62.
Briggs, B.H. (1956). J. Atmos. Terr. Phys. 8, 171.
Hoffmann, H.J., Fechtig, H., Grün, E. and Kissel, J. (1975). Planet. Space Sci. 23, 215.
Kleiber, J. (1888). Astron. Nachr. 118, 345.
Kresák, L. and Vozárová, M. (1953). Bull. Astron. Inst. Czech. 4, 128.
McIntosh, B.A. (1971). In: Evolutionary and Physical Properties of Meteoroids, IAU Colloquium 13 = NASA SP-319, 193.
McCrosky, R.E. (1957). Bull. Astron. Inst. Czech. 8, 1.
Millman, P.M. (1936). J. Roy. Astron. Soc. Can. 30, 338.
Poole, L.M.G. (1965). Bull. Astron. Inst. Czech. 16, 364.
Porubčan, V. (1968). Bull. Astron. Inst. Czech. 19, 316.
Porubčan, V. (1974). Bull. Astron. Inst. Czech. 25, 353.
Savrukhin, A.P. (1951). Astron. Cirk. Kazan, 121, 9.
Shain, C.A. and Kerr, F.J. (1955). J. Atmos. Terr. Phys. 6, 280.

4.4
THE HELIOCENTRIC DISTRIBUTION OF THE METEOR BODIES AT THE VICINITY OF THE EARTH'S ORBIT

V.V. Andreev, O.I. Belkovich, V.S. Tokhtas'ev
Engelhardt Astronomical Observatory, Kazan/U.S.S.R.

A determination of velocity distributions of sporadic meteor bodies and distributions of their radiants over the celestial sphere corrected for the motion and attraction of the earth is one of the main problem of the meteor astronomy. An analytical model of these distributions has certain interest.

There are now three traditional ground-based methods of meteor observations: visual, photographic and radar ones. All these methods yield strongly different apparent velocity and radiant distributions. The differences on the one hand may be interpreted as due to their own distinctive features for different observed mass ranges. On the other hand the differences may be caused by the methods of observations.

All methods of observations deal with different sides of the meteor phenomenon. The intensity of light by visual observations, the linear electron density by radar, the intensity of a meteor track at the plate in photographic observations are registered. Thus one has to obtain the distributions for particles over certain mass ranges in order to compare results of observations.

We have assumed the following form of the relation of the ionisation coefficient ß with the velocity:

$$ß = ß_o (v - v_o)^{3.5}.$$

Similar relations exist for the luminous efficiency for visual and photographic meteors:

$$\tau = \tau_o' (v - v_o)^{0.35}, \qquad (v \leq 16 \text{ km/sec}),$$
$$\tau = \tau_o'' (v + v_o)^{-1}, \qquad (v \geq 16 \text{ km/sec}).$$

v_o = the threshold velocity, below this velocity ionisation and luminosity are practically negligible.
$v_o = 8,8$ km/sec, $ß_o = 8,4 \times 10^{-7}$; $\tau_o' = 1,15 \times 10^{-13}$ and $\tau_o'' = 5,7 \times 10^{-12}$.
The geocentric velocity and flux density distributions over the celestial sphere for photographic and radar meteors were derived taking into account the above mentioned modified relations. No significant differences in the distributions were found in the results.

The heliocentric distributions of velocities and radiants were derived from geocentric ones. The formula of astronomical selection has been used in the derivation which is different from Öpik's one. The radiant density in the antapex direction exceeds the density in the apex direction by approximately 10^3 times.

4.5 FIREBALLS AS AN ATMOSPHERIC SOURCE OF METEORITIC DUST

Zdeněk Ceplecha
Astronomical Institute
Czechoslovak Academy of Sciences
251 65 Ondřejov Observatory

This note emphasizes the importance of very bright fireballs as a powerful source of meteoritic dust in the upper atmosphere and on the Earth's surface. Recently we obtained a direct observational evidence when we photographed the Šumava Fireball, an object of -22 absolute stellar magnitude, on Dec. 4, 1974 at $17^h 57.5^m$ UT. This is the brightest object on the European Network photographic records during almost 12 years of systematic fireball observations. We obtained 4 all-sky-camera records of the Šumava Fireball from two Czech and one German station of the Network. The most important combination for the trajectory computations proved to be station No. 5 at Ondřejov (Astronomical Institute of the Czechoslovak Academy of Sciences) and station No. 65 at Bernau (Max-Planck-Institut für Kernphysik, Heidelberg). The results are given in Table 1. At Ondřejov we also obtained 4 spectral records of the Šumava Fireball; the most detailed record has a dispersion of 5 Å/mm.

The fireball began the luminous trajectory at a height of 99 km with a velocity of 27 km/s. A very steep increase of brightness took place starting at 82 km height. The maximum brightness of -22 stellar absolute magnitude was achieved at 73 km height. The maximum was a flare of long duration: 14 additional shorter flares were observed: four of them were extremely intense, occurring at heights of 73, 68, 65 and 61 km. The meteor ended its luminous trajectory at 55 km height with a small flare and with almost unchanged initial velocity. The trajectory was inclined by $27°$ to the horizon of the terminal point. The 90 km long luminous trajectory was traversed in 3.4 seconds.

Using the luminous efficiency given by Ayers et al. (1970) and by McCrosky (1973), the initial mass was computed as 200 metric tons. This might seem to be an excessively large value, but we note that the measured energy radiated within the panchromatic bandpass (from 3600 to 6600 Å) is 10^{19} erg (300 000 kWh). If the entire kinetic energy of the meteoroid would be radiated within the panchromatic bandpass, 3 metric tons are still needed to produce the Šumava Fireball. Taking the experimentally measured value of the luminous efficiency (between 1 and 2 % for this velocity), we see that 200 tons were necessary to produce the fireball. Owing to an insignificant change of velocity and a very high termination height the terminal mass had to be virtually zero. Any significant mass at a height of 55 km cannot move with velocity higher than 20 km/s without ablation being observed. The entire mass has ablated during the short interval of 3.4 seconds. The structure of the body was very fragile and it resembles the weakest known meteoroids of the Draconid shower. A striking difference between two photographic fireballs with the same velocity is demonstrated in Table 2, which compares the Šumava Fireball with a fireball photographed within the Prairie Network by McCrosky (1973), PN 40660 B. (This is the closest fireball in velocity and in inclination to the surface, which we could find among all PN and EN photographic records). The ratio of brightness is almost 5 orders of magnitude, still the terminal point of the fainter PN-fireball is 26 km lower!

Ceplecha and McCrosky (1976) have recently finished a thorough study of PN-fireballs. Three groups according to the structure of meteoroids were found: group I has the structural strength and density of ordinary chondrites; group II belongs to the carbonaceous chondrites; group III consists of weak material of cometary origin. Group III contains meteoroids of type IIIA, the "stronger" cometary material, and of type IIIB, the "Draconid type" cometary material. The number of fireballs observed in each group is about the same, i.e. about 1/3 of all observed fireballs are of type I, 1/3 of type II and 1/3 of type III. The Šumava Fireball is of type IIIB and the Prairie Network fireball 40660 B is of type I. The Šumava Fireball probably belongs to the meteor shower of Northern Chi Orionids. (Table 1).

The spectral records of the Šumava Fireball do not differ from spectra of fireballs of about -10 stellar magnitude. The emission line spectrum contains FeI, CaI, CaII, MgI, NaI, CrI, MnI, AlI (preliminary indentification). There is some hope that the 5 Å/mm spectrum might permit a study of the line profiles, namely for the brightest feature, the sodium D-doublet. Iron is the most important radiator: more than 2/3 of the observed lines belong to it; multiplet 15 of FeI is the strongest one, not much fainter than the sodium D-lines.

Let us now study the contribution of such big bodies to the Earth's environment. The total area-time coverage of the European Network is about 4×10^{23} cm^{2}s up to now; (this is approximately equivalent to a 1 day coverage of the entire Earth's surface). If N is the number of bodies per cm^2 per second having 200 metric tons and more, we have $\log N \approx -23.6$. If we extrapolate the data given by McCrosky (1968;

Fig. 2 and 3) almost two orders towards bigger masses, we find log N ≈ -24.3, and doing the same in the energy plot, we find log N ≈ -24.5. This is a relatively good order of magnitude agreement showing that the large number of fireballs given by McCrosky (1968) having masses of up to 5 metric tons is still observed for objects with masses of 200 tons. It is important to emphasize that the Šumava Fireball is not a statistical freak: within the EN coverage we have photographed 4 objects of -17 magnitude, 2 objects of -18 magnitude and 1 object of -19 magnitude. These statistics are well in accordance with fireball data published by McCrosky (1968; Fig. 1).

If we adopt log N ≈ -24.0, we arrive at an impact rate of 1 kg/s for the whole Earth's surface from bodies of 200 tons and more, and at about 2 days between successive fireballs belonging to bodies of 200 tons or larger hitting the Earth's atmosphere. This infall produces a practically continuous source of fine dust between heights of 100 and 50 km. If we assume that the 200 tons will be converted into grains of 10^{-15} g, then one such particle per cm^2 is available every 5 seconds (2000 particles of 10^{-15} g/m^2s). If we assume that the fireball body is converted into particles of 10^{-6} g, we have one such particle per m^2 every six days. These values are close to the observed numbers of meteoritic dust particles in the Earth's atmosphere, which are known to be significantly higher (by orders of magnitude) than the numbers observed at 1 AU free space.

I think that there is enough evidence (obtained from photographic records of fireballs within the PN- and EN-network) that fireballs are the most important source of meteoritic dust in the Earth's atmosphere (and on the Earth's surface). It is concluded that bodies of dimensions from tens of centimeters to tens of meters, i.e. with masses from 10^5 to 10^9 grams are decisive in producing the atmospheric dust of meteoritic origin.

The physical details of the conversion of a large fireball mass into dust particles are beyond the scope of this contribution. Apart from direct fragmentation into fine dust, a significant fraction of the mass has to evaporate, since we observe most of the fireball light as the emission lines of hot gas. Subsequent condensation may play an important role in forming the tiny dust particles.

I am very much indebted to Dr. T. Kirsten, Mr. H. Haag and Mr. G. Hauth from the Max-Planck-Institut für Kernphysik, Heidelberg, for supervising and performing the observational work in Germany and for sending me the films exposed at German stations of the EN. I thank Mrs. M. Ježková and Mr. J. Boček for their help with the film measurements and computations.

REFERENCES

Ayers, W.G., McCrosky, R.E., Shao, C.-Y.: 1970, Smithson.Astrophys. Spec.Rep. 317.

Ceplecha, Z., McCrosky, R.E.: 1975, Fireball End Heights: A Diagnostic for the Structure of Meteoritic Material, in prep. for JGR

Lindblad, B.A.: 1971, Smithson.Contr.Astrophys., No. 12.

McCrosky, R.E.: 1968, Smithson.Astrophys.Obs.Spec.Rep., No. 280.

McCrosky, R.E.: 1973, private communication.

Table 1.: Šumava Fireball (EN 041274)

date	Dec 4, 1974	initial mass	2×10^8 g
time UT	17^h 57.5^m	terminal mass	virtually zero
max. brightness	-22 abs.stel.mag.	radiant α_R app.	$75°$
initial velocity	27 km/s	δ_R	$28°$
beginning height	99 km	shower (Lindblad)	Northern χ Orionids
λ	$14°38'24"$ E.Gr.	orbit a	2.0 a.u.
φ	$49°16'38"$	e	0.76
end height	55 km	q	0.47 a.u.
λ	$13°31'20"$ E.Gr.	Q	3.5 a.u.
φ	$49°08'10"$	ω	$282°$
inclination to the surface	$27°$	Ω 1950.0	$251.9°$
length	90 km	i	$2°$
duration	3.4 s	π	$174°$
		spectral records	FeI, CaI, CaII, MgI, NaI, CrI, MnI, AlI

Table 2.

fireball	EN 041274 Šumava	PN 40660 B
initial velocity	27 km/s	26.5 km/s
$\cos Z_R$	0.46	0.60
beginning height	99 km	93 km
terminal height	55 km	29 km
max. brightness	-22 abs.mag.	-10 abs.mag.
initial mass	200 000 kg	8 kg
terminal velocity	$\approx$ 26 km/s	6.5 km/s
terminal mass	$\approx$ 0	30 g
probable structure	very weak cometary material	compact stone
type	IIIB (Draconid type)	I (ordinary-chondrite type)

4.6   INTERPLANETARY DUST IN THE VICINITY OF THE EARTH

G.M. Teptin
Kazan State University/U.S.S.R.

The characteristics of the flux density of the interplanetary small particles in the vicinity of the Earth are considered. From the results of the analysis of the characteristics of the suspended dusty particles in the turbulent upper atmosphere the density of flux is obtained. The radar meteor measurements of the turbulent intensity in the upper atmosphere allow to evaluate the flux density of the interplanetary dust as a function of the magnitude of the particles at the intervals of the sizes from $\sim 1\mu$ to $\sim 0.05\mu$. The comparison with the results of different experiments shows a good agreement.

4.7   METEOR RADAR RATES AND THE SOLAR CYCLE

B.A. Lindblad
Lund Observatory, Lund/Sweden

Long-term variations in meteor radar rates, echo amplitudes, and meteor end-point heights have been observed. These variations appear to be controlled to a large extent by the solar cycle of activity. Exceptionally high echo rates in 1953, 1963 and 1972 coincide approximately with solar cycle minima. The long-term variations are explained in terms of a solar controlled change in the atmospheric density gradient at the meteor ablation level.

To be published in "Nature".

4.8 **EVOLUTION AND DETECTABILITY OF INTERPLANETARY DUST STREAMS**

Ľubor Kresák
Astronomical Institute
Slovak Academy of Sciences
Bratislava, Czechoslovakia

Sudden enhancements in responses recorded by micrometeoroid detectors flown on spacecrafts have been repeatedly attributed to encounters with streams of cometary debris similar to, or identical with, the meteor streams known from ground-based observations. For measurements made in the Earth-Moon environment, spacecraft effects, atmospheric fragmentation of larger particles, and possibly lunar ejecta can be misinterpreted as interplanetary streams. For deep space observations it is necessary to inquire whether a compact dust stream can persist under the dispersive and destructive effects which increase rapidly with decreasing particle size.

The reduction of solar attraction by direct radiation pressure is believed to produce a cutoff in the mass of the particles moving in circumsolar orbits - except for submicron grains where light scattering reduces the effect of radiation pressure. The size range of particles swept out of the solar system depends on the critical value of the radiation parameter $\beta = 0.93 \times 10^{-4} m^{-1/3} \varrho^{-2/3}$, where m is particle mass in grams and ϱ particle density in g cm^{-3}. There is a wide-spread misconception that the cutoff is governed by the condition $\beta = 1$ representing an equilibrium between gravitation and radiation forces. This applies only to fictitious particles of zero angular momentum. On the other hand, the critical value of β corresponding to emissions at perihelion (Harwit, 1963) is exceeded appreciably if the dust is released at a larger distance from the Sun.

The initial semimajor axis of the orbit of a dust particle released at a negligible velocity from a parent orbit of semimajor axis a_0 and eccentricity e_0 at a heliocentric distance r is

$$a = (1 - \beta) r a_0 (r - 2 \beta a_0)^{-1} \qquad (1)$$

Consequently, the critical value of β for a parabolic escape varies between $\frac{1}{2}(1 - e_0)$, if the particle is emitted at perihelion, and $\frac{1}{2}(1 + e_0)$, if it is emitted at the aphelion of the parent comet. While the dust emission is obviously most abundant near perihelion, each comet spends most of a revolution in remote parts of its orbit, the contribution from which may not be disregarded for short-period comets. In fact, we have clearcut observational evidence of strong cometary activity beyond the orbit of Jupiter, ranging from sudden outbursts, to splitting of the nuclei, and to the presence of conspicuous tails, including indirect evidence from the tracing backwards of tail structures observed near the Sun. A rough model assuming that the mass loss is proportional to the radiative energy input, i.e. to the rate of change of true anomaly, was used by us to compute the variation of the mass population index for several comet-meteor associations.

The result is a break in the mass distribution function, with a progressive depletion beginning at $m = 6.5 \times 10^{-12} Q^{-2}(1 - e_0)^{-3}$, e.g. at $m = 1.8 \times 10^{-9} Q^{-2}$ for P/Encke or $1.0 \times 10^{-7} Q^{-2}$ for P/Swift-Tuttle, the parent comet of the Perseids. In a mass range which is about three orders of magnitude for short-period comets, a linear log n / log m relation would only change its slope, until a final cutoff is produced by the absence of emission activity beyond a certain distance from the Sun. This limit is always above, but may be close to, $m = 0.8 \times 10^{-12} Q^{-2}$, i.e. in the region where scattering effects begin. Hence Soberman's conjecture (1971) that there is no cutoff at all may apply for some particle compositions, provided that the cometary emissions occur at large distances from the Sun. Maintaining fine dust within the solar system requires that the ratio a_0/r not be excessively large, so as to keep $\underline{a}$ positive for sufficiently high values of β. For comets of long period, and especially those coming from Oort's Cloud, the limited heliocentric region of activity moves both the break and the cutoff of the mass distribution curve into the size range of ordinary meteors. Hence no fine dust released from these comets remains in elliptic orbits, and a passage through the comet's tail is the only possibility of detecting a dust stream. For a nearly circular orbit of the parent body where the escape condition $\beta > \frac{1}{2}$ used by Zook and Berg (1975) applies the break and the cutoff coincide.

It is essential that the post-separation Poynting-Robertson spiralling begins from a starting orbit determined by the direct radiation pressure, with $a > a_0$, $\beta < \frac{1}{2} r a_0^{-1}$. This sets a definite lower limit of Poynting-Robertson lifetime, depending on the orbital elements of the parent body. The shortest time of inspiralling into the Sun is 1.4×10^4 yr for perihelion emissions from P/Encke, 1.9×10^3 yr for aphelion emissions from P/Encke, and as high as 1.0×10^6 yr for perihelion emissions from P/Swift-Tuttle. The role of the Poynting-Robertson effect seems to have been overestimated in most of the current comparisons of the competitive evolutionary processes. Due to the initial blowing off by direct radiation pressure, the time for complete inspiralling may become comparable with the destruction lifetime, and a considerable proportion of the particles may not survive.

Now, each reduction of the particle size by fragmentation or abrasion reinforces the radiative repulsion. For a particle of radius s eroded at a heliocentric distance r we have

$$d a = - \beta (1 - \beta)^{-1} (2 a r^{-1} - 1) a s^{-1} d s \qquad (2)$$

which may be compared with the inspiralling rate determined by Wyatt and Whipple (1950). The solar wind sputtering rates, as estimated by Wehner et al. (1963) and by Ashworth and McDonnell (1974) are entirely inadequate to compensate for the Poynting-Robertson drag. Rotational splitting (Radzievskij, 1954; Paddack, 1969) would require the bursting speed to be attained every 10^3 yr for the ejecta from P/Encke. The impact erosion rates estimated by Whipple (1967) would be sufficient to convert the inspiralling into a prevailing drift outwards for cometary debris of loose structure. The abundance of shower associations among ordinary meteors, combined with the rate of disintegration of periodic comets and the rate of perturbational dispersion of meteor streams, lends strong support to a high destruction rate, irrespective of the exact mechanism (Kresák, 1968). However, the model of McDonnell and Ashworth (1972) predicts a steep drop of the erosion rate below $m \sim 10^{-5}$g, which would permit inspiralling for most of the cometary dust starting from short-period orbits.

Just as the heliocentric distance of the point of separation determines the efficiency of the radiation pressure in changing the size of the starting orbit, the distribution of the eroding medium governs its efficiency in offsetting the inspiralling motion. For example, for P/Encke the same erosion rate is about 150 times more effective if applied at perihelion than at aphelion. Apart from the dispersion by planetary perturbations, which is accelerated by the rapid radiative dispersion in mean anomaly, it appears that compact streams of dust particles released from normal comets can never pass inside their orbits, as claimed by Alexander et al. (1970) for P/Encke or by

Berg and Gerloff (1971) for the two comet associations already criticized by Levin and Simonenko (1972).

It has been definitely established by optical and radio observations that all permanent meteor showers become less prominent with respect to the sporadic background as the particle mass decreases (Millman, 1970). There are three reasons to expect that this trend is maintained, or even increased, in the range of smaller dust particles detected on spacecraft : the depletion of smaller sizes by hyperbolic escape of the dust component emitted near perihelion; increasing revolution periods with decreasing size, which reduces their encounter frequency relative to their numbers; and a size-dependent effect of abrasion which makes the relative mass loss increase with decreasing mass. Therefore, showers of cometary dust should not be detectable in deep space, unless the probe enters a comet tail which still contains fresh hyperbolic ejecta. Such encounters, however, should be rare.

Recent in situ measurements of impact velocities and directions seem to be at variance with the identification of normal comets as a principal source of interplanetary dust. In this connection, two types of hypothetical parent bodies, unconfirmed by direct observation, deserve attention. If Harwit's (1967) interplanetary boulders revolve in orbits of low eccentricity in the region of terrestrial planets, a low total energy would protect even very fine dust from being ejected and from being destroyed before arriving close to the Sun. During the inspiralling phase the particles would cover a broad size range, and exhibit low impact velocities on space probes, perhaps 10 km s^{-1} and less at $r = 1$. At the terminal phase of escape, following a partial vaporization (Belton, 1967), the mass range should be relatively narrow, with geocentric velocities near 40 km s^{-1}. The erosion and light-scattering effects would act selectively according to the particle composition. A definite concentration to the plane of ecliptic would follow from a general correlation between inclinations and eccentricities of the parent bodies.

Another intriguing possibility is the existence of intermediate boulder-type products of disintegration in cometary orbits of very small perihelion distance. A unique example is the Kreutz group of sungrazing comets, where progressive disintegration of the nuclei seems to have produced at least one hundred separate comets (Kresák, 1966). The orbital ellipse of this system of comets may be occupied by numerous sizeable fragments which, at each sungrazing passage, can liberate dust without building up visible comas and tails. The dust would move away from the Sun at velocities exceeding 100 km s^{-1} at $r = 1$. Since the Kreutz group is apparently of recent origin (Marsden, 1967), one can speculate that there are similar older streams of cometary fragments which no longer contain active comets. These might represent a significant source of high-velocity particles with fluxes variable both in time and space. These streams may fail to show an ecliptical concentration, but may contribute to collisional interaction with the low-inclination low-eccentricity component. The orbits of the dust particles released at small heliocentric distances would be nearly the same as those of the particles of solar origin suggested by Hemenway et al. (1973), but the two sources would be distinguishable if reliable mass and velocity measurements were made.

REFERENCES

Alexander, W.M., Arthur, C.W., Corbin, J.D. and Bohn, J.L. (1970). Space Research 10, 252.
Ashworth, D.G. and McDonnell, J.A.M. (1974). Space Research 14, 723.
Belton, M.J.S. (1967). In : The Zodiacal Light and the Interplanetary Medium. NASA SP-150, p. 301.
Berg, O.E. and Gerloff, U. (1971). Space Research 11, 225.
Harwit, M. (1963). J. Geophys. Res. 68, 2171.
Harwit, M. (1967). In : The Zodiacal Light and the Interplanetary Medium. NASA SP-150, p. 307.
Hemenway, C.L., Erkes, J.W., Greenberg, J.M., Hallgren, D.S. and Schmalberger, D.C. (1973). Space Research 13, 1121.
Kresák, L. (1966). Bull. Astron. Inst. Czech. 17, 188.
Kresák, L. (1968). In : Physics and Dynamics of Meteors. IAU Symp. 33, p. 391.
Levin, B.Yu. and Simonenko, A.N. (1972). Kosmicheskie Issledovaniya 10, 113.
Marsden, B.G. (1967). Astron. J. 72, 1170.
McDonnell, J.A.M. and Ashworth, D.G. (1972). Space Research 12, 333.
Millman, P.M. (1970). Space Research 10, 260.
Paddack, S.J. (1969). J. Geophys. Res. 74, 4379.
Radzievskij, V.V. (1954). Dokl. Akad. Nauk SSSR 97, 49.
Soberman, R.K. (1971). Rev. Geophys. Space Phys. 9, 239.
Wehner, G.K., KenKnight, C.E. and Rosenburg, D.L. (1963). Planet. Space Sci. 11, 885.
Whipple, F.L. (1967). In : The Zodiacal Light and the Interplanetary Medium. NASA SP-150, p. 409.
Wyatt, S.P. and Whipple, F.L. (1950). Astrophys. J. 111, 134.
Zook, H.A. and Berg, O.E. (1975). Planet. Space Sci. 23, 183.

4.9 ON THE STRUCTURE OF HYPERBOLIC INTERPLANETARY DUST STREAMS

L. Kresák and E.M. Pittich
Astronomical Institute
Slovak Academy of Sciences
Bratislava, Czechoslovakia

The only type of concentration of cometary dust with a reasonable probability of being detected by deep-space probes, are the dust tails emanating from passing comets. Essentially all the dust released from long-period comets leaves the solar system on hyperbolic orbits, because the radiation pressure limit is high (corresponding to centimetre-sized grains emitted in the vicinity of the Earth's orbit by comets of the Oort's Cloud). For the short-period comets the dynamical conditions for retention of emitted particles within the solar system are much more favourable, but those which remain in circumsolar orbits tend to disperse rather rapidly (Kresák, 1976a).

The Earth and its artificial satellites can encounter the dust debris only from comets whose perihelion distance is less than 1 A.U. On the average two such comets are observed per year. The average frequency of perihelion passages is 1.4 yr^{-1} for long-period comets ($P > 200$ yr), 0.1 yr^{-1} for the intermediate group of the Comet Halley type (20 yr $< P <$ 200 yr), 0.4 yr^{-1} for comets of the Jupiter family (4 yr $< P <$ 20 yr), and 0.3 yr^{-1} for Comet Encke. The statistics of the long-period comets is less complete, but the bias is not particularly serious for $q < 1$ (Kresák, 1975b). The short-period comets tend to liberate much less dust per revolution and their average absolute brightness is much lower. The probability of penetrating the tail of a short-period comet is increased by their concentration to the plane of ecliptic, but their low inclinations at the same time imply lower impact velocities and fluxes, and hence higher requirements on the minimum detectable concentration of particles. Taking all this into account it can be inferred that the prospects of detecting dust debris from comets are much better for nearly parabolic comets.

The detectability of the direct dust tails has already been discussed, and predictions provided, by Poultney (1972, 1974). He concluded that each passage of a Pioneer 8/9 detector would only be expected to yield one detected particle, but that other observable effects would be produced by the injection of the dust into the atmosphere.

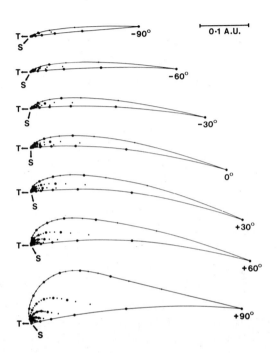

In order to get a clearer insight into the geometry and detectability of the dust tails, some model computations have been performed by the authors. The above figure shows the evolution of a dust stream produced by low-velocity emissions from a comet moving in a parabolic orbit of q = 0.5. The dots indicate the crossing points of selected syndynes and synchrones. The upper boundary is the syndyne occupied by particles of mass $m = 10^{-8}$g, assuming spherical shape and Whipple's (1967) mean density of photographic cometary meteors $\varrho = 0.44$ g cm^{-3}. For Verniani's (1973) value $\varrho = 0.8$ g cm^{-3}, obtained for faint radio meteors, this corresponds to $m = 3 \times 10^{-9}$g; for Ceplecha's (1967) value $\varrho = 1.4$ g cm^{-3} to $m = 10^{-9}$g; for stone of $\varrho = 3.2$ g cm^{-3} to $m = 2 \times 10^{-10}$ g; and for iron of $\varrho = 7.8$ g cm^{-3} to $m = 3 \times 10^{-11}$ g. Each following syndyne, proceeding downwards, refers to particle masses one order of magnitude higher, up to 10^4 times the first value. The lower boundary is the synchrone formed by dust emission at true anomaly $v = -150°$, i.e. at a heliocentric distance of 7.5 AU. Proceeding to the left, the synchrone step is $+10°$ in v, with solid circles marking each

30°. If the amount of dust liberated from the nucleus is proportional to the radiative energy input, it is also proportional to dv/dt, and equal amounts are confined between each neighbouring pair of synchrones at a given time. If the mass distribution of the particles is dn $\propto$ m^{-2}dm, a relation which is nearly satisfied for all permanent meteor showers in the visual range (Kresáková, 1966), each zone between two neighbouring syndynes includes an equal total mass, but a total number of particles differing by a factor of 10 increasing upwards. Instantaneous structures of the tail are plotted for seven values of v ($r \leq 1$), covering a period of 77.5 days; these are indicated on the right. The direction of motion of the comet T, defining the horizontal axis of the plot, and the direction to the Sun S are also shown.

In spite of a number of simplifications involved, some interesting inferences are possible. Since the dust tail occupies a thin layer, the point of its crossing by the detector determines uniquely the size (syndyne) and time of release (synchrone) of the particles encountered. For this reason no information on the mass distribution or time dependence of the emissions can be gathered unless the probe moves in the orbiting plane of the comet. The curvature and width of the dust tail rather than its length tends to increase during the apparition. The area occupied by the dust increases faster than the mass input, so that for the post-perihelion arc the probability of encountering the tail is greater but, on the average, the flux is the same. The sunward boundary of the tail is determined by the heliocentric distance at which the comet begins to emit the dust particles; the outer boundary by the smallest particle mass which can be detected. Since the spread area varies nearly as $m^{-2/3}$, the probability of encounter would tend to increase sharply with instrument sensitivity. On the other hand, if the thickness of the layer is proportional to the distance from the nucleus (due to the orthogonal component of the low emission velocities), the loci of uniform spatial concentration would nearly coincide with the synchrones in our model size distribution. An instrument with higher sensitivity for exploring the outer regions of the tail would detect relatively more sporadic particles which might mask the presence of the tail. Statistically, the encounter with the earlier emissions would be more likely. This implies that post-perihelion intercepts of the early pre-perihelion emissions, i.e. material from the time prior to a full development of the coma, has the highest probability of being recorded.

It should be emphasized that a single passage through a comet tail would not yield a reliable estimate of the total mass of the dust released, since emissions in bursts may produce an irregular synchrone pattern. In general, the probability of detecting a dust stream is low and depends on the sensitivity of the instrument. Unless a probe is intentionally launched into a comet tail, only very small particles should be detectable at chance encounters, and the flux enhancement should mostly vanish in the sporadic background. A passage through the region of large particles and high particle concentrations would require a close approach to the comet and the crossing of its orbital plane at a suitable position angle, which would occur very rarely.

REFERENCES

Ceplecha, Z. (1967). In : Meteor Orbits and Dust, Smithson. Contrib. Astrophys. 11 = NASA SP-135, p. 35.
Kresák, L. (1976a). This Volume.
Kresák, L. (1975b). Bull. Astron. Inst. Czech. 26, 92.
Kresáková, M. (1966). Contrib. Astron. Obs. Skalnaté Pleso 3, 75.
Poultney, S.K. (1972). Space Research 12, 403.
Poultney, S.K. (1974). Space Research 14, 707.
Verniani, F. (1973). J. Geophys. Res. 78, 8429.
Whipple, F.L. (1967). In : The Zodiacal Light and the Interplanetary Medium. NASA SP-150, p. 409.

4.10
EXPECTED DISTRIBUTION OF SOME OF THE ORBITAL ELEMENTS OF INTERSTELLAR PARTICLES IN THE SOLAR SYSTEM

O.I. Belkovich and I.N. Potapov
Engelhardt Astronomical Observatory, Kazan/U.S.S.R.

A two-dimensions distribution $p(e,q)$ of eccentricities e and perihelion distances q can be derived by means of the formula for the probability transformation:

$$p(e,q) = p(v,\alpha) \cdot \begin{vmatrix} \frac{\partial v}{\partial e} & \frac{\partial \alpha}{\partial e} \\ \frac{\partial v}{\partial q} & \frac{\partial \alpha}{\partial q} \end{vmatrix} \quad (1)$$

where v is the heliocentric velocity at infinity, α is the impact parameter. Assuming v and α are independent and $p(v) = C_1$, $p(\alpha) = C_2 \alpha^2$, we have

$$p(v,\alpha) = p(v)p(\alpha) = C_1 C_2 \alpha^2. \quad (2)$$

Here C_1 and C_2 are constants.

The well-known relations of celestial mechanics give

$$(3) \quad v^2 = \frac{e-1}{q}, \qquad \alpha^2 = \frac{e+1}{e-1}, \quad (4)$$

where q is in AU and v is in units of the earth's velocity.

From eq. (1) taking into account for eqs. (2) - (4) one can derive:

$$p(e,q) = \frac{C_1 C_2 \, q^{3/2} \, e(e+1)^{1/2}}{(e-1)^2} \quad (5)$$

Distributions $p(e)$ and $p(q)$ were derived from the integration of eq. (5):

$$p(e) = \begin{cases} C_3 e (e^2-1)^{1/2}, & (1 \leq e \leq e_o) \\ C_4 e (e^2-1)^{1/2} \left[\left(\frac{e_m - 1}{e-1} \right)^{5/2} - 1 \right] & (e_o \leq e \leq e_m) \end{cases} \quad (6)$$

e_o and e_m are found from the minimum and maximum values of the velocities and maximum size of the region near the sun that can be observed. In the real case $e_o \sim 1.0001$.

$$p(q) = C_5 q^{1/2}, \quad (7)$$

C_3, C_4 and C_5 are constants.

One can see from eq. (6) there is a strong concentration of the parameter e near 1.

Similar results were obtained for some other forms of the velocity distributions $p(v)$.

5 DYNAMICS AND EVOLUTION

5.1
SOURCES OF INTERPLANETARY DUST

Fred L. Whipple
Center for Astrophysics
Harvard College Observatory and Smithsonian Astrophysical Observatory
Cambridge, Massachusetts 02138

ABSTRACT

Attention is centered on cosmic dust measures made by sensors on Pioneers 8 and 9 in Earth-like orbits. The conclusion follows Zook and Berg that the particles are largely "β-meteoroids," interplanetary impact debris expelled by solar radiation pressure. An analysis of periodic comet orbits and comets observed during the missions failed to yield correlations, except possibly for debris from Comet Encke. A treatment of β-meteoroids from this stream is presented.

The solar-centered maximum impact rate is considered and indicates that formation of the observed β-meteoroids occurs more at solar distances greater than 0.5 AU rather than less.

The interesting variation of observed pulse heights with impact direction is analyzed and qualitatively explained in terms of a narrow distribution function of particle masses among the β-meteoroids with a maximum $\Delta \log N/\Delta \log m$ near or below 10^{-13} gm, in general agreement with lunar microcratering data. The mean pulse-height appears to increase when the mass sampling veers towards a minimum in the normal distribution at greater masses.

The observations appear consistent with the author's early conclusion that the interplanetary particles are largely of cometary origin, collision being the major destructive process.

INTRODUCTION

Investigations by Dohnanyi (1970, 1972, 1973) confirm the author's conclusion (Whipple, 1955, 1967) that meteoroids in the interplanetary complex (IPC) derive primarily from cometary debris and are mostly destroyed by mutual collisions. During the subsequent several years remarkable progress has been made in measuring the occurrence of smaller dust particles in the IPC by means of space probes and studies of microcraters on lunar material. Solids of small masses in the range 10^{-14}-10^{-12} grams produced by collisions move in unstable orbits because of their high surface-to-area ratio and consequent acceleration away from the Sun by radiation pressure. Thus their lifetimes in the solar system are extremely short compared to 10^4-10^5 years for the larger average particles in the IPC.

In this paper I concentrate on the cosmic dust measures made by the sensors on Pioneers 8 and 9 in Earth-like orbits about the Sun reported on by Berg and Grün (1973) (BG, hereafter) and McDonnell, Berg and Richardson (1975) (MBR, hereafter). They give data on 319 front-film-grid cosmic dust impacts over a period of seven years, combined for Pioneers 8 and 9. These sensors turn rapidly in the plane of the ecliptic and show a preponderance of impact events originating from particles in orbits with apparent radiants in the solar direction (see Fig. 1). In careful analyses based on both observational and theoretical material, BG and also Grün, Berg and Dohnanyi (1973) show that these events are not noise and are not caused by solar activity but measure true impacts of cosmic dust. The limiting impact function of mass, m, and velocity, v, of impact is determined by ground calibration to be $mv^{2.6} = 1.1 \times 10^4$ (cgs) (BG Fig. 5). The pulse height is logarithmically compressed in data transmission to $\log mv^{2.6}$.

The event rate in Fig. 1 increases rapidly from the antapex ($\phi = 180°$) direction of the space probe motion with respect to the Sun until a maximum is reached just past the solar direction at $\phi = 90°$. The rate decreases monotonically through the apex ($\phi = 0°$) to the antisun ($\phi = 270°$) direction and remains very low around to the antapex. Zook and Berg (1975) (ZB, hereafter) present a thorough analysis of this event frequency distribution as a function of spacecraft orientation, on the basis that the small particles are produced by collisions within the Earth's orbit among the IPC particles, and move outward in hyperbolic orbits at relatively high acceleration by solar radiation pressure. Since the masses required by the calibrations of BG at reasonable interplanetary velocities are consistent with particles in unstable orbits, the arguments by ZB appear to be entirely satisfactory in principle. Were the bodies in stable Earth-crossing orbits the numbers and velocities impacting from the general anti-solar direction should be about equal to those from the solar direction. The observations vehemently deny this expectation and support their premise. ZB call these particles beta-meteoroids because they are produced by the outward acceleration of solar radiation at sizeable values of beta, the ratio of solar light pressure to solar gravity, a quantity which is constant for a given particle as a function of solar distance because of the inverse square laws involved.

In the following sections I shall deal with four aspects of these observations (1) an effort to find direct correlations with comets or cometary streams of debris, (2) possible correlations with Comet Encke debris, (3) problems of the direction of impact, and (4) a discussion of the intriguing variation of pulse height with impact direction (see Fig. 1).

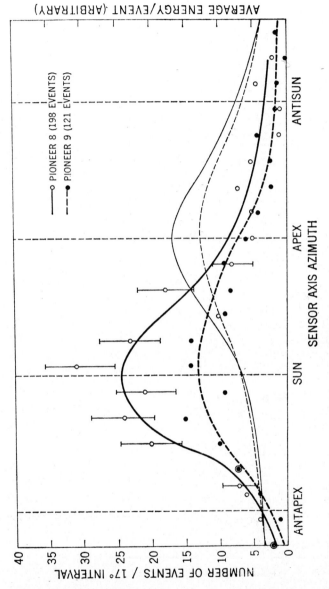

Fig. 1. The Azimuthal Distributions of 319 Front Film-Grid Events (coincidence) during 7 yrs. of Observation from Pioneers 8 and 9 (from Zook and Berg, 1975).

A SEARCH FOR COMETARY SOURCES IN BETA METEOROIDS

MBR find (Fig. 2) that the overall frequency of events for Pioneers 8 and 9, over seven years, peak sharply at heliocentric longitude $\lambda = 9°$ (~Oct. 1 in Earth position) with a deep minimum near $\lambda = 90°$, (~Dec. 20) a broad maximum near $\lambda = 190°$ (~Apr. 1) and another minimum near $\lambda = 280°$ (~July 1). It seemed worthwile to search among the periodic comets and among comets observed concurrently with the space experiments, for orbital characteristics in which the interaction of the associated meteoric streams with interplanetary debris might produce systematic collision areas that could lead to beta-meteoroids observed by Pioneers 8 and 9.

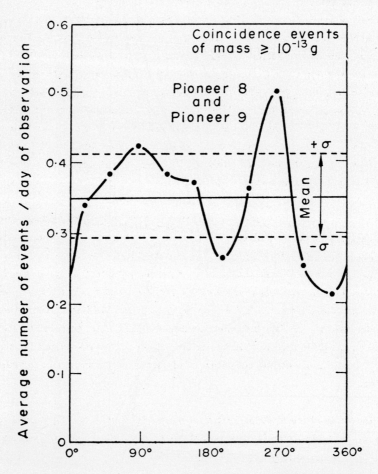

Fig. 2. FFG Events from Pioneers 8 and 9 as a Function of Heliocentric Position, 0° Occurring Jan. 1, 1968 (from McDonnell, Berg and Richardson, 1975).

I studied the orbits of 24 periodic comets with perihelion distance, $q < 1.1$ AU, plotting the positions of the orbital intersections with the plane of the ecliptic as well as the general motions. I calibrated the expected collisional effects in terms of the absolute magnitudes of the comets and their velocities near the plane of the ecliptic and at perihelion. No direct correlation appeared except for the case of Comet Encke, which will be discussed below. I made similar studies of the ten comets with perihelion distance less than 1 AU observed in the interval from late 1968 to the middle of 1972, comparing the actual positions of Pioneers 8 and 9 in their orbits about the Sun with those of the cometary orbits. Again no correlations of predictable event frequencies with actual concurrent comets appeared except for P/Encke.

DEBRIS FROM PERIODIC COMET ENCKE (TAURID METEORS)

There is strong evidence that P/Encke has been a major contributor to the IPC over several thousands of years (Whipple, 1940, 1955). An elliptical torus of meteoroids, with $q \sim 0.34$ AU and aphelion distance, Q, ~ 4 (+0.5-2.0) AU, is produced by the differential rate of nodal regression produced by Jupiter perturbations on P/Encke debris in orbits of different Qs. The lines of orbital apsides move slowly: Near perihelion the particles happen to have their maximum inclination to the ecliptic because of the nature of the perturbations. The inclination, i, varies cyclically from $\sim 4°$ to $\sim 16°$. Thus collisions among the P/Encke meteoroids and with other debris of the IPC should occur most frequently near perihelion to produce β-meteoroids. Perihelion occurs over a considerable range from $\lambda \sim 110°$ to $\sim 160°$, centering at $\lambda \sim 130°$. The latter is the zero point (vertical) in Fig. 3, the orbit of P/Encke being the lower ellipse section plotted.

The several curves plotted in Fig. 3 represent the motions of β-meteoroids after their production by collision near perihelion and their subsequent ejection by radiation pressure at values of β from zero (P/Encke orbit) to 100, as indicated by the numbers indicated in the diagram. Values of β above 5 must certainly be infrequent, but the plot of such motions adds to the esthetics of Fig. 3. The relative velocities of impact with the Earth and approximately with Pioneers 8 and 9 are indicated just within the Earth's orbit in Fig. 3, the unit of velocity being the Earth's orbital velocity 29.76 km/sec (see also Table 1). Actual impact velocities would generally be slightly smaller than calculated because I have not included an uncertain term varying $\sim \cos i$, representing reduction in velocity by collisions among the particles near perihelion.

Because the nodes of the P/Encke particles are not far from q at large values of i, the mutual collisions might lead to a fair fraction of particles with low i, which could intercept the Earth's orbit.

Of particular interest is the direction of impact, ϕ, on Pioneers 8 and 9 as a function of β for P/Encke β-meteoroids produced at perihelion. From Table I it appears that β-meteoroids produced by P/Encke would generally impact at $\phi < 90°$ (Sun direction towards apex) by a moderate angle. Particles produced at a solar radial distance r > 0.34 AU would impact with still smaller values of ϕ and with lower velocities.

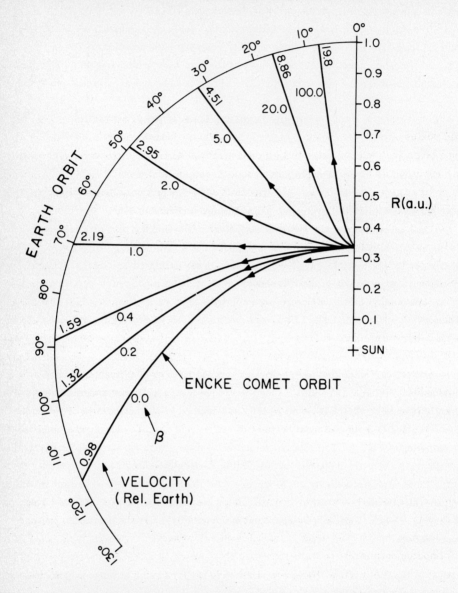

Fig. 3. Calculated Orbits of Particles Released from Periodic Comet Encke at Perihelion (on vertical line) with various Values of β (light-pressure/gravity) and Intercepting the Earth's Orbit with Relative Velocities as indicated (unit 29.76 km/sec).

Impacts with Pioneers 8 and 9 would occur at $\lambda = 110°$ ($\pm 20°$) + $90°$ ($\pm 30°$) for P/Encke β-meteoroids. This would center near the broad peak occurring about Apr. 1 in Earth orbit ($90°$ in Fig. 2).

Table I

Impact Directions and Velocities
for P/Encke β-meteoroids

β	0.0	0.2	0.4	1.0	2.0	5.0
φ (deg)	78	81	82	85	86	87
v (km/sec)	29	39	47	65	88	134

Otherwise I find that most comet-stream produced β-meteoroids would tend to occur near the broad minimum in Fig. 2 (positions 270° to 30°). I conclude that most comets do not provide much collisional activity in the IPC, and must represent minor fluctuations of particle density. Perhaps this is not surprising in view of the long time constant of 10^4-10^5 years for average collisional destruction in the IPC and the low space density in cometary meteoroid streams. Sekanina (1975) has identified and studied 275 streams among 19,698 radio-meteor orbits observed by the Radio Meteor Project at Havana, Illinois, Dec. 2, 1968 to Dec. 4, 1969. The masses involved are the order of magnitude of 10^{-4} to 10^{-3} g. (Cook et al., 1972) not much larger than the maximum of the logarithmic mass distribution function (Whipple, 1967), and therefore of a size to contribute significantly to collisions in the IPC. For only 10 of the 275 streams does Sekanina find the space density equal to or greater than the sporadic space density, and this for the very center of the streams, the most concentrated volumes. For only the Geminids and the Scorpiids-Sagittariids does the central space density exceed 10 times the sporadic space density. Two observed Comet Encke streams (Taurids and Southern Arietids) show central space densities twice that of the sporadic. Near perihelion their space densities must be many times greater. Considering the expected large volume of the Encke Comet stream and its expected higher concentration near perihelion, collisions there might well lead to a significant number of β-meteoroids observable from Earth's orbit.

PROBLEMS OF THE DIRECTION OF IMPACT

Zook and Berg have carried out a most thorough and careful study of the β-meteoroids and the Pioneer 8 and 9 impacts. I wish only to add a few details in this and the following section. The fact that the directions of impact center so strongly in the solar direction (φ = 90°) imposes constraints on the heliocentric distances at which the collisions occur, as ZB point out. I wish to amplify their point, holding to the constraint that φ = 90° and considering typical original orbits.

For a body moving about the Sun, subject only to the radial force of gravity the angular momentum, h, is a constant given by

$$r^2 \frac{d\theta}{dt} = h = [GM\, q(1+e)]^{1/2} , \qquad (1)$$

where θ is the true anomaly in the orbit, G the constant of gravitation, M the mass of the Sun, q the perihelion distance, and e the orbital eccentricity.

For a β-meteoroid released at zero velocity, the value of h remains constant, which means that the velocity perpendicular to the radius vector, $rd\theta/dt$, remains equal to h/r, regardless of β. For such a β-meteoroid to impact the spacecraft in Earth orbit at $\phi = 90$, the original value of h_0 must equal that of the Earth, h_e, given by

$$h_0/1 \text{ AU} = (GM/1 \text{ AU})^{1/2} = h_e/1 \text{ AU} \quad , \tag{2}$$

the Earth's velocity in circular orbit. If we start with orbits of varying q and Q (aphelion distance) with i = 0 and release β-meteoroids with zero velocity at q, then $1 + e = 2Q/(q+Q)$ and the constraint on q and Q of Eq. 1 and 2 becomes

$$q = Q/(2Q - 1) \quad . \tag{3}$$

The curve in Fig. 4 represents this relation between q and Q for β-meteoroids released at q and impacting a sensor in Earth-orbit from the direction of the Sun at $\phi = 90°$. Even for a parabolic initial orbit the minimum q is 0.5 AU. It is easily shown that release at solar distances above the curve in Fig. 1 will require larger values of q for the initial orbit, while a correction for finite inclination to the ecliptic will also increase the constrained q above the curve. The values of h for Pioneers 8 and 9 are $1.02 \, h_e$ and $0.93 \, h_e$, respectively. Thus the higher e and smaller q of Pioneer 9 does not much affect these arguments, nor the observations in Figs. 1 and 2 as compared to those of Pioneer 8.

Now it is true that the angle of acceptance of the Pioneer sensors extends to ±60° with reduced probability with angle (Grün, Berg and Dohnanyi, 1973), but the near symmetry around $\phi = 90°$ in Fig. 1, indicates a preponderance of impact directions very close to $\phi = 90°$ and the necessity of a significant fraction towards the antapex, where ϕ is considerably greater than 90°. It is easy to show that the latter impacts must come from bodies in orbits of high e, low i, and q more nearly 1 AU.

Thus I conclude that the β-meteoroids in large measure are produced in orbits with q > 0.5 AU, rather than from orbits with much smaller q, as proposed by ZB. Without access to the original detailed data, the further development of this argument here seems fruitless.

Note that Fig. 4 presents the velocities of impact for $\beta = 0.99$ in terms of the Earth's orbital velocity. For $\phi \gtrsim 90°$, the velocities of impact are not high, and will continue to decrease to less than 15 km/sec as ϕ increases from 90° to 180° at the antapex.

The inclusion of interstellar meteoroids could, of course, redirect this argument if they preponderately occur at $90° < \phi < 180°$.

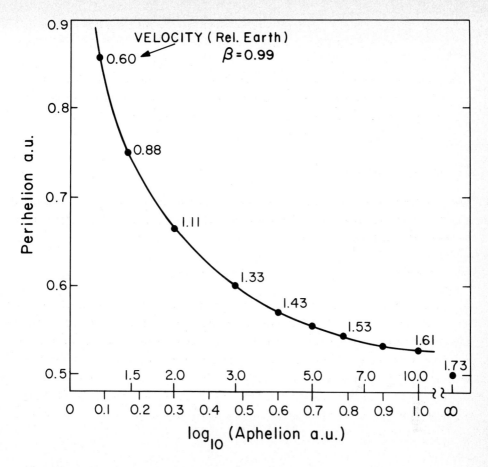

Fig. 4. The Relationship between Perihelion and Aphelion of Orbits such that β-meteoroids Released at Perihelion would Encounter the Earth Apparently from the Solar Direction (φ = 90°). The Unit of Velocity is 29.76 km/sec, for β = 0.99.

Hemenway (Hemenway et al., 1972) particles of high β and high velocity radially from the Sun would impact with φ's somewhat less than 90°.

THE VARIATION OF THE MEAN PULSE HEIGHT

A most significant observation of the Pioneer 8 and 9 FFG events by BG (Fig. 1), is the variation of the mean pulse height with the direction of impact. The mean pulse height, H, approximately proportional to energy ($mv^{2.6}$), increases steadily from the antapex to a maximum at the apex, being nearly symmetrical about the apex direction of impact. Note that the system averages the <u>logarithm</u> of H, not the linear quantity. For a uniform distribution of particle masses in the particle source, ⟨log H⟩ should be little dependent on the particle mass.

Suppose the cumulative number, N, of particle masses follows an inverse mass, m, power law over a considerable range in m:

$$N = a/m^b ,\qquad(4)$$

where a is a constant and b > 0.

The measured pulse height, H, for a velocity, v, of impact is given by

$$H = H_0 \, mv^{2.6} ,\qquad(5)$$

where $H_0^{-1} = 1.1 \times 10^4$ cgs (BG, Fig. 5).

Unit pulse height, the limit registered by Pioneers 8 and 9, will occur at a given velocity v_1, and mass m_1, given by $m_1 = 1.1 \times 10^4/v_1^{2.6}$ cgs. Thus at the given velocity for particles of mass, m,

$$H = m/m_1 ,\qquad(6)$$

and

$$\langle \ln H \rangle = \int_{m_m}^{m_1} \ln H \, \frac{dN}{dm} \, dm \bigg/ \int_{m_m}^{m_1} \frac{dN}{dm} \, dm \qquad(7)$$

where m_m is the maximum mass of the distribution function.

From equations 4, 6, and 7, setting $x_i = \ln m_i$ and $\epsilon = m_1/m_m$, we find

$$\langle \ln H \rangle = -\ln m_1 - bm_1^b (1-\epsilon^b)^{-1} \int_{x_m}^{x_1} x e^{-bx} \, dx ,\qquad(8)$$

and, finally, after integration and substitution of m_i in the limits

$$\langle \ln H \rangle = \frac{1}{b} - \epsilon^b \ln \epsilon \, (1-\epsilon^b)^{-1} .\qquad(9)$$

In the case of a considerable range in the mass distribution function $\epsilon \ll 1$, the second term in Eq. 6 becomes negligible, and $\langle \ln H \rangle \to 1/b$. Varying v changes m_1 inversely as $v^{2.6}$ but does not change $\langle \ln H \rangle$, unless v is so small that $m_1/m_m = \epsilon$ becomes appreciable.

In view of these considerations the large variation in $\langle \log H \rangle$ observed by BG demands a special distribution function of mass. The variation cannot be accounted for by a changing velocity in a uniform distribution function over a large range in m. As we shall see, this is a powerful additional argument for the "β-meteoroid" theory of ZB, following BG.

Let us now look to the studies of the mass distribution of particles in the 10^{-12} gm range as measured from lunar micro-cratering and beautifully summarized by Hörz et al. (1975). This notable consortium of scientists find strong evidence for a deep minimum in the $\Delta \log N/\Delta \log m$ slope of the mass distribution function near 10^{-8} gm, and a relatively sharp maximum in the neighborhood of 10^{-11}-10^{-13} gm (their Fig. 15).

To illustrate the effect on $\langle \log H \rangle$ of such a concentration of particle sizes in a narrow mass range, I have adopted an extreme artificial distribution of mass in the mathematical form

$$\frac{\Delta \log N}{\Delta \log m} = -\frac{1.1}{1 + 10 \log (m/m_0)} , \qquad (10)$$

where $\log m_0 = -13$.

Numerical integrations then lead to various functions on a $\log_{10}$ scale in terms of $\log m$ as shown in Fig. 5. The $\Delta \log N/\Delta \log m$ curve is Eq. 10 from which the $\log N$ curve is derived, beginning with $N = 1$ at $\log m = -11.5$. Other $\log N$ curves beginning at $\log m = -12.0$ and -12.5 are not shown to avoid confusion, but would be displaced vertically below the $\log N$ curve shown. The three $\langle \log H \rangle$ curves of the log (pulse height) assume the distribution function gives $N = 1$ at the three masses, $\log m = -11.5$, -12.0 and -12.5. The quantity $\langle \log H \rangle$ applies to a chosen value of m_1 on the abscissa.

The $\langle \log H \rangle$ curves of Fig. 5 show how the pulse height, averaged logarithmically, can vary with the critical value of m_1 for unit pulse height. A shift from a starting velocity v to a higher velocity will shift m_1 to a lower value and hence to the left in Fig. 5. Such a shift might either increase or decrease $\langle \log H \rangle$, depending upon the value of m_1 compared to m_0, and upon the starting value of m for log N. Usually, however, for a more physically acceptable mass distribution function the change should not be so marked.

To explain the pulse heights observed by BG (Fig. 1), I suggest that the growth in the pulse height from antapex to apex arises from β-meteoroids of increasing impact velocity, representing a decrease in mass following some $\langle \log H \rangle$ curve less steep than those of Fig. 5, but suggesting particles more massive than the peak of the $\Delta \log N/\Delta \log m$ curve. As the velocity increases beyond $\phi = 90°$, the Sun impact direction, $\langle \log H \rangle$ continues its increase. The rapid rise in $\langle \log H \rangle$ and rapid decrease in N near $\phi = 60°$, some 30° towards the apex from the Sun direction, must then arise from the superposition of another mass distribution function, not necessarily with a velocity increase, possibly a decrease, as suggested by the HEOS velocities by Hoffmann et al. (1974, 1975). Probably this is the tail end of the ordinary meteoroid distribution with a small value of b as indicated by the dip in the $\Delta \log N/\Delta \log m$ curve of Hörz et al. Towards the antisun direction, ϕ decreasing through 270°, the velocities are again lower, the masses higher and most of the material comes from the normal meteoroid mass distribution, β-meteoroids not making a significant contribution.

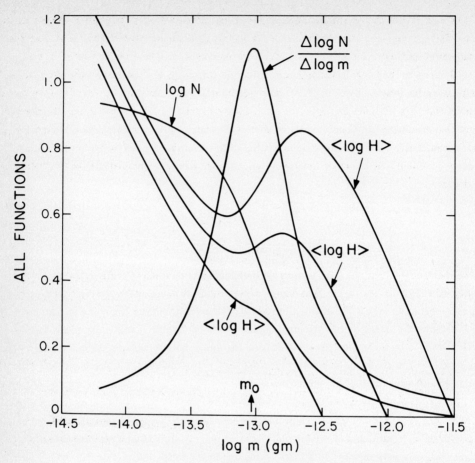

Fig. 5. log N (Cumulative) and Mean log H (Pulse Height)
Developed from $\dfrac{\Delta \log N}{\Delta \log m} = \dfrac{-1.1}{1 + 10 \log (m/m_0)}$
⟨log H⟩ Curves begin at log m (gm) = -11.5, -12.0 and -12.5.

This qualitative explanation falls short of a detailed explanation of the ⟨log H⟩ variation with direction, but strongly supports the β-meteoroid explanation by demanding a narrow special mass-distribution function for these particles, as would be expected from the limited range in mass for particles that can be greatly accelerated by solar radiation pressure.

From the Pioneer 8 and 9 FFG measures Rhee, Berg and Richardson (1974) find no significant change in particle concentrations with heliocentric distance between 0.76 and 1.08 AU and no concentration at 1 AU. From the same data Zook (1975) finds the spatial concentration of the parent bodies increasing with heliocentric distance. He develops a theory for the production rate and radial distance from the source of β-meteoroids.

The definitive treatment of the Pioneer 8 and 9 data should include not only the frequency of impact with heliocentric distance and possibly longitude, but also include the important data of impact direction and pulse height. Possibly meteor streams may also play a role in these distribution functions. Note that for β-meteoroids the velocity will probably increase with decreasing mass within the mass-distribution function. The $\langle \log H \rangle$ curves in Fig. 5 were derived on the assumption that for a given m_1 at H = 1, the velocity of the whole mass-distribution function is given by the value applicable to m_1. Thus among actual β-meteoroids a more complicated relationship will be found among $\langle \log H \rangle$, m_1 and the mass-distribution function. The value of $\langle \log H \rangle$ will undoubtedly not increase so rapidly with decreasing m_1 and increasing velocity as indicated in Fig. 5. Note also that the actual meteoroids will possess a velocity-distribution function at each mass, further complicating the theoretical analysis.

REFERENCES

Berg, O. E. and Grün, E., 1973, Space Research XIII, Akademie-Verlag, Berlin, 1047-1055.

Cook, A. F., Williams, J. T. and Shao, C.-Y., 1972, IN Meteor Research Program, pp. 83-89. Final Report Contract NSR 09-015-33, NASA CR-2109.

Dohnanyi, J. S., 1970, J. Geophys. Res., 75, 3468-3493.

Dohnanyi, J. S., 1972, Icarus, 17, 1-48.

Dohnanyi, J. S., 1973, Evolutionary and Physical Properties of Meteoroids. ed. Hemenway, Millman and Cooks, Wash. DC, NASA, Sp. Rep. 319, 363-374.

Grün, E., Berg, O. E., and Dohnanyi, J. S., 1973, Space Research XIII, Akademie-Verlag, Berlin, 1056-1062.

Hemenway, C. L., Hallgren, D. S. and Schmalberger, D. C., 1972, Nature, 238, 256.

Hoffmann, H. J., Fechtig, H., Grün, E. and Kissel, J., 1974, XVII COSPAR at Sao Paulo, Brazil.

Hoffmann, H. J., Fechtig, H., Grün, E. and Kissel, J., 1975, Planet. and Spac. Sci., 23, 215-224.

Hörz, F., Brownlee, D. E., Fechtig, H., Hartung, J., Morrison, D. A., Neukum, G., Schneider, E., Vedder, J. F., and Gault, D. E., 1975, Planet. and Spac. Sci., 151-172.

McDonnell, J. A. M., Berg, O. E., and Richardson, F. F., 1975, Planet. and Spac. Sci., 23, 205-214.

Rhee, J. W., Berg, O. E. and Richardson, F. F., 1974, Geophys. Res. Let., 1, No. 8, 345-346.

Sekanina, Z., 1975, CFA Preprint 313.

Whipple, F. L., 1940, Proc. Amer. Phil. Soc., 83, 711-745.

Whipple, F. L., 1955, Ap. J., 121, 750-770.

Whipple, F. L., 1967, "The Zodiacal Light and the Interplanetary Medium," NASA Sp. Rep. 150, Wash. DC, 409-426.

Zook, H. A., 1975, Private Communication.

Zook, H. A., and Berg, O. E., 1975, Planet. and Spac. Sci., 23, 183-203.

5.2   DYNAMICS OF INTERPLANETARY DUST AND RELATED TOPICS

Jan Trulsen
The Auroral Observatory
University of Tromsø
Tromsø, Norway

Abstract

The problem of the effects of mutual collisions for the dynamics of interplanetary dust particles and grains is reviewed. Collisions are shown to give a rather characteristic dynamical signature, the importance of these effects depending mainly on the mean free collision time and the degree of inelasticity. Although a few attempts to look for collisional effects in the solar system have been made, rather much work remains to be done before the problem is fully understood.

Introduction

The dynamics of small interplanetary bodies, grains and dust - in short the meteoritic complex - depend on the type and strength of perturbing forces acting. For the smallest, charged dust particles the Lorentz force will be decisive. For particles in the micron to the millimeter size range radiation pressure with the associated Poynting-Robertson effect and solar wind pressure are important. For larger bodies planetary perturbations have to be taken into account. The dynamical effects of these forces have been discussed to a considerable extent in the literature [1]. Recent progress has particularly been made regarding the radiation pressure effects through the application of Mie scattering theory.

The types of forces mentioned above are examples of continuously acting forces. Of a quite different nature is the impulsive force resulting from mutual collisions between members of the meteoritic complex. Such collisions have previously mostly been taken into account in order to explain the size spectrum of this complex [2]. Such studies concentrate on the fragmentational properties of high velocity impacts. The dynamical aspects of the collision process are grossly neglected by simply introducing the average impact velocity as a parameter in the model.

Since the dynamical effects of mutual collisions have been given little regard we shall here concentrate on this aspect, reviewing the model studies that have been done so far. Subsequently, the importance of the obtained results for the present day solar system will be discussed. This topic has only been discussed to a limited extent yet. Many questions remain unanswered. We do, however, hope to point out some of the characteristic dynamical signatures of mutual collisions so that the cause can be recognized when one is facing the effect.

The dynamical importance of mutual collisions was proposed by Alfvén [3], predicting that particles moving in neighbouring Kepler orbits around a central body would tend to be collisionally focused into a stream - a jetstream. This hypothesis constitutes a cornerstone of the cosmological theory of Alfvén and Arrhenius [4], jetstreams being considered the parent structure in which accreation of planets and satellites from smaller grains took place.

Insight into the physics of jetstreams can be gained by studying a simple model. Consider a circular jetstream with mass very much smaller than that of the central body so that selfgravitational effects are negligible. Let the mean free collision time in the stream be long compared to the orbital period. An individual grain in the stream will between collisions follow a Kepler orbit. Now compare the velocity components of this grain as it crosses the symmetry plane of the stream with the velocity $\sqrt{\mu/r}$ of an observer in a circular motion at the same distance. To lowest order in eccentricity e and inclination i the result is:

$$v_r = \sqrt{\frac{\mu}{a}}\, e \sin E$$

$$v_z = \sqrt{\frac{\mu}{a}}\, i \cos (E - \omega) \qquad (1)$$

$$v_\varphi - \sqrt{\frac{\mu}{r}} = \frac{1}{2}\sqrt{\frac{\mu}{a}}\, e \cos E .$$

Here a is the semimajor axis, E the eccentric anomaly and ω the
argument of pericentrum. As is easily seen the contribution from each
grain to the velocity spread in the stream in the azimuthal direction
is systematically down by a factor 2 relative to the contribution in the
radial direction. The velocity distribution in the stream is therefore
necessarily strongly non-Maxwellian.

From the kinetic theory of gases it is well known that elastic
collisions will tend to Maxwellize the velocity distribution. This
menas - as seen from (1) - that a balance between the average eccentricity and inclination of the individual orbits in the stream should be
established. However, the deficiency of the velocity spread in the
azimuthal direction compared to the radial direction can only be decreased by steadily increasing the average eccentricity. This means
that more and more grains are put onto hyperbolic orbits through collisions.

The conclusion is therefore that the stream configuration can
only be maintained in the presence of collisions with a sufficient degree
of inelasticity. The energy lost through collisions can in principle
be taken from two sources, from the potential energy of the stream resulting in a shrinking of the stream towards the central body, or from
the thermal motions in the stream. If thermal motions constitute the
main source of energy the mean eccentricity and inclination of the individual orbits in the stream must decrease with collisions. This is
equivalent to a focusing of the stream into its plane of symmetry while
the individual orbits at the same time are becoming more and more circular. In addition to this effect Alfvén also predicted the existence
of a corresponding radial focusing in the sence that also the radial
thickness of the stream should decrease with collisions under suitable
conditions.

From this qualitative discussion of basic jetstream physics
- which can also be repeated for eccentric jetstreams - we then proceed
to the more quantitative model studies.

Analytical model studies

The model studies that have been performed so far have all restricted themselves to the simple situation of a stream of negligible
total mass, consisting of grains of equal size, subject only to the

perturbing force due to mutual collisions and with a mean free collision time long compared to the typical orbital period. The Boltzmann equation:

$$\frac{\partial f}{\partial t} + \underline{v} \cdot \frac{\partial f}{\partial \underline{r}} + \underline{a} \cdot \frac{\partial f}{\partial \underline{v}} = I(f,f) \tag{2}$$

constitutes the natural framework for a quantitative discussion of this system. Here $f(\underline{r},\underline{v},t)$ is the distribution function of the stream particles in phase space, $\underline{a}$ is the acceleration of an individual grain due to the gravitational field of the central body. The nonlinear Boltzmann collision operator can be given in the general form [5]:

$$I(f,f) = \int d\underline{g}\, d\Omega\, \frac{d\sigma}{d\Omega}\, |\underline{g}|\{f(\underline{v} + \tfrac{1}{2}\underline{g}' - \tfrac{1}{2}\underline{g})f(\underline{v} + \tfrac{1}{2}\underline{g}' + \tfrac{1}{2}\underline{g})$$
$$- f(\underline{v})f(\underline{v} + \underline{g})\}\,, \tag{3}$$

where $\underline{g}$ and $\underline{g}'$ are the relative velocity vectors of two colliding particles before and after the collision, Ω is the scattering solid angle and $d\sigma/d\Omega$ is the scattering cross section as a function of relative velocity. The degree of inelasticity is determined by the functional relationship between $\underline{g}$ and $\underline{g}'$. For completely inelastic collisions $\underline{g}' = 0$ while $|\underline{g}'| = |\underline{g}|$ for elastic collisions. A more specific form of the collision operator for a particular collision model is given in [6].

Two types of expansion procedures for the Boltzmann equation have been attempted, both based on the assumption of a mean free collision time long compared to the orbital period. This author [6] makes use of a power series expansion in time of the distribution function. The collisional induced deviation $\delta f(\underline{r},\underline{v},t)$ of the distribution function from an initial state $f_o(\underline{r},\underline{v})$, chosen as a function of $\underline{r}$ and $\underline{v}$ only through constants of motion of the two-body problem, is for small enough times t given by:

$$\delta f(\underline{r},\underline{v},t) = t\left(\frac{\partial f_o}{\partial t}\right) + \frac{1}{2} t^2 \left(\frac{\partial^2 f_o}{\partial t^2}\right) + \ldots \,. \tag{4}$$

The time derivatives are determined from the Boltzmann equation. It can now be shown that the collisional induced change in the distribution function gives rise to an additional mass flux in the stream:

$$\delta(n\underline{U}) \equiv \int d\underline{v}\, \underline{v}\, \delta f = -\frac{1}{2} t^2\, \nabla \cdot \int d\underline{v}\, \underline{v}\, \underline{v}\, I(f_o, f_o) \,. \tag{5}$$

By calculating the direction of the flux vector $\delta(n\underline{U})$ at different points of a cross-section of the stream an idea of the initial dispersal or focusing of the stream can be gained.

The obvious weakness of this approach, as clearly demonstrated by the numerical simulations to be described below, is that (5) might only portray initial transients in the stream due to "improper" starting conditions. It is, however, felt that combined with the experience gained from the numerical simulations, the method could give valuable information of the dependence of stream dynamics on the choice of specific collision models. Such a study has not yet been done.

The other expansion approach to the Boltzmann equation by Baxter and Thompson [5] directs itself to the question of the existence of the radial focusing mechanism, as predicted by Alfvén. These authors consider both three and two dimensional streams — the stream particles being constrained to move in the same plane in the latter case. Since their results are similar for the two cases, we restrict ourselves to the simpler two-dimensional case.

Circular jetstreams are considered. It is assumed that the distribution function depending on $\underline{r}$ and $\underline{v}$ only through angular momentum L and eccentricity e, $f(L,e^2)$, remains a slowly varying function of L. Making use of a Taylor series expansion of f in L the authors then derive an equation for the evolution of the angular momentum distribution:

$$h(L) \equiv \int dr\, d\varphi\, dv_r\, f(L, e^2(r, v_r, L)) , \qquad (6)$$

which with the additional assumption $f(L,e^2) = h(L)\psi(e^2)$ takes the simple form:

$$\frac{\partial h}{\partial t} = D(L) \frac{\partial^2}{\partial L^2} h^2(L,t) . \qquad (7)$$

The diffusion coefficient is given as:

$$D(L) = \frac{\pi}{4} L^4 \int dg_r\, dg_\varphi\, de^2\, de'^2\, \frac{\partial(r, v_r)}{\partial(e^2, e'^2)} \psi(e^2)\psi(e'^2)$$
$$\sigma(\underline{g})|\underline{g}|[(g_r^2 - g_\varphi^2)\beta(\underline{g}) - g_\varphi^2 \alpha(\underline{g})] . \qquad (8)$$

Here $\alpha(\underline{g})$ and $\beta(\underline{g})$ are the average fractional energy loss and average fractional energy deflected in a collision with relative velocity $\underline{g} = (g_r, g_\varphi)$. These quantities are therefore measures of how much the relative velocity vector shrinks and twists in an average collision, respectively. Further, e and e' are the eccentricities of particles at (r, v_r, L) and $(r, v_r + g_r, L + r g_\varphi)$ respectively, while ψ, σ and the Jacobian determinant are all positive quantities.

Since the angular momentum distribution (6) is closely related to particle density in the stream, (7) tells that if $D(L) < 0$ then the density at density maxima where $\partial^2 n^2/\partial L^2 < 0$ will increase, leading to still more enhanced maxima. This phenomenon for which the term "negative diffusion" was coined would be the conjectured radial focusing. From (8) it is seen that an increasing inelasticity decreases $D(L)$ and therefore promotes such a focusing. The effect of energy deflection depends on the explicit form of $\beta(\underline{g})$ and scattering cross-section. One would normally expect a decreased energy deflection to promote radial focusing of the stream.

This conclusion is in qualitative agreement with the results of numerical simulations. The beauty of the result is, however, spoiled by the fact that the applied series expansion is valid only if the distribution function remains not only a slowly varying function of angular momentum but an even more slowly varying function of eccentricity. The latter requirement is clearly too restrictive.

Numerical simulation studies

Numerical simulations have by far given the best insight into stream dynamics yet [7]. Results obtained indicate that it is natural to divide the dynamical evolution of streams into two separate stages. The first of these seems to be almost independent of the particular choice of collision model. Starting from an arbitrarily prescribed initial state a rapid "thermalization" takes place during the first 1 - 2 mean free collision times. During this time the particle distribution adjusts itself such that an approximate balance between the mean square velocities in the radial and polar directions in the stream is set up. Since the mean square of these velocity components are determined by the distributions of eccentricities and inclinations, the width of these distributions adjust to each other during the first transient evolutionary stage.

This is demonstrated in figure 1 for four different simulations starting from the same initial state but with different degrees of inelasticity. The specific collision model employed will be described below. The width of the distributions of eccentricities and inclinations are represented by the average values of these quantities, <e> and <i>. In the present case we clearly has a rather "unbalanced" initial state. This type of effect is also predicted from the time expanded Boltzmann equation [6].

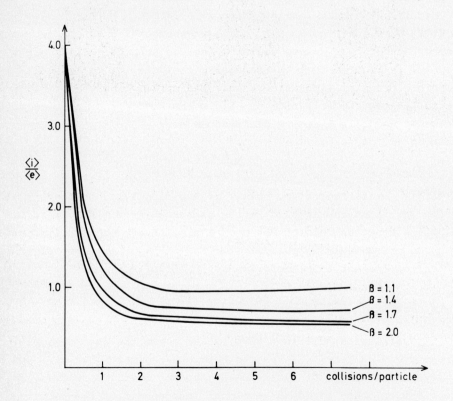

Figure 1. Evolution of the ratio of mean values of inclination and eccentricity for different degrees of inelasticity, starting from the same initial state.

A more detailed analysis shows that not only the width but also the shape of the distribution of inclinations and eccentricities are adjusted during the first stage. This is demonstrated in figure 2 for a circular jetstream. Shown are particle number histograms as a function of eccentricity at three different times for a particular simulation. Starting from a uniform distribution of eccentricities in the interval $e = 0.1$ to 0.3 the distribution can be approximated quite well with a Rayleigh distribution after only one mean free collision time. The same conclusion applies to the distribution of inclinations.

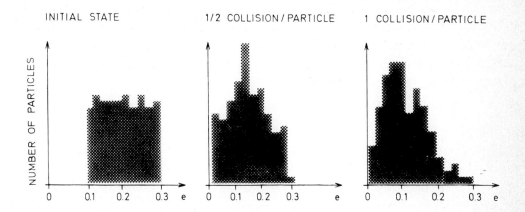

Figure 2. Evolution of the distribution of eccentricities.

A simple physical explanation of this fact can be given along the following lines. Consider a particle with specified orbital elements. In the first approximation the motion of this particle can be considered as a circular motion superimposed two harmonic oscillations, one in the radial direction with amplitude ae and one in the polar direction with amplitude ai. A large number of such oscillators interact via interparticle collisions. One would expect this system of interacting oscillators to evolve towards a state a equilibrium with Rayleigh distributed oscillator amplitudes. This would again give rise to Rayleigh distributed eccentricities and inclinations:

$$F(e) \sim e \exp(-\alpha e^2) \tag{9}$$

and

$$F(i) \sim i \exp(-\delta i^2) . \tag{10}$$

The result applies to circular jetstreams, that is for streams having an isotropic distribution of individual pericentrum vectors $\underline{p}$, $\langle \underline{p} \rangle = 0$. For elliptic jetstreams, $\langle \underline{p} \rangle = \underline{\eta} \neq 0$, the distribution of eccentricities takes the alternative form:

$$F(e) \sim e I_0(2\alpha\eta e)\exp(-\alpha(e^2 + \eta^2)) , \tag{11}$$

where $I_o(z)$ is the modified Bessel function of zero order. This is a Rice distribution. It is known from the theory of random signals to be the resulting amplitude distribution if a narrow-band gaussian noise is superimposed on a deterministic sinusoidal signal. Certain physical similarities exist between this system and our elliptic stream. For small degrees of anisotrophy, $|<\underline{P}>|/<|\underline{P}|> \ll 1$, the Rice distribution reduces to a Rayleigh distribution. In the opposite case, $|<\underline{P}>|/<|\underline{P}|> \approx 1$, a narrow gaussian distribution centered at $e = \eta$ results.

The properties of jetstream dynamics discussed so far seems to be essentially independent of the choice of collision model. For the subsequent and more slowly evolving stage this is not so. Here the degree of inelasticity and the amount of energy deflection in a typical collision seems to be of decisive importance. It is therefore necessary at this point to describe in some detail the different collision models that have been used. Both two and three-dimensional simulations were performed. The latter ones were all done with what will be called the β-model. It is described in terms of one parameter β by which help the degree of inelasticity can be varied. The pre- and post-collisional relative velocities $\underline{g}$ and $\underline{g}'$ are related by:

$$\underline{g}' = \underline{g} - \beta \underline{g} \cdot \underline{k}\underline{k} , \qquad (12)$$

where $\underline{k}$ is the unit impact vector, parallel to the line connecting the centres of two colliding particles at impact. The component of the relative velocity normal to $\underline{k}$ is left unchanged while the parallel component is reversed and diminished, β taking values in the range (1, 2). Elastic collisions correspond to $\beta = 2$. For $\beta = 1$ half the kinetic energy in the centre of mass system will be lost in an average collision.

The two-dimensional simulations were performed with different collision models. In addition to the β-model results for the snowflake model will also be reviewed. In this model the components of the post-collisional relative velocity parallel and perpendicular to the pre-collisional relative velocity is given by:

$$\begin{aligned} g'_{\parallel} &= g(1 - C \cos \theta) \\ g'_{\perp} &= B\, g \sin \theta \cos \theta \end{aligned} \qquad (13)$$

where θ is the angle between $\underline{g}$ and $\underline{k}$. By varying the parameters

C and B the amount of energy lost and deflected in collisions can be varied, respectively.

The simulations indicate that the evolution of the distributions of eccentricities and inclinations during the second stage depends mainly on the degree of inelasticity. When the inelasticity is sufficient, that is, when 30 - 40 per cent of the kinetic energy in the centre of mass system of two colliding particles is lost in an average collision, the widths of these distributions decrease with time. This means that the orbits of the individual particles will evolve towards circular orbits in the symmetry plane of the stream. If this requirement is not fulfilled the opposite evolution takes place. The average eccentricity and inclination then increase with time, more and more particles being put onto hyperbolic orbits and literally being "kicked off" from the central body. This is demonstrated in figure 3 which refers to the same set of simulations as discussed in figure 1. The β-values 1.1, 1.4, 1.7 and 2.0 correspond to average energy losses of 50, 40, 25 and zero per cent respectively. The initial increase in eccentricity is due to the "thermalization process" described above. After only 5 elastic collisions per particle an appreciable fraction of these particles belong to the tail of the distribution of eccentricities extending beyond $e = 1$ and are therefore "kicked off". It is important to note that any amount of energy loss will not bring about a focusing of the stream in its plane of symmetry. The inelasticity has to exceed a certain limit before the inherent tendency of elastic collisions to Maxwellize the velocity distribution is overcome. This limit is not exceeded for $\beta = 1.7$. The

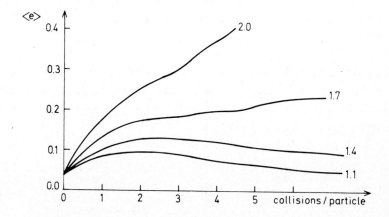

Figure 3. Evolution of the mean value of eccentricity for different degrees of inelasticity.

energy lost in this case is taken from the potential energy, the average value of 1/a increasing with time.

The final effect to be discussed is that of the radial focusing of the stream. This effect seems to depend critically on the average amount of energy deflected. The smaller the deflection of the relative velocity vector in an average collision, the better are the chances that a radial focusing will take place. With the β-model no radial focusing was observed in the three or two-dimensional simulations. The two-dimensional simulations with the snowflake model do, however, show this effect under favourable conditions, reductions in the radial width of the stream with a factor 2 to 3 having been observed. This does, however, require a rather small amount of energy deflected in the average collision. This is demonstrated in figure 4, where the spread of the

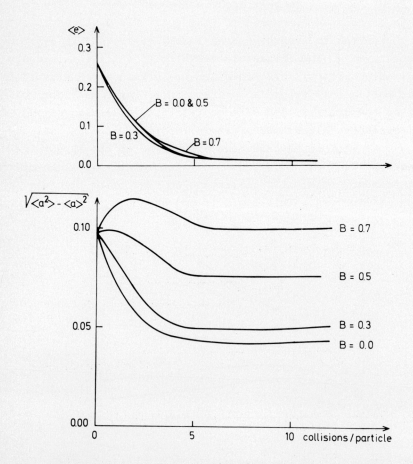

Figure 4. Radial focusing for different amounts of energy deflection, starting from the same initial state.

individual semi-major axis has been taken as a measure of the radial width of the stream.

The radial focusing seems to require that a certain minimum width condition for the stream be satisfied. There thus seems to exist a maximal allowed radial density gradient in the stream for a given spread of eccentricities. This effect is demonstrated in figure 5 which refers to a set of two-dimensional simulations, keeping the collision model and the initial distribution of eccentricities unchanged, but varying the initial radial width of the stream.

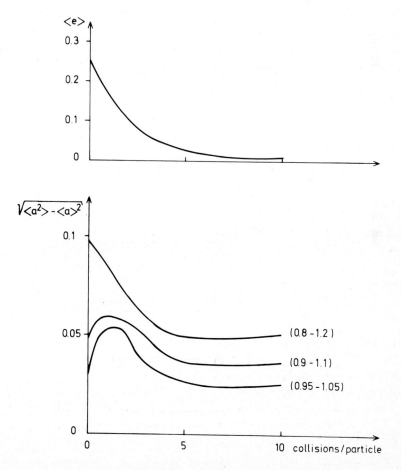

Figure 5. Radial focusing in streams of different initial radial thickness.

Finally, any radial focusing requires a sufficient spread of eccentricities. It is through this spread that the necessary coupling between particles in the stream is at all possible. When the average eccentricity has decreased to almost zero no more effect is seen in figures 4 and 5.

Dynamical effects of collisions in the present solar system

The studies reviewed above are all restricted to rather idealized stream models. Only streams of same sized spherical particles have been considered. Thus possible characteristic effects due to a broad particle size spectrum are not known. Particle spin has not been taken into account.

The model studies have mainly considered number conserving collisions. Fragmentational collisions are, however, expected to be of importance - at least in other connections [2] - and possibly also accreational collisions. The latter ones will necessarily be completely inelastic. Laboratory studies on hypervelocity impacts indicate that about one half of the kinetic energy is spent on crushing and heating the rock [8]. This would bring fragmentational collisions in line with the most inelastic collisions studied in the numerical simulations. Another facet of fragmentational collisions is that dispersing a given mass in smaller particles will increase the collision frequency and therefore tend to enhance the dynamical importance of collisions.

Contributions to the dynamics of the meteoritic complex come from different sources as noted in the introduction. A simple way of estimating the relative importance of these sources is to compare their different characteristic time scales. For collisions the relevant time scale is clearly the mean free collision time.

With these introductory remarks in mind we next turn to a discussion of possible dynamical effects of collisions in our present day solar system. This discussion will necessarily have to be rather sketchy and partly speculative.

It is expected that the thermalization effect of collisions would be one of their most characteristic dynamical signatures. A search for such characteristica can be made in the visual asteroid population. The ratio of the mean values of inclination and eccentricity of the numbered asteroids turns out to be $<i>/<e> \approx 1$. Further, in figure 6 the number of these asteroids as functions of inclination and eccentricity are plotted together with the Rayleigh distributions adjusted to the corresponding mean value. Inclinations are here taken relative to the ecliptic plane. A more correct procedure would require inclinations relative to the symmetry plane of the asteroid population. Even if the actual distributions do not fit the Rayleigh distributions to within the 95 per cent fractile of a χ^2-test, similarities are clearly seen. It is tempting to speculate that we are here observing an effect of collisions in the asteroid population.

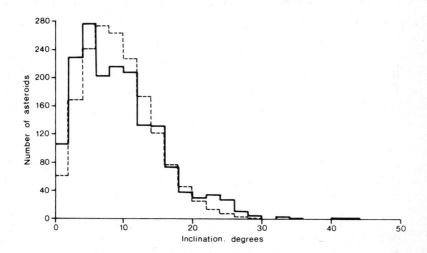

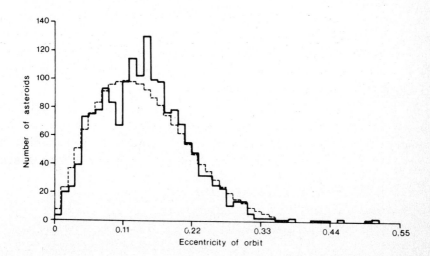

Figure 6. Distributions of inclinations and eccentricities for the numbered main belt asteroids, together with the corresponding Rayleigh distributions.

A characteristic property of the Poynting-Robertson effect for the interplanetary dust population [1] is that the individual orbital planes remain unchanged while the average eccentricity decreases. This process would eventually lead to a collisionally "unstable" dust population. If the mean free collision time in this population is comparable to Poynting-Robertson lifetimes, a thermalization process is expected to take place whereby a finite average eccentricity is maintained at the expense of a decreasing average inclination. Thus, since the Poynting-Robertson effect only constitutes one of at least two competing processes in the real system, conclusions from existing Poynting-Robertson effect calculations should be taken with caution. Whether any decrease in average inclination with decreasing solar distance has been observed is not known to this author.

The simulation results have also been applied to the Saturnian ring problem [9]. Whatever the origin of the rings their particles must at present or in the past have suffered mutual collisions. From the mere existence of the rings it could then be concluded that these particles should have a rather high degree of inelasticity. Secondly, ring lifetime arguments put an upper limit to the possible ring thickness. This follows because increased thickness leads to increased collision frequency and thereby increased rate of energy dissipation. This question can be studied using the results from the numerical simulations. From a given ring thickness the average inclination of the individual particle orbits in the ring is determined. This in turn determines the corresponding average eccentricity. With Rayleigh distributed inclinations and eccentricities we thus see that the ring thickness uniquely determines the velocity distribution in the ring. In this way the energy dissipation rate can be estimated. The conclusion is reached that the ring thickness is at least an order of magnitude less than the 1 - 2 km thickness often inferred from optical observations [10].

The applications discussed so far all refer to what was previously denoted as circular jetstreams. We, however, also observe elliptic stream configurations in the solar system, such as meteor streams and asteroidal jetstreams. The latter ones [3], [11] are represented by clusterings of visual asteroids in similar orbits. Although a rigid analysis of the statistical significance of such clusterings has not yet been attempted, the probability of a random appearance at our times of the major asteroidal jetstreams seems sufficiently small that a physical mechanism able to hold the streams together against the secular perturbations due to Jupiter, should be sought. According to Alfvén's hypothesis, this mechanism is the collisional focusing effect, the visual jetstream asteroids being taken as some kind of Brownian particles in a background collisionally dominated stream of subvisual asteroids. Attempts of looking for the collisional thermalization process in any of these proposed elliptic streams have not been made. Estimates of mean free collision times obviously do not exist. Upper limits to such times set by optical brightness constraints also seems to be lacking.

For meteor streams Mendis [12] has made an attempt to estimate the relative importance of different perturbing forces. Planetary perturbations will give rise to a secular variation of the individual particle orbits in the stream. This will give rise to slowly changing and revolving orbits with typical times for a complete revolution of the order of 10^5 years. This time in itself is of little interest since a revolution of the whole meteor stream is of no relevance in the present connection. Mendis does, however, estimate the timescale for differential perturbations to bring individual grains out of the main bulk of the stream to be of the same order of magnitude. The timescale for the orbit of a 100 μm dust grain to shrink by an amount comparable to the radial thickness of typical meteor streams by the Poynting-Robertson effect is also found to be of the same order of magnitude. On the other hand Mendis comes out with 10^4 years as a typical mean free collision time for a meteor stream of mass 10^{18} g, thickness 10^7 km, semi-major axis 3 a.u., consisting of particles of average size 100 μm and having an average internal velocity of about 100 m/s. This is certainly an indication that we are in the right ballpark and that important collisional contributions to the dynamical evolution of meteor streams are expected. More detailed studies of this aspect are needed.

A slight warning should be raised at this point. Discussions based entirely on the concept of mean free collision time will not always suffice to settle the problem. Thus, to establish the collisional effects on planetary induced longitudinal focusing in meteor streams [13], a more complete study would be necessary.

Conclusion

So far model studies of the dynamical importance of mutual collisions have only treated rather idealized situations. Dynamical effects of fragmentational collisions and a broad particle size spectrum have not been dealt with. The interaction between a stream and a background gas or sporadic meteor population has not been given any detailed discussion. Nevertheless, studies performed have already revealed characteristic dynamical signatures expected from mutual collisions.

Only limited effort has been made to apply these results to the real solar system. Evaluations of mean free collision times and expected degree of inelasticity are required. It is hoped that such questions will be given consideration in the future and that the effects of collisions on the dynamics of our solar system will no more simply be overlooked but ascribed their proper weights.

References

[1]: Wyatt, S.P. and F.L. Whipple   (1950)   Astrophys. J. 111, 134

Kaiser, C.G.   (1970)   Astrophys. J. 159, 77

Lamy, Ph.L.   (1974)   Astron. and Astrophys. 33, 191

Lamy, Ph.L.   (1974)   Astron. and Astrophys. 35, 197

Mukai, T. and S. Mukai   (1973)   Publ. Astron. Soc. Japan 25, 481

Mukai, T., T. Yamamoto, H. Hasegawa, A. Fujiwara and C. Koike   (1974)   Publ. Astron. Soc. Japan 25, 445

Carpenter, D.C. and R.R. Pastusek   (1967)   Planet. Space Sci. 15, 593

[2]: Dohnanyi, J.S.   (1971)   in "Physical Studies of Minor Planets", T. Gehrels (ed.), NASA SP-267

[3]: Alfvén, H.   (1969)   Astrophys. Space Sci. 4, 84

[4]: Alfvén, H. and G. Arrhenius   (1970)   Astrophys. Space Sci. 8, 338

Alfvén, H. and G. Arrhenius   (1970)   Astrophys. Space Sci. 9, 3

Alfvén, H. and G. Arrhenius   (1973)   Astrophys. Space Sci. 21, 117

Alfvén, H. and G. Arrhenius   (1974)   Astrophys. Space Sci. 29, 63

Ip, W.-H.   (1974)   Astrophys. Space Sci. 21, 57

[5]:	Baxter, D.C. and W.B. Thompson	(1971)	in "Physical Studies of Minor Planets", T. Gehrels (ed.), NASA SP-267
	Baxter, D.C. and W.B. Thompson	(1973)	Astrophys. J. <u>183</u>, 323
[6]:	Trulsen, J.	(1971)	Astrophys. Space Sci. <u>12</u>, 329
[7]:	Trulsen, J.	(1972)	Astrophys. Space Sci. <u>18</u>, 3
[8]:	Anders, E.	(1972)	in "From Plasma to Planets", A. Elvius (ed.), Almquist and Wiksell, discussion p. 193
[9]:	Trulsen, J.	(1972)	Astrophys. Space Sci. <u>17</u>, 330
[10]:	Bobrov, M.S.	(1970)	in "Surfaces and Interiors of Planets and Satellites", A. Dollfys (ed.), Acad. Press
[11]:	Arnold, J.R.	(1969)	Astron. J. <u>74</u>, 10
	Danielsson, L.	(1969)	Astrophys. Space Sci. <u>5</u>, 53
	Danielsson, L.	(1971)	in "Physical Studies of Minor Planets", T. Gehrels (ed.), NASA SP-267
	Lindblad, B.A.	(1972)	in "From Plasma to Planets", A. Elvius, Almquist and Wiksell
	Lindblad, B.A. and R.B. Southworth	(1971)	in "Physical Studies of Minor Planets", T. Gehrels (ed.), NASA SP-267
[12]:	Mendis, D.A.	(1973)	Astrophys. Space Sci. <u>20</u>, 165
[13]:	Alfvén, H.	(1972)	in "The Motion, Evolution of Orbits and Origin of Comets", Chebotarev et al. (eds.), IAU, D. Reidel Publ.
	Trulsen, J.	(1972)	in "The Motion, Evolution of Orbits and Origin of Comets", Chebotarev et al. (eds.), IAU, D. Reidel Publ.

5.3

MODELING OF THE ORBITAL EVOLUTION OF VAPORIZING DUST PARTICLES NEAR THE SUN

Zdenek Sekanina
Center for Astrophysics
Harvard College Observatory and Smithsonian Astrophysical Observatory
Cambridge, Massachusetts 02138, U.S.A.

The Poynting-Robertson (P-R) effect (Robertson, 1937, Wyatt and Whipple, 1950), assisted by a pseudo P-R effect due to the sputtering (Whipple, 1955, 1967), is known to cause small dust particles in interplanetary space to spiral toward the sun. Evaporation from the surface of such particles thus increases progressively with time and their size is being reduced accordingly. When the rate of evaporation is no longer negligibly low, it induces on the particle a measurable dynamical effect, which is associated with the implied variations in the magnitude of solar radiation pressure relative to solar attraction. By gradually reducing solar attraction, the particle evaporation tends to increase the orbit dimensions, thus acting against P-R. The P-R inward spiraling, far exceeding the dynamical effect from evaporation at larger heliocentric distances, slows gradually down as the particle approaches the sun, and virtually ceases when the critical distance is reached, where the two forces approximately balance each other. Then, typically, the perihelion distance stabilizes, while the eccentricity starts increasing very rapidly until the particle is swept out of the solar system. This, in brief, is the orbital evolution of a vaporizing particle in the absence of other potentially important but rather poorly known processes, such as particle collisions, rotational bursting, electric charging and interactions with the solar wind and with the interplanetary magnetic field.

If we assume that the particle is rapidly rotating, spherical in shape, of a uniform density ρ, and that most of the solar energy it absorbs is spent on reradiation, the linear vaporization rate of the particle $\dot{a}$ is given, in the first approximation, by

$$\dot{a} = (A/\rho) \exp[B(1-r^{\frac{1}{2}})],$$

where A is the normalized (to 1 AU from the sun) vaporization flux from the particle (mass per unit surface area per unit of time), r is the heliocentric distance (in AU), $B = 1.81 \, L \, (\varepsilon/\kappa)^{\frac{1}{4}}$, L is the latent heat of vaporization (in kcal mole^{-1}), κ is the absorptivity of the particle's surface for solar radiation and ε its emissivity for reradiation. In the following the expression $L \, (\varepsilon/\kappa)^{\frac{1}{4}}$ will be termed the effective latent heat of vaporization.

The calculations based on this model of particle evaporation indicate that a particle, whose pre-evaporation orbit was circular, is expelled from the solar system on a hyperbolic orbit as soon as radiation pressure attains about 0.8 of solar attraction. The expulsion limit, however, is lower for elongated pre-evaporation orbits; it amounts, for example, only to 0.5 - 0.6 of solar attraction for the pre-evaporation eccentricity of 0.2. Nevertheless, purely dielectric particles, some of which may never be subject to radiation pressure exceeding 0.5 of solar gravity, could perhaps, under certain circumstances, vaporize off completely near the sun. However, this possibility is not here pursued further, as it is considered rather untypical.

In order to determine quantitatively the relation between the initial (pre-evaporation) and final (at expulsion) physical and dynamical characteristics of the vaporizing particles, we computed a total of 64 runs, varying the initial particle size and density, the effective latent heat of vaporization and the normalized vaporization flux, the scattering efficiency for radiation pressure and the eccentricity of the initial orbit. We arrived at the following basic conclusions.

The <u>final particle size</u> is on the order of magnitude of 0.1 micron. It is essentially independent of the initial particle size, the effective latent heat of vaporization and the normalized vaporization flux. It is directly proportional to the scattering efficiency for radiation pressure, inversely proportional to the particle density and it increases with increasing eccentricity of the initial orbit.

The <u>perihelion distance of the final orbit</u> varies in inverse proportion to an approximately 2.2 power of the effective latent heat, from less than 0.1 AU above 70 kcal mole^{-1} to more than 1 AU below 25 kcal mole^{-1}, and it decreases with decreasing normalized vaporization flux.

It is, however, virtually independent of the initial particle size, its density and the scattering efficiency, and only slightly dependent on the initial eccentricity.

The heliocentric velocity at expulsion can lie anywhere between the parabolic limit and a maximum hyperbolic velocity, which is determined by the final perihelion distance, the two vaporization constants, the scattering efficiency, the particle density and the final particle size. However, since the particles are strongly affected by radiation pressure, their velocity of escape from the solar system and the maximum hyperbolic velocity are, at 1 AU from the sun, typically less than 20 and 30 km s^{-1}, respectively. The velocities are somewhat higher than indicated for more eccentric initial orbits and for materials of the effective latent heat of vaporization exceeding 100 kcal mole^{-1}.

The maximum intercept velocity at expulsion, i.e., the maximum velocity of an expelled particle relative to the earth at the encounter, increases with increasing effective latent heat of vaporization, but attains no more than 40 km s^{-1} at 100 kcal mole^{-1}.

The maximum intercept angle at expulsion, i.e., the maximum angle toward the sun subtended by the direction from which the particle intercepts the earth and by the earth's apex direction, also increases with increasing effective latent heat, reaching about 50° at 100 kcal mole^{-1}.

Finally, the expulsion lifetime of a vaporizing particle, measured by the span of time from the onset of appreciable evaporation to expulsion, decreases from some 1000 years at the effective latent heat of 30 kcal mole^{-1} to about 10 years at 100 kcal mole^{-1}. The lifetime increases somewhat with the initial particle size and the particle density.

This research was supported by a grant NGR 09-015-159 from the National Aeronautics and Space Administration.

References

Robertson, H. P. (1937), Mon. Not. Roy. Astron. Soc., 97, 423.
Whipple, F. L. (1955), Astrophys. J., 121, 750.
Whipple, F. L. (1967), in "The Zodiacal Light and the Interplanetary Medium", Weinberg, J.L., Ed., NASA SP-150, Washington, D.C., p. 409.
Wyatt, S. P., and Whipple, F. L. (1950), Astrophys. J., 111, 134.

ORBITAL EVOLUTION OF CIRCUM-SOLAR DUST GRAINS

Ph. Lamy
Laboratoire d'Astronomie Spatiale, Marseille

Abstract. The orbital evolution of circum-solar dust grains is obtained by numerical integration of the equations of motion which includes the grains' interactions with the solar radiation field and the solar wind. Our past solution (Lamy, 1974) is improved by avoiding a classical approximation for the Poynting-Robertson term and leads to an important revision of the orbital behaviour. Results are presented for obsidian grains whose inward spiraling is stopped by the effect of sublimation.

1. INTRODUCTION

The dynamics of circum-solar dust grains which experience sublimation was first investigated by Lamy (1974), hereafter referred as Paper I, who gave a qualitative description of the orbital behaviour of grains of silicates (obsidian and andesite) and of iron. Independently, Mukai et al. (1974) considered the case of obsidian and graphite. The trajectories were obtained by numerical integrations showing that the inward spiraling of interplanetary grains under the Poynting-Robertson and corpuscular drags are either reduced or counterbalanced by the effect of the net increase of the radiation force caused by the decrease of the grain's radii when sublimating. We show here that the classical approximation for the Poynting-Robertson drag introduced by Wyatt and Whipple (1950)-in which the cross-sections for absorption and radiation pressure are taken to be equal - completely breaks down for sufficiently small grains. We reconsider the case of obsidian to illustrate this point. Our solution for the interaction with the solar radiation field and the solar wind and the corresponding approximations are the same as in Paper I except for the temperature distribution which has been improved. The reader is referred to this article for further details.

2. EQUATIONS OF MOTION

The interaction with the solar radiation field is responsible for the radiation pressure force (with the associated Poynting-Robertson drag) and the thermal equilibrium of grains. The corresponding equations are given in Paper I but the numerical solution was improved. As a consequence, we found the temperature distribution law (in °K)

$$T_g(R) = \exp\left[7.389 \, (\ln R)^{-0.1945}\right]$$

with R expressed in solar radii, to be best suited to obsidian grains of radius $s \simeq 1 \, \mu m$ close to the Sun. The temperature controls the sublimation rate as

$$\frac{ds}{dt} \, (\mu m/sec) = 408 \, \frac{p}{\delta} \, \sqrt{67/T_g}$$

via the vapor pressure p, expressed in tor and given by Centolanzi and Chapman (1966):

$$\log p = 10.915 - 24928.3/T_g$$

$\delta = 2.37 \, gm/cm^3$ is the density of obsidian.

The direct impacts of solar wind protons and α-particles onto interplanetary grains give rise to a force

$$\vec{F} = \pi \, s^2 \, |\vec{w} - \vec{V}| \, (\vec{w} - \vec{V}) \, \sum_{p,\alpha} m_i \, n_i$$

(Baines et al., 1965) whose tangential component F_f (the corpuscular drag), alone, is of importance. Here, $\vec{w}$ denotes the solar wind velocity, $\vec{V}$ the grain's orbital velocity and m_i and n_i, the mass and number density of each particle species. The plasma or Coulomb drag was neglected as in Paper I but the sputtering was included since it was found that it is not negligible (Lamy, 1975). Possible electromagnetic intereactions were not considered.

r and f being the usual polar coordinates, the equations of motion can be written:

$$\ddot{r} - r \, \dot{f}^2 = - \frac{GM_\odot}{r^2} + \frac{3}{4 \, s \, \delta \, c} \, \frac{\Omega}{\pi} \, I_p$$

$$- \frac{3}{4 \, s \, c} \, \frac{\dot{r}}{c} \, \frac{\Omega}{\pi} \, (I_p + I_a)$$

$$\frac{1}{r} \frac{d}{dt} (r^2 \, \dot{f}) = - \frac{3}{4 \, s \, \delta} \, \frac{r \, \dot{f}}{c^2} \, \frac{\Omega}{\pi} \, I_a + \frac{F_f}{\pi \, s^2}$$

$$\frac{ds}{dt} = \dot{s}_{sublimation} + \dot{s}_{sputtering}$$

Ω is the solid angle subtented by the sun at the location of the grain, c is the velocity of light and I_p et I_a are two integrals:

$$I_p = \int Q_{pr}(\lambda) \, F_\Theta(\lambda) \, d\lambda$$

$$I_a = \int Q_{abs}(\lambda) \, F_\Theta(\lambda) \, d\lambda$$

where the Q's are the efficiency factors for radiation pressure and absorption as given by Mie theory and $F_\Theta(\lambda)$ is the monochromatic emissive power of the sun. Following the treatment of Wyatt and Whipple (1950), these two integrals have been often taken equal in the litterature, thus implying that $Q_{pr}(\lambda) = Q_{abs}(\lambda)$. This has already been pointed out by Mukai et al. (1974) who used the correct equations. In order to show that the above approximation is invalid **for micronic grains, the two integrals are** plotted over a large range of values of s (Fig. 1).

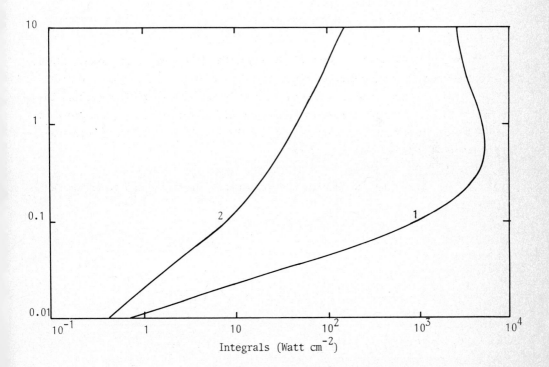

Fig. 1. The integrals for radiation pressure (1) and absorption (2)

For s ≃ 1 μm, the difference reaches two order of magnitudes. Although this does not appear in the above equations, the Poynting-Robertson terms were corrected according to Guess (1968) to take into account the solid angle effect of the sun.

After eliminating the time and switching to the true anomaly as independent variable, the equations were numerically integrated using the Runge-Kutta fourth order method. The program contained a model of the solar wind derived from the measurements of Koutchmy (1972). The integrals I_p and I_a were computed by linear interpolation between two of ten points on a log-log scale. In this present study, the initial conditions include circular orbits only.

3. RESULTS AND DISCUSSION

A simultaneous plot of the aphelion and perihelion distances and of the grain's radius versus the number of revolution as already used by Lamy (1975) was found to best illustrate the orbital evolution.

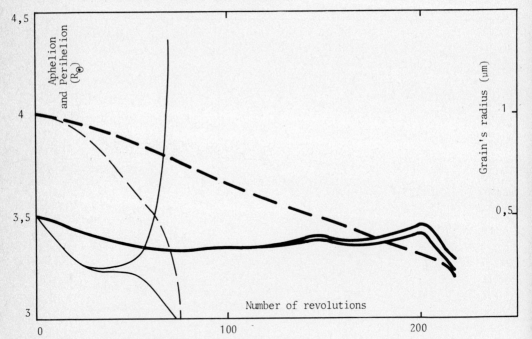

Fig. 2. Orbital evolution of an obsidian grain ($s_o = 1 \mu m$) starting from an initially circular orbit, 3.5 $R_\odot$ in radius: aphelion and perihelion distances (solid lines) and the grain's (broken line) versus the number of revolution. The Thin lines correspond to the approximation $Q_{abs} = Q_{pr}$ for the Poynting-Robertson drag, and the thick lines, to the exact solution.

Figure 2 shows the result for an obsidian grain of initial radius 1 μm starting on a circular orbit of radius 3.5 $R_\odot$. It is in complete agreement with the theoretical study based on perturbation theory given in Paper I: sublimation stops the inward spiraling and causes the orbit to expand, oscillate and to become slightly elliptic. As stated in Paper I, grains with radius smaller than 0.2 μm have such a short lifetime that they play a negligible role. The solution based on the Wyatt and Whipple approximation is presented on the same graph; since the Poynting-Robertson drag is overestimated, it controls much strongly the orbital evolution - which is drastically different - and shortens the lifetime of the grain. As a final example, the new result for an obsidian grain of initial radius 0.5 μm starting on a circular orbit of radius 3.5 $R_\odot$ is presented in figure 3.

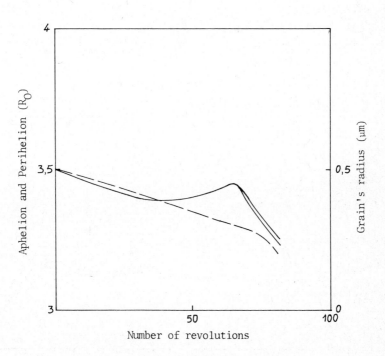

Fig. 3. Orbital evolution of an obisidian grain (s_o = 0.5 μm) starting from an initially circular orbit, 3.5 $R_\odot$ in radius: aphelion and perihelion distances (solid lines) and the grain's radius (broken line) versus the number of revolution.

REFERENCES

Baines,M.J., Williams,I.P., Asebiomo,A.S. 1965, Monthly Notices Roy. Astron. Soc. 130, 63.

Centolanzi,F.J. and Chapman,D.R. 1966, J. Geophys. Res. 71, 1735.

Guess,A.W. 1962, Astrophys. J. 135, 855.

Koutchmy,S. 1972, Solar Phys. 24, 373.

Lamy,Ph.L. 1974, Astron. Astrophys. 33, 191.

Lamy,Ph.L. 1975, Doctoral Thesis, Cornell University.

Mukai,T., Yamamoto,T., Hasegawa,H., Fujiwara,A. and Koike,C. 1974, Publ. Astron. Soc. Japan 26, 445.

Wyatt,S.P., Whipple,F.L. 1950, Astrophys. J. 111, 842.

5.5

TEMPERATURE DISTRIBUTION AND LIFETIME
OF INTERPLANETARY ICE GRAINS

Ph. Lamy and M.F. Jousselme
Laboratoire d'Astronomie Spatiale, Marseille

Abstract. An improved solution of the temperature distribution of interplanetary ice grains is presented using the refractive index measured at 100° K. The efficiency factors for absorption are obtained from Mie theory and the calculation is carried out for micronic and submicronic grains at 50, 100 and 150° K; corresponding lifetimes are given.

1. INTRODUCTION

The problem of the temperature distribution of interplanetary dust grains has been recently investigated in detail by Mukai and Mukai (1973) and Lamy (1974). The main improvement of these studies consists in using Mie theory to compute the efficiency factors for absorption Q_{abs}, taking into account the dependence of the complex indices of refraction on wavelength.

The case of ice presents particular problems: as shown by the aforementioned authors, ice grains cannot survive at temperature larger than 150° K approximately. We need therefore to consider typical temperatures of 50 and 100° K and use the refractive index $m(\lambda)$ measured at these temperatures. The absorbed energy is consequently radiated in the far infrared so that $m(\lambda)$ and the efficiency factor are required over a very large region. Past computations were based on the compilation of the refractive index at 273° K performed by Irvine and Pollack (1968). We present here an improved solution to the problem by using the refractive index measured at 100° K. The computational procedure and the vapor pressure formula have also been revised and should lead to an increased accuracy. The equations for the temperature and the sublimation rate will be the same as those given by Lamy (1974) and the reader is referred to this article for a complete discussion of the assumptions and techniques of solution. In particular, grains are assumed spherical for simplicity.

2. TEMPERATURE DISTRIBUTION

a) Equation: the equation relating the heliocentric distance R and the temperature T_g of a grain of radius s as given by Lamy (1974) is:

$$\left[\frac{R_\odot}{R}\right]^2 \int_{0.2}^{15} Q_{abs}\, F_\odot(\lambda)\, d\lambda = 4\left[\int_{1}^{300} Q_{abs}\, B(\lambda, T_g) d\lambda + \frac{dE}{dt} L_s(T_g)\right]$$

where $F_\odot(\lambda)$ is the monochromatic emissive power of the Sun, $R_\odot$ its radius; dE/dt is the mass sublimation rate of ice at temperature T_g and $L_s(T_g)$ its latent heat of sublimation at the same temperature. $B(\lambda,T_g)$ is Planck's function:

$$B(\lambda,T_g) = 2\pi h c^2 \lambda^{-5} \left[\exp\frac{hc}{k\lambda T_g} - 1\right]^{-1}$$

b) Numerical solution: the efficiency factor Q_{abs} is a function of wavelength λ, of the complex refractive index of ice $m(\lambda) = n(\lambda) - ik(\lambda)$ and of the grain's radius s and was computed, using the rigorous Mie theory, at 192 values of wavelength, between 0.2 and 300 μm, for each of the following 10 values of s: 0.01, 0.05, 0.1, 0.2, 0.3, 0.4, 0.5, 1 and 10 μm. For the spectral region 1.25 - 300 μm, we obtained $m(\lambda)$ from the experimental results of Bertie, et al. (1969) for hexagonal ice at 100° K (measurement of absorbance and Kramers-Krönig analysis). Below 1.25 μm, results are almost non-existant. For n, the value for 0.579 μm as reported in the Smithsonian Tables (1969) allows to scale the refractive index of water as given by Allen (1973) since ($n_{water} - n_{ice}$) appears to be nearly constant and equals to 0.23 approximately. Our selected values are slightly larger than those retained by Isobe (1971), the difference amounting to 0.04 approximately. For k, the compilation of Irvine and Pollack (1968) shows a similar behavior for ice and water below 0.95 μm. Therefore, we bridged the results of Bertie et al. (1969) and those for water down to 0.2 μm. The error resulting from this approximation is quite negligible since k is very small in this region. The discussion of Irvine and Pollack (1968) shows that the temperature effect is extremely small below 1 μm (where there is no absorption band); therefore, we consider that our data for $m(\lambda)$ are relevant to a temperature of 100° K and certainly apply with good accuracy in the range 50-150° K. Table 1 gives the steps used for the calculation of the integrals whose limits (in μm) appear in the above equation. The solar spectrum in the region 0.2 - 4 μm is obtained from the compilation of Labs and Neckel (1968). Beyond, a black-body model is suitable and we used a temperature of 5500° K in the 4 - 7.5 μm region and 5000° K in the 7.5 - 15. μm region. The calculation of the sublimation term L_s dE/dt will be presented in the third section.

c) Results: figure 1 shows the heliocentric distance as function of the grain's radius for three different temperatures, 50, 100 and 150° K. We see that the previous results are substantially modified as ice grains at a given temperature come now closer to the Sun.

Spectral domain (μm)	Step (μm)
0.2 - 0.4	0.01
0.4 - 0.7	0.02
0.7 - 1	0.05
1 - 6	0.1
6 - 15	0.25
15 - 30	0.5
30 - 100	5
100 - 300	10

Table 1. Steps of integration

3. SUBLIMATION OF ICE GRAINS

The mass sublimation rate in gm sec^{-1} cm^{-2} is (Lamy, 1974):

$$\frac{dE}{dt} = 4.08 \times 10^{-2} \, p \, \sqrt{18/T}$$

The vapor pressure p (expressed in tor) was recently reconsidered by Jancso et al. (1970); their proposed formula is likely to be the most reliable now available:

$$\log p \, (\text{Tor}) = -2481.604/T + 3.57 \log T - 3.097 \times 10^{-3} \, T - 1.76 \times 10^{-7} \, T^2 + 1.902$$

The latent heat of sublimation may now be calculated using the Clausius-Clapeyron equation. The lifetime of ice grains may be obtained directly from dE/dt (see Lamy, 1974). Taking an initial radius of 1 μm, we found a lifetime of 6.4×10^{13} sec for a grain at 100° K (located at 3.02 AU) and of 1.4×10^5 sec for a grain at 150° K (1.3 AU). In this latter case this is only 3×10^{-3} of the period of the corresponding keplerian orbit.

ACKNOWLEDGMENT

We are grateful to J. Pollack for drawing our attention to several references on the optical properties of ice.

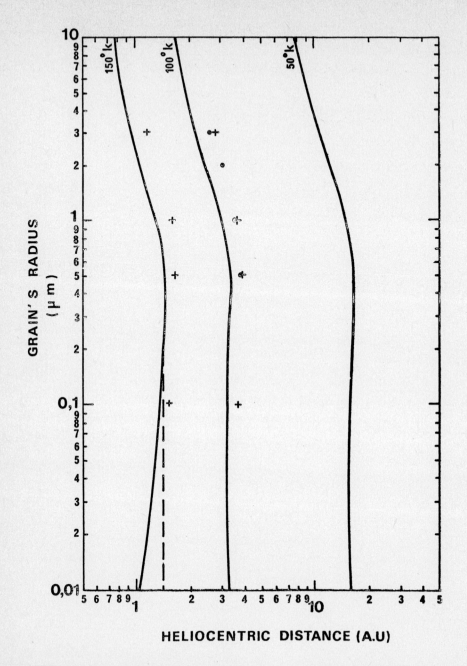

Fig. 1. Temperature distribution of interplanetary ice grains as function of their radius. The broken line corresponds to the result without the sublimation term. The dots are the results of Mukai (1973) and the crosses, those of Lamy (1974).

REFERENCES

Allen,C.W. 1973, Astrophysical Quantities, third Edition, The Athlone Press, London.

Irvine,W.M. and Pollack,J.B. 1968, Icarus $\underline{8}$, 324.

Isobe,S. 1971, Ann. Tokyo Astron. Obs. $\underline{12}$, 263.

Jancso,G., Pupezin,J. and Van Hook,W.A. 1970, J. Phys. Chem. $\underline{74}$, 2984.

Labs,D. and Neckel,H. 1968, Z. Astrophys. $\underline{69}$, 1.

Lamy,Ph.L. 1974, Astron. Astrophys. $\underline{35}$, 197.

Mukai,T. and Mukai,S. 1973, Publ. Astron. Soc. Japan $\underline{25}$, 481.

RADIAL DISTRIBUTION OF METEORIC PARTICLES IN INTERPLANETARY SPACE

John W. Rhee

Rose-Hulman Institute of Technology
Terre Haute, Indiana 47803

and

Goddard Space Flight Center
Greenbelt, Maryland 20771

An attempt has been made to derive heliocentric distribution of meteoric particles in interplanetary space. Poynting-Robertson effect, collision process, and cometary dust injection have been included in this study. Three different radial distribution functions for different sizes have been derived. The number density appears to depend on particle size.

A series of important experiments has been carried out to obtain some information about the heliocentric distribution of cosmic dust particles which produce zodiacal light (Alexander, 1962; Alexander et al., 1965; Rhee et al., 1974; Humes et al., 1974). Recently Southworth and Sekanina (1973) studied the distribution of meteoroids in the solar system from the orbits of approximately 20,000 radar meteors observed in 1969. They reached the surprising new conclusion that the space density in ecliptic plane was minimum near $0.7 \sim 0.8$ AU and maximum in the asteroid belt, between 2 and 3 AU. The zodiacal light data obtained by Hanner et al., (1974) from Pioneer 10 drops off quite rapidly between 2.4 AU and 3.2 AU. There appears to be no zodiacal light beyond 3.3 AU. In the following we shall attempt to derive heliocentric distribution functions on a number of simple and plausible physical arguments.

(a) <u>Poynting-Robertson Effect</u>

It is a well-known fact that meteoric particles lose angular momentum in interplanetary space and spiral into the sun. If the Poynting-Robertson effect operates alone in interplanetary space, the particle number density N must satisfy

$$\left(\frac{dN}{dr}\right)_{P-R} = \left(\frac{dN}{dt}\right)_{P-R} \left(\frac{dt}{dr}\right)_{P-R} = -\left(\frac{N}{r}\right) \tag{1}$$

which has a solution of $N \sim r^{-1}$ and r is the distance.

(b) <u>Inter-Particle Collision Process</u>

Southworth and Sekanina has shown that due to collisions N must satisfy

$$\left(\frac{dN}{dr}\right)_c = \left(\frac{dN}{dt}\right)_c \left(\frac{dt}{dr}\right)_{P-R} = AN^2\sqrt{r} \tag{2}$$

where A is a constant. The exact solution of Equation 2 is $N \sim r^{-1.5}$.

(c) Cometary Particle Injection

Since Poynting-Robertson effect and collision theory cannot explain the recent observational data from Pioneer 10, it is proposed that direct cometary injection will have to be taken into account. This might be done as

$$\left(\frac{dN}{dr}\right)_i = \left(\frac{dN}{dt}\right)_i \left(\frac{dt}{dr}\right)_{p-R} \quad (3)$$

where $\left(\frac{dN}{dt}\right)_i$ represents cometary particle injection rate. The dust emission rate from comets is usually assumed to vary with r as $r^{-\alpha}$ where α usually ranges from 0.5 to 5 (Sekanina, 1974). It is not clear whether cometary brightness is directly related to particle injection rate or not. Cometary brightness depends also on r as $r^{-\beta}$ where β typically ranges from 2 to 6. It is quite possible that most of dust emitted by comets does not remain in the solar system due to solar radiation pressure. At any rate Equation 3 can be written as

$$\left(\frac{dN}{dr}\right)_i = -\beta r^{-p} \quad (4)$$

where β and p are constants.

(d) Southworth-Sekanina Equation

Following Southworth and Sekanina (1973), we shall expand $\frac{dN}{dr}$ in terms of $\left(\frac{dr}{dt}\right)_{p-R}$ as follows:

$$\frac{dN}{dr} = \sum_{i=1}^{3} \left(\frac{dN}{dt}\right)_i \left(\frac{dt}{dr}\right)_{p-R} = -\frac{N}{r} + AN^2\sqrt{r} - \beta r^{-p} \quad (5)$$

Defining $y = N^{-1}$, Equation 5 can be rewritten as

$$\frac{dy}{dr} - \frac{y}{r} = -A\sqrt{r} + \beta y^2 r^{-p} \quad (6)$$

The following three approximate specific solutions of Equation 6 have been derived

Case (a): $\quad p = 0, \quad y = A_1 r - A_2 r^{3/2} + A_3 r^4$

$$\frac{N}{N_0} = \frac{0.23}{r - 0.87 r^{3/2} + 0.1 r^4} \quad (7)$$

Case (b): $\quad p = 1, \quad y = A_4 r - A_5 r^{3/2} + A_6 r^3$

$$\frac{N}{N_0} = \frac{0.25}{r - 0.8 r^{3/2} + 0.05 r^3} \quad (8)$$

Case (c): $\quad p = 2, \quad y = A_7 r - A_8 r^{3/2} + A_9 r^2$

$$\frac{N}{N_0} = \frac{0.58}{r - 0.45 r^{3/2} + 0.03 r^2} \quad (9)$$

where N_0 is the number density at 1 AU and $A_1 \ldots A_9$ are constants. These three cases represent cometary injection rates of r^{-1}, r^{-2}, and r^{-3}, respectively. A cometary injection rate of r^{-4} leads to an unphysical situation and must be discarded under this approximation.

Case (a) has been fitted with Pioneer 8 and Pioneer 9 data (Rhee et al., 1974) and the normalized number density is represented by Equation 7. Equation 7 is also shown in Figure 1. An examination of Figure 1 shows that for $r < 0.4$ AU, $N \sim r^{-1}$ and for $r > 2$ AU, $N \sim r^{-4}$. This simple model predicts that most of the interplanetary dust particles is concentrated within 1 AU in the solar system. The dust density appears to vary little between 0.7 AU and 1.1 AU. The dust density at 2 AU is 20% of that at 1 AU and at 3.3 AU it is only about 2% of that at 1 AU.

Case (b) is a very interesting model since it shows one minimum and one maximum. This case has been fitted with the radar meteor data (Southworth and Sekanina, 1973) and the result is shown as Equation 8 and in Figure 2. This model again demonstrates that for $r < 0.3$ AU, $N \sim r^{-1}$ and for $r > 3$ AU, $N \sim r^{-3}$. The space density is minimum near 0.8 AU and maximum at 2.4 AU.

The last case has been compared with the Pioneer 10 penetration data and is shown in Figure 3. Here again for small r, $N \sim r^{-1}$. Between 1 AU and 3 AU the number density

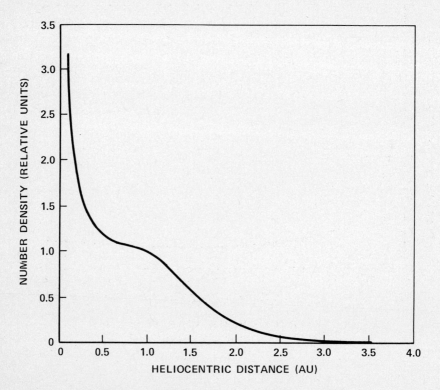

Figure 1. Radial Distribution of Zodiacal Dust

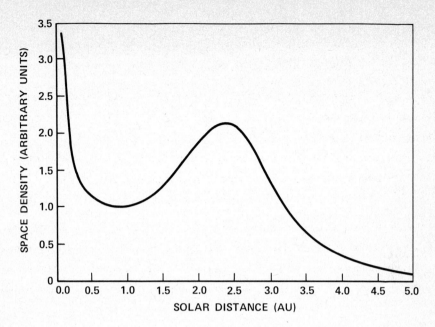

Figure 2. Space Density of Radar Meteors

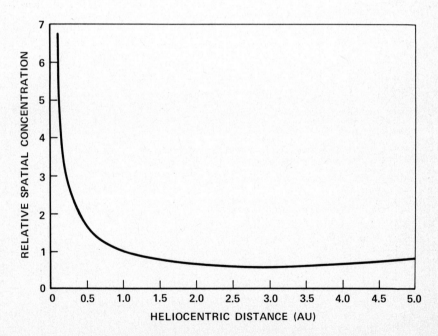

Figure 3. Spatial Concentration of Meteoroids of Mass $10^{-9} \sim 10^{-10}$ g (normalized to that at 1 AU)

decreases very slowly and then starts increasing at 3 AU. The spatial concentration at 3 AU is about 62% of that at 1 AU and at 5 AU it is only 80% of that at 1 AU. Thus this model predicts one minimum and no maximum in interplanetary space. It is to be noted that this model cannot explain the kind of data gap observed from the Pioneer 10 penetration measurements.

For zodiacal dust particles the spatial concentration is maximum near the sun and is essentially zero beyond 3.3 AU. The space density of larger particles such as radar meteors shows a minimum near 0.8 AU and a maximum near 2.5 AU. For medium size particles of mass 10^{-12} kg and larger the number density has a minimum at 3 AU even though the spatial variation between 1 AU and 5 AU is no more than 50% of that at 1 AU. In all of these cases the concentration is inversely proportional to the radial distance between 0.1 AU and 0.5 AU.

REFERENCES

Alexander, W. M., Cosmic Dust, Science 138, 1098-1099, 1962.

Alexander, W. M., McCracken, C. W., and Bohn, J. L., Zodiacal Dust, Science 149, 1240, 1965.

Hanner, M. S., Weinberg, J. L., DeShields II, L. M., Green, B. A., and Toler, G. N., Zodiacal Light and the Asteroid Belt: The View from Pioneer 10, J. Geophys. Res. 79, 3671-3675, 1974.

Humes, D. H., Alverez, J. M., O'Neal, R. L., and Kinard, W. H., The Interplanetary and Near-Jupiter Meteoroid Environment, J. Geophys. Res. 79, 3677-3684, 1974.

Rhee, J. W., Berg, O. E., and Richardson, F. F., Heliocentric Distribution of Cosmic Dust Intercepted by Pioneer 8 and 9, Geophys. Res. Letters, 1, 345-346, 1974.

Sekanina, Z., On the Nature of the Antitail of Comet Kohoutek 1973f, Preprint No. 106, Center for Astrophysics, Cambridge, Mass., May, 1974.

Southworth, R. B. and Sekanina, Z., Physical and Dynamical Studies of Meteors, 38, NASA CR-2316, Oct. 1973.

5.7   ROTATIONAL BURSTING OF INTERPLANETARY DUST PARTICLES

Stephen J. Paddack
Goddard Space Flight Center, Greenbelt, Maryland 20771

John W. Rhee
(On sabbatical leave from Rose-Hulman Institute of Technology, Terre Haute, Indiana 47803)
Goddard Space Flight Center, Greenbelt, Maryland 20771
and
University of Maryland, College Park, Maryland 20740

Abstract. Solar radiation pressure can cause rotational bursting and eventual elimination from the solar system of small asymmetric interplanetary particles by a windmill effect. The life span determined by this process for stony meteoritic material or tektite glass with radii of 1 cm is on the order of 10^5 years. Same size material which contains iron, nickel or aluminum, with properties such that it is subject to 5 percent of the amount of spin damping as pure metals, can be removed from the solar system on the order of 10^6 years by this process. Ordinary chondritic material, despite its high resistivity, is subject to a type of magnetic spin damping, in addition to the normal spin damping, with the consequent result that this type material cannot be removed from the solar system by this process. This depletion mechanism appears to work faster than the traditional Poynting-Robertson effect by approximately two orders of magnitude for the nonmetallic particles and one order of magnitude for the metallic particles.

Introduction. Kresák (1968) has drawn attention to a puzzling deficiency in the numbers of small meteors. From the laws of celestial mechanics he notes that the breakup time of meteor showers is on the order of 10^3 to 10^5 years. After the breakup of the showers, the meteors continue to exist as individuals in individual orbits (sporadic meteors) until they are destroyed. Because there are only about five times as many sporadic meteors as shower meteors, the lifetime of meteors as solid objects must have a range of 5×10^3 to 5×10^5 years. But lifetime determined by the Poynting-Robertson effect is about 10^7 years for a 1 cm radius particle. Hence, some other more effective mechanism of destruction must exist. We here suggest that it is rotational bursting.

Radzievskii (1954) suggested that solar radiation pressure could cause rotational bursting of asteroids and meteoroids having irregular albedos over their entire surfaces. As an example, he calculated that a 1 cm granite cube at 0.4 au from the sun would reach bursting speed in about 1000 years.

Paddack (1969, 1973) demonstrated that rotational bursting could be caused by the interaction of solar radiation pressure and irregular surface geometry irrespective of variations in albedo. In rotational bursting, smaller particles are produced with each burst. Subsequently, they will reach a size where the gravitational attraction to the sun is balanced by the repelling effect of radiation pressure also some particles will reach hyperbolic speeds. Once this point is reached, the particles can no longer remain in the solar system and are removed from it.

On the basis of experimentation, it was conservatively estimated that nonmagnetic meteoroids and tektites in heliocentric orbits reach bursting speed in about 60,000 years. These analyses did not take into account magnetic spin damping effects for tektites because of their high electrical resistivity (Hoyte et al., 1965) and low magnetic permeability. Most meteoritic dust is stony (Öpik, 1958; Banderman, 1968) but also contains a metallic component which may cause them to be subject to spin damping. Some stony meteorites have been found to contain between a few tenths of a percent and 20 percent metal (Wood, 1963). Damping has been investigated since (Paddack and Rhee, 1975) and the following is an extension of that analysis.

Analytical Procedure. An irregularly shaped metallic body can also reach rotational bursting speed by this windmill effect. Applying the experimentally determined asymmetry factor, ℓ, which is 0.05 percent of the maximum dimension of the body (Paddack, 1969, 1973), and using radiation pressure, P_r, and total projected area, A, the effective torque, N_1, due to radiation pressure, is given by

$$N_1 = 0.1 \, P_r \, A \ell. \tag{1}$$

That the angular speed of a conductor is reduced in an external magnetic field is a well-known phenomenon. The magnitude of the retarding torque for a rotating sphere having angular velocity ω and effective radius b is given by Wilson (1961).

$$N_2 = -\frac{2\pi B^2 b^5 \omega}{15\rho} \tag{2}$$

where ρ is the resistivity and B is the interplanetary magnetic field intensity.

For material with randomly distributed metal particles, such as exhibited by some chondrites (Wood, 1963), there is a magnetic spin damping effect related to, and in addition to, normal magnetic spin damping. In this case, each small particle of metal imbedded in the rotating small celestial body tends to retard its spin. The intergrated effect of these metallic particles can cause significant spin damping. The imbedded metal particle torque, N_3, is given by

$$N_3 = -\left[\frac{2\pi B^2 c^5 \omega}{15\rho_c}\right] \frac{M}{m_c} \tag{3}$$

where c is the average radius of the imbedded metallic particles, ρ_c is the resistivity of the imbedded metallic particles, M is the total mass of the metal in the small celestial body, and m_c is mass of an average single imbedded metallic particle.

Damping due to induced currents was also considered but it was found to be insignificantly small. Consequently, it is possible to express the rotational equation of motion for small interplanetary particles, including the damping effects due to interplanetary magnetic field, as

$$I\frac{d\omega}{dt} = N_1 + N_2 + N_3 \qquad (4)$$

where I is the moment of inertia. The solution of Equation 4 is

$$\omega = \frac{\alpha}{\beta}(1 - e^{-\beta t}) \qquad (5)$$

where $\alpha = N_1/I$, and $\beta = -(N_2 + N_3)/I\omega$.

Equation 5 shows that the angular motion is stabilized under the damping effect and its upper limit is α/β.

According to Timoshenko (1942), the stress, σ, developed in a rotating solid disk of radius b is given by

$$\sigma = \frac{(3+\mu)}{8} \delta b^2 \omega^2 \qquad (6)$$

where δ is the density and μ is the Poisson ratio. The time needed for elimination from the solar system by radiation pressure as a consequence of successive bursting can be estimated by multiplying the time to the first burst by a factor at 2.7 (Paddack, 1973).

The time to elimination from the solar system for certain kinds of materials has been calculated by use of Equations 5 and 6 and tabulated in Table 1. The following assumptions were used: heliocentric circular orbits at one astronomical unit, radiation pressure of 4.5×10^{-5} dyn cm^{-2}, 5 gamma for B (the interplanetary field intensity (Ness, 1965)), standard resistivity for the metals, $3 \times 10^{+14}$ ohm-cm for tektites (Hoyte et al., 1965), 15 ohm-cm for meteoroids (Wood, 1963), and velocity of 30 km sec^{-1}. Handbook values were used for the tensile strength of metals, 6.9×10^8 dyn cm^{-2} for tektites (Centolanzi, 1969), and 5.0×10^7 dyn cm^{-2} for meteoroids (Öpik, 1958). Handbook values were used for the densities of metals, 2.4 g cm^{-3} for tektites (Chao, 1963), and 3.6 g cm^{-3} for meteoroids (Wood, 1963). In the same table, those particles that reach equilibrium below bursting angular speed are indicated as NB (no bursting).

In the case of chondritic meteoroids, metallic particles are assumed to be randomly distributed in them. These imbedded metallic particles are assumed to be iron spherules with radii varying from 25 to 250 μm. The percentage of iron in the chondrites was varied from 0.5 to 5 percent (Wood, 1963). For the only elimination time shown in the table (actually time to equilibrium in this case), a percentage of iron of 4 percent was assumed, and the radii of the imbedded particles was assumed to be 25 μm.

Table 1

Elimination Time for Different Materials as a Function of Particle Radius

Radius (cm)	Al		Fe		Ni		Tektites	Chondritic Meteoroids	Stony Meteor (nonmetallic)
	100%	5%	100%	5%	100%	5%			
10^{-4}	1.52×10^2	1.52×10^2	6.25×10^2	6.25×10^2	1.49×10^2	1.49×10^2	6.43×10^1	**	2.12×10^1
10^{-3}	1.52×10^3	1.52×10^3	6.26×10^3	6.25×10^3	1.49×10^3	1.49×10^3	6.43×10^2	**	2.12×10^2
10^{-2}	1.53×10^4	1.52×10^4	6.30×10^4	6.26×10^4	1.50×10^4	1.49×10^4	6.43×10^3	NB*	2.12×10^3
10^{-1}	1.64×10^5	1.52×10^5	6.76×10^5	6.28×10^5	1.51×10^5	1.49×10^5	6.43×10^4	NB*	2.12×10^4
1	NB*	1.58×10^6	NB*	6.49×10^6	1.65×10^6	1.50×10^6	6.43×10^5	1.03×10^6	2.12×10^5

*NB means no bursting.
**No calculation done because imbedded metallic particle size is equal to or greater than size of small celestial body.

Summary. Solar radiation pressure can cause rotational bursting of certain small asymmetric interplanetary particles and eliminate them from the solar system. Particle lifetimes of between a few tens of years and on the order of 10^6 years are predicted for particles in heliocentric orbits at 1 au. The lifetimes vary as a function of the amount of spin damping to which the particles are subject, their external geometry, their chemical composition, and their physical structure. Particles subject to substantial spin damping, such as some pure metals, or chondritic material with metallic particles imbedded in it, may become stabilized before reaching the bursting point.

For nonmetallic particles, such as tektites or other metal-free silicate material, the retarding torques are so small that there is no effective spin stabilization, and consequently these can be eliminated from the solar system inbetween a few tens of years and in about 60,000 years for particles with radii between 10^{-4} and 1 cm in heliocentric orbits at 1 au.

This depletion mechanism works faster than the traditional Poynting-Robertson effect by approximately one order of magnitude for metallic particles, which are not stabilized by spin damping, and about two orders of magnitude for nonmetallic particles.

The percentage of pure nickel-iron-type meteors, which are about gram size, has been found to be only about 1 or 2 percent. Also, observational evidence indicates silicate rather than metallic material for the composition of dust in comets (Millman, 1974). Therefore, a great percentage of the particulate material in interplanetary space should be affected by this process.

For those particles with substantial amounts of metal, the distribution of the metallic component can have a significant effect on spin damping. Any gaps in the conducting path can appreciably lower the effectiveness of normal magnetic spin damping. The spin damping effect of particles of metal imbedded in the small celestial body, however, can be substantial, as was shown for chondritic meteoroids. Despite the spin damping that can occur, rotational bursting appears to be a reasonably important process to be considered in the study of the lifetime and population of small bodies in the solar system. It is also reasonable to expect those particles subject to substantial spin damping to be older particles, because they would be eliminated or destroyed by slower acting mechanisms.

Acknowledgments. The authors would like to thank Drs. John A. O'Keefe and David P. Stern for helpful discussions.

References

Banderman, Lothar W., Physical Properties and Dynamics of Interplanetary Dust, Ph.D Thesis, University of Maryland, 1968, p. 189.

Centolanzi, Frank J., Maximum Tektite Size as Limited by Thermal Stress and Aerodynamic Loads, J. Geophys. Res. 74, pp. 6725-6736, 1969.

Chao, E. C. T., The Petrographic and Chemical Characteristics of Tektites, in Tektites, edited by J. A. O'Keefe, pp. 51-94, University of Chicago Press, Chicago, Ill. 1963.

Hoyte, A., F. Senftle, and P. Wirtz, Electrical Resistivity and Viscosity of Tektite Glass, J. Geophys. Res., 70, pp. 1985-1994, 1965.

Kresák, L., Structure and Evolution of Meteor Streams, in Physics and Dynamics of Meteors, edited by L. Kresák and P. M. Millman, pp. 391-403, D. Reidel Pub. Co., Dordrecht, Holland, 1968.

Millman, Peter M., Dust in the Solar System, Smithsonian Contributions to Astrophysics, (in press), 1974, pp. 17-18.

Ness, Norman F., The Interplanetary Medium, in Introduction to Space Science, edited by Hess, Wilmot N., pp. 323-346, Gordon and Breach, New York, 1965.

Öpik, Ernst J., Physics of Meteor Flight in the Atmosphere, pp. 24-26, Interscience Pub. Inc., New York, 1958.

Paddack, Stephen J., Rotational Bursting of Small Celestial Bodies: Effects of Radiation Pressure, J. Geophys. Res., 74, pp. 4379-4381, 1969.

Paddack, Stephen J., Rotational Bursting of Small Celestial Bodies: Effects of Radiation Pressure, Ph.D. Thesis, Catholic University of America, Washington, D.C., 1973.

Paddack, Stephen J. and Rhee, John W., Rotational Bursting of Interplanetary Dust Particles, Geophys. Res. Letters, 2, pp. 365-367, 1975.

Radzievskii, V. V., A Mechanism of the Disintegration of Asteroids and Meteorites, Dokl. Akad. Nauk SSSR, 97, pp. 49-52, 1954.

Timoshenko, S., Strength of Materials, p. 249, D. Van Nostrand, Princeton, N.J., 1942.

Wilson, R. H., Rotational Magnetohydrodynamics and Steering of Space Vehicles, NASA TN 566, Goddard Space Flight Center, 1961.

Wood, J. A., Physics and Chemistry of Meteorites, in The Moon, Meteorites, and Comets, edited by B. M. Middlehurst and G. P. Kuiper, pp. 327-401, University of Chicago Press, Chicago, Illinois, 1963.

5.8 LUNAR EJECTA IN HELIOCENTRIC SPACE

W.M. Alexander and M.A. Richards
Baylor University, USA

Studies of the parameters of micron and submicron ejecta particles from laboratory hypervelocity impact experiments have been accomplished using a light-gas-gun to accelerate milligram particles to velocities of 4 km/s onto a basalt like rock target.

Preliminary results give a mass-distribution for the micro-ejecta particles with velocities ≥ 3 km/s relative to the target. From these results, it is seen that less than 0.03 % of the mass of the 4 km/s primary impacting particle leaves the impact crater with velocities equal to or greater than "lunar" escape velocity.

It is shown that over 80 % of these particles escape the earth-moon gravitational sphere of influence and enter heliocentric space with 1 AU perihelions. Ejecta particles with mass less than 0.75 pg leave the interplanetary system due to radiation pressure. Particles with masses less than 1.47 pg are probably perturbed by Jupiter. Thus, ejecta particles with masses greater than 1.47 pg are injected into an extremely small volume of heliocentric space which is symmetrical to the ecliptic plane. The major force acting on these size particles in this volume is the Poynting-Robertson force. The dwell-time of these particles in the above space is found to be between 600 and 1000 years. Calculations for the resulting spatial density of lunar ejecta in this volume of heliocentric space shows that the spatial density is of the same order of magnitude as the in-situ measurements of Pioneers 8 and 9 and Mariner IV.

5.9 RADIATION PRESSURE ON INTERPLANETARY DUST PARTICLES

G. Schwehm

Bereich Extraterrestrische Physik, Universität Bochum, F R G

The force acting on an interplanetary dust particle due to solar radiation pressure at a distance R from the sun is given by

$$F_{rad} = \frac{\pi r^2}{c} \frac{R_o^2}{R^2} \int_0^\infty Q_{pr}[m(\lambda), x(\lambda)] \, s_\lambda \, d\lambda$$

with $R_o = 1$ AU, r the radius of the particle, $m(\lambda)$ its complex refractive index, c the velocity of light, λ the wavelength, s_λ the solar flux outside the earth's atmosphere per unit area and wavelength range. The function Q_{pr} is the efficiency factor of the radiation pressure as given by Mie-theory, which depends on the refractive index m and the size parameter x of the particle defined as the ratio of the circumference of the particle to the wavelength.

In figure 1 $Q_{pr}(m,x)$ is plotted versus size parameter x to give an impression of the properties of this quantity.

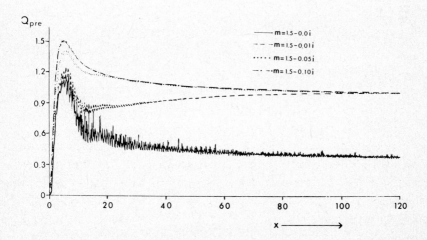

<u>Fig. 1:</u> Q_{pr} for different imaginary parts of the index of refraction; real part fixed ($\lambda = 0.53$ μ).

It is more convenient to use the quantity ß defined as the ratio of the force due to radiation pressure F_{rad} to the force of gravitational attraction F_{grav}

$$ß = \frac{F_{rad}}{F_{grav}} = \frac{A}{\rho \cdot a} \int_0^\infty Q_{pr} \, s_\lambda \, d\lambda$$

which is independent of the distance of the particle from the sun.

Here ρ is the density of the dust particle and the constant $A = 3R_o^2/4cfM_\odot$; f is the gravitational constant, and $M_\odot$ the mass of the sun.

The integral was evaluated using a computer program, which has been written to compute scattering functions of core-mantle particles (Giese, Schwehm, Zerull 1973) based on the theory of Güttler(1952). In this paper, however, only homogeneous spheres will be considered, because of the amount of data then to be taken into account a thorough discussion of the ß values for core-mantle particles would be beyond the scope of this short contribution and will be published elswhere. For the solar radiation flux, values given by Labs and Neckel(1970) have been used.

ß has been evaluated for different materials regarding the wavelength dependence of their complex index of refraction. The integration was carried out for every value of the radius over the whole range of wavelength from 0.2 μ to 50(100)μ.

In fig. 2 ß is plotted versus particle radius r for Obsidian and Andesite. The shape

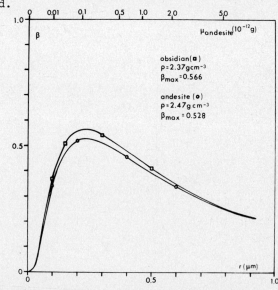

Fig. 2: ß values for Obsidian (□) and Andesite (o); index of refraction after Pollack et al. (1973).

of both curves is very similar with the maximum of ß being about
0.53 for Andesite.

The values have been checked with those given by Lamy (1974), which
show a peak value of ß for Andesite to be 0.59. These small
discrepencies are due to the fact that the values for the solar
radiation used in this paper are more accurate, because for
$\lambda > 0.6$ µ Lamy used a black body approximation. Even if the bulk
of the contribution to the integral for the radiation pressure
comes from the region 0.2 to 0.6 µ the approximation of the
radiation flux for $\lambda > 0.6$ µ causes some inaccuracies.

The most particular fact is that both curves remain smaller than
ß = 1 for spherical particles consisting of these nearly pure
dielectric materials, where Obsidian is a volcanic glass and
Andesite represents stony meteorites. These two models of dust
grains will in no case, even with a slightly higher absorption
meet the requirement of ß values in the range of $0.9 \leq ß \leq 1.1$
found for dust particles of the comet Kohoutek (1973 f)
by Grün et al. (1976).

A better grain model
to meet this
requirements would be
for example silicate
particles with the
values of the
refractive index
tabulated in a paper
by Isobe (1975). The
ß values for silicate
particles are plotted
in figure 3.

The lower curve
represents the ß
values of a purely
dielectric quartz
particle. This curve
has been computed to
check the values with
Gindilis et al. (1969)

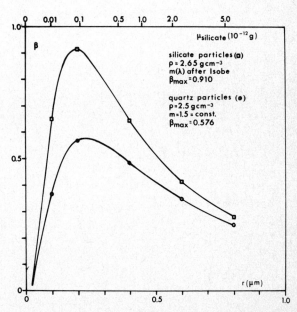

Fig. 3: ß values for silicate particles
(□); refractive index after Isobe
(1974) and for quartz particles
(o) with constant indec of
refraction m = 1.5.

where good agreement was found besides the fact that they found their values for the Q_{pr} by interpolation of tabulated values, whereas in this paper in every step of the calculation the exact value of the efficiency factor was used.

The values of the wavelength dependend refractive indices to be found in the literature are in most cases based on measurements of nearly pure crystalline substances, which are not likely to be very good examples of interplanetary dust particles. The real dust particles in space are effected by collisions and sputtering which will change the surface properties very much. Local impurities and radiation damage will change the optical properties of the particle and will lead in the case of radiation damage to a much higher absorption of radiation as was found during sun simulation tests in the Apollo program. To show how the shape of the curves is changed by an increasing absorption in the grain, I have plotted several curves for a fixed real part of the refractive index and variable imaginary part (figure 4). In order to simplify the matter this was done with the refractive index kept constant over the whole range of wavelengths. The corresponding efficiency factors are shown in fig.1.

The shape of the curve for the dielectric particles is similar to that for Obsidian and Andesite particles (fig.2). With increasing imaginary part of the refractive index, i.e. increasing absorption, the maximum shifts towards smaller radii and the peak becomes very sharp and shows very high values for ß.

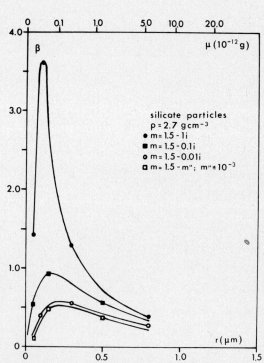

Fig. 4: ß values for different imaginary parts of the refractive index, real part fixed; $\rho = 2.7$ g·cm^{-3}.

This looks very similar to the features of the curve for iron computed with constant index of refraction m = 1.27 - 1.37i as it is often found in the literature.

As already mentioned all the calculation till now have been based on the idealistic assumption of spherical particles and on refractive indices of crystalline substances. Therefore it would be most valuable to learn more about the dynamical behaviour of irregular shaped particles and to get more and more reliable information on the complex indices of refraction for both the visible and infrared region for more realistic materials.

Another problem I would like to indicate is the density of the interplanetary dust grains. If we assume a much lower density than that used in this and most of the other calculations the ß values would be much higher and this would have a very strong influence on the dynamics of the grains.

This research was supported by a grant from the Bundesministerium für Forschung und Technologie (Project HELIOS, E 10 Z).

Literature

Giese, R.H., Schwehm, G., Zerull, R. 1974, Grundlagenuntersuchungen zur Interpretation extraterrestrischer Zodiakallichtmessungen und Lichtstreuung von Staubpartikeln verschiedener Formen, BMFT-FB W 74-10.

Gindilis, L.M., Divari, N.B., Reznova, L.V. 1969, Sov.Astr.A.J. 13, 114.

Grün, E., Kissel, J., Hoffmann, H.-J. 1976, Dust emission from Comet Kohoutek (1973f) at large distances from the sun, this volume.

Güttler, A. 1952, Ann.Phys. 11, 65.

Isobe, S. 1975, Ann. Tokyo Astron.Obs., Second Series, 14, 141.

Labs, D., Neckel, H. 1970, Solar Phys. 15, 79.

Lamy, Ph.L. 1974, Astron. and Astrophys. 35, 192.

Pollack, J.B., Toon, O.B., Khare, B.N. 1973, Icarus 19, 372.

ARE INTERPLANETARY GRAINS CRYSTALLINE?

S. Drapatz and K.W. Michel
Max-Planck-Institut für Physik und Astrophysik
Institut für extraterrestrische Physik
8046 Garching, W.-Germany

<u>Abstract</u>: The optical properties of interplanetary grains depend not only on chemical composition, size, and shape but also on lattice defects. It is argued that practically all sources for the zodiacal dust cloud yield grains with highly disturbed crystal structure. Healing of these defects can occur when the grains are heated in the vicinity of the sun ($\leq$ 0.1 AU) and various orbits are considered, on which annealed grains with nearly perfect crystal structure can return to larger heliocentric distances to reveal the sharp optical features of cold well-ordered crystals. However, we do not find any processes, which produce healed particles of sufficient number to affect the properties of the general zodiacal cloud. Therefore, the optical properties of interplanetary grains are determined by a high degree of lattice perturbation.

1. Introduction

As is well known, there can exist a great difference in the absorption coefficients ($\alpha = 4\pi k \tilde{\nu}$) of the same material if the degree of lattice perturbation is different, as can be seen for silicates (Fig.1).

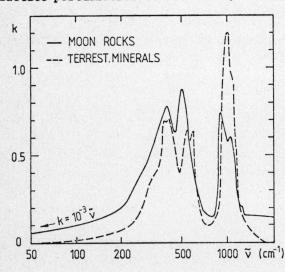

<u>Fig. 1</u>: Extinction coefficient k of silicates with high (moon rocks) and low (terrestrial minerals) degree of lattice perturbation. Averaged moon rock data shown here correspond to samples with low plagioclase content (PERRY et al. 1972). Terrestrial mineral data are taken from POLLACK et al. (1973) and HUFFMAN (1975).

So, for the thermal emission of grains and correlated temperature distribution the question is important, whether grains are well ordered or disturbed, e.g. by radiation damage. Their structure is also important in the investigation of the linear and circular polarization. To investigate the structure of grains the knowledge of processes is necessary that create or destroy crystal structures, viz. which transform well-ordered crystalline structures into disturbed lattices with defects, and vice versa. If material exists in an amorphous phase annealing is possible by recrystallization, i.e. defect-free grains grow within the old deformed ones if roughly $T > \frac{1}{2} T_{melt}$ (Wigley 1971) or if the material melts and cools slowly ($\Delta T/\Delta t < 1$ K/h, SCHOTT 1971). In addition a disturbed lattice may be partly healed by diffusion of point defects (vacancies and interstitials) or impurities to the surface (at $T >$ some $100°$ K). On the other hand radiation damage destroys crystal structures by generation of point defects inside the crystal and sputtering at its surface. Crystals with high defect concentration, amorphous grains and highly heated grains have similarly smeared out and diffuse optical properties, whereas well-ordered crystals show often distinct, sharp absorption and emission features. Hence, close to the sun we expect disordered grains. Perturbed grains can gain sharp bands if they return from the solar vicinity to larger heliocentric distance where they cool. As opposed to well-ordered crystals, the optical properties of perturbed grains show little dependence on temperature.

The question whether the interplanetary grains are well ordered now depends on the answer to the following problems:
- What are the sources of interplanetary grains and do they supply crystalline or highly perturbed material?
- What are the orbits of the grains and can one of the mentioned healing processes occur, while the grain moves on its orbit?

2. The sources of interplanetary grains

The possible sources of interplanetary grains are "primary" ones: interstellar medium, asteroidal belt, comets, primordial debris, and "secondary" sources, which change the size and orbit parameters of primaries or generate secondaries: gravitational force of planets (esp. Jupiter, VAGHI 1973), rotational bursting, collisional break-up (ZOOK and BERG 1975), evaporation (LAMY 1974), impact on planets or their satellites (mercury, moon).

The crystal structure of primary particles of all sources is probably highly perturbed. The interstellar and primordial debris (high orbital inclination) grains were exposed to cosmic rays, which cause high defect concentration (ratio of density of point defects to density of atoms in the grain $n_d/n_o > 0.01$ for some 10^6 y exposure time to interstellar 2 MeV-protons with a flux of 10^2 cm^{-2} sec^{-1}). The comets probably have been formed at temperatures 250 < T < 400 K (Biermann 1975) in the presolar nebula at distances beyond Jupiter's orbit. When orbiting around the sun at R > 4 AU, the internal temperature does not rise beyond $T = 280/\sqrt{R} < 140$ K (no defect healing). As far as asteroidal grains are concerned, only the inner regions of larger (> 1 μ) grains may be crystalline, in spite of the solar wind sputtering and surface erosion, since the penetration depth for 1 keV-protons is low ($\leq$ 0.1 μ in quartz, Hines 1960). The question arises, is there an appreciable fraction of these particles moving around the sun on elliptic orbits with eccentricities high enough that they approach the sun and are heated above $T = \frac{1}{2} T_m$ (corresponding to perihelion distances of $q \approx 0.1$ AU) and that they still have semi-major axis a > 0.5 AU to contribute noticeably to the interplanetary dust.

a) <u>Interstellar particles:</u> An interstellar particle (particle radius a_s, velocity $v_\infty = 20$ km/sec outside the heliosphere) will be captured into an elliptic orbit, if the change Δe of the eccentricity e of the initially hyperbolic orbit during one revolution due to radiation pressure is $\Delta e > e - 1$. Particles with $a_s \leq 0.1$ μ, where solar wind pressure and Lorentz force are dominant (ELSÄSSER 1967) are not considered. The formula below gives perihelion distances q for the particles of maximum impact parameter for capture (= 0.1 AU, at particle sizes $a_s = 0.7$ μ)

$$q \approx R_\odot \left(\frac{20 \text{ km sec}^{-1}}{v_\infty}\right)^{-4/3}$$

For defect annealing and to avoid vaporization q has to be in the range $10 R_\odot < q < 20 R_\odot$, which seems to rule out this source.

b) <u>Comets:</u> The perihelion distance of bright periodic comets (see e.g. compilation by ALLEN 1973) is $q = a(1-e) > 0.3$ AU. Since the change of the eccentricity $\frac{de}{dt}$ is much faster than change of the semi-major axis $\frac{da}{dt}$, these particles will not move in highly eccentric orbits ($q \leq 0.1$ AU), if no perturbation of the orbit other than Poynting-Robertson effect occurs. In addition it has been emphasized by HARWIT (1963) that most cometary dust grains will leave the solar system in hyperbolic orbits.

c) <u>Asteroids:</u> If particles are generated by asteroidal collisions and grinding, their relative velocities (some km sec^{-1}, G.W. WETHERILL 1968) will be smaller than the orbit velocity of asteroids ($\approx 1.8 \cdot 10^6$ cm sec^{-1} at 2.8 AU). Then the eccentricity of the particle orbit is roughly

$$e = \left[\left(1 - \frac{v_\perp^2 r}{\mu}\right)^2 + \left(\frac{v_\| v_\perp r}{\mu}\right)^2\right]^{1/2} \approx 1 - \frac{v_\perp^2 r}{\mu}$$

where $\mu = GM_\odot - \alpha c$ the usual solar gravitational constant decreased by radiation pressure and $v_\perp$, $v_\parallel$ are the particle velocities relative to the sun-asteroid vector r. In order to obtain perihelion distances q 0.1AU a serious perturbation of the orbit has to occur by other than radiation pressure forces which is very unlikely as has also been shown by DOHNANYI (1976).

3. The change of particle orbits and sizes

Particles of asteroidal or cometary origin may be deflected into Mars-crossing orbits and then by Mars into Earth-crossing orbits etc. But even these multi-step processes yield eccentricities too low for $q < 0.1$ AU to occur (WETHERILL 1967). Also "secondary" particles can be generated by impact of primaries on moon or mercury. Take the moon as an example. If lunar material (density ρ_m) is hit by meteorites (density ρ'_m, velocity $w \approx 30$ km sec^{-1}) then the velocity u of emitted particles (DRAPATZ and MICHEL 1974) is

$$u = w(\sqrt{\tfrac{\rho m}{\rho m'}} + 1)^{-1/2} \approx \tfrac{w}{\sqrt{2}}$$

Neglecting in a rough approximation the perturbation of the particle's orbit by the earth and the relative velocity of earth and moon with respect to the earth's orbiting velocity, one finds that for particle radii $a_s \geq 1$ µ the fraction of ejected particles that can become crystalline is $f \approx 0.1$, but considerably smaller for smaller w (see also GAULT 1964).

Inside $r = 20$ $R_\odot$ particles' mass (and hence orbit) may be drastically changed by evaporation of particles or collisional break-up. While the first process (investigated by LAMY 1974) leads to disappearance of particles, the second process might be important and has been considered in another context by ZOOK and BERG (1975). On the basis of their derivation particles with masses $1 - 10^{-5}$ g suffer a catastrophic collision at $r \leq 0.2$ AU (where relative velocities of particles and spatial densities are quite high). The collision debris are small enough for radiation pressure to change the orbit significantly. Particles spiral inside an orbit R for a maximum ratio β_{max} of radiation force to gravitational forces

$$\frac{\beta_{max} - 0.5}{(r_i/R)^2} \frac{\beta_{max} - 1}{r_i/R} = \frac{1}{2}$$

if the meteorites' initial circular orbits have radius r_i, and the debris has zero velocity relative to one of the colliding bodies. Therefore, a considerable fraction of the smaller particles can reach larger heliocentric distances (e.g. $R = 0.6$ AU for $r_i = 0.1$ AU and 0.5 µ particles).

Nevertheless, the contribution of these particles to the mass of the general zodiacal cloud is negligible.

4. Summary

We do not find any processes in interplanetary space which return grains, whose defect concentration has been reduced in the solar vicinity, to larger heliocentric distance with sufficient efficiency, to affect the properties of the general zodiacal cloud. Thus, the imaginary part of the refractive index (related to the extinction coefficient by $\alpha = 4\pi k \tilde{\nu}$) shown for disturbed lunar rocks in Fig. 1 might be recommended also for interplanetary grains. Unfortunately, optical properties of radiation damaged silicates have not yet been determined for the visible spectrum. The extinction coefficients, known for well-ordered terrestrial materials, would imply an absorption of less than 1 % of the incident solar radiation for particles of size $a_s = \lambda/2\pi$, whereas much higher absorption and as consequence higher grain temperatures are expected for more realistic perturbed crystals. Finally it should be mentioned that lattice defects are also important for interstellar grains (DRAPATZ and MICHEL, 1975).

5. References

Allen, C.W.A., Astrophysical Quantities, Athlone Press (1973)
Biermann, L., private communication (1975)
Dohnanyi, J.S., this volume, 1976
Drapatz, S., and K.W. Michel, Z. Naturf. **29a**, 870 (1974)
Drapatz, S., and K.W. Michel, Workshop on Interstellar Grains, Gregynog Hall, Wales (1975)
Elsässer, H., NASA SP-150 p. 287 (1967)
Gault, D.E., et al., NASA TN D-1767 (1963)
Harwit, M., J. Geophys. Res. **68**, 2171 (1963)
Hines, R.L., Phys. Rev. **120**, 1626 (1960)
Huffman, D.R., private communication (1975)
Lamy, Ph.L., Astr. Astrophys. **33**, 191 (1974)
Perry, C.H., et al., The Moon **4**, 315 (1972)
Pollack, J.B., et al., Icarus **19**, 372 (1973)
Schott, "Schott-Zerodur", Mainz (1971)
Wetherill, G.W., J. Geophys. Res. **72**, 2429 (1967)
Wigley, D.A., "Mechanical Properties of Material at Low Temperatures" Plenum Press N.Y. (1971)
Vaghi, S., Astron. and Astrophys. **24**, 107 (1973)
Zook, H.A., and O.E. Berg, Planet. Space Sci. **23**, 183 (1975)

5.11

A TECHNIQUE FOR MEASURING THE INTERSTELLAR COMPONENT OF COSMIC DUST

Daniel A. Tomandl
Department of Astronomy
University of Washington
Seattle, Washington USA 98195

That there is an interstellar (IS) component in cosmic dust has been demonstrated by Pioneers 8 and 9 (Wolf, Rhee, and Berg, 1975). The Pioneer spacecrafts distinguished the IS from interplanetary (IP) dust by measuring particle velocity and direction. Unfortunately, detectors that are capable of measuring a particle's velocity and direction are so restricted in their sensitive area and/or solid angle that their event rate is very low. Pioneers 8 and 9, for example, detected 1-5 IS particles out of 20 events in 7 spacecraft-years of operation (Wolf, Rhee, and Berg, 1976). More events can be obtained, however, if one uses a detector that only measures the direction of travel. The direction alone can be sufficient to distinguish between IS and IP dust--at least on a statistical basis. For example, if most IP dust travels in directions near the ecliptic plane, then an IS flux from out of the plane should be detectable. This paper will examine the use of direction alone in detecting IS particles.

There are two basic types of space probes that are available to look for IS dust by its direction: 1) spacecraft with particle detectors (for example, Pioneers 8 and 9), and 2) the moon. The Pioneers have detected more than 800 particles which did not have enough energy to make a velocity measurement possible but for which there is approximate direction information (the front film data). Lunar samples can be used to look for particle travel directions by examining the microcraters formed upon impact. Only those samples whose lunar orientation was documented can be used. To determine the impact direction more accurately than to within 1 hemisphere, one can examine the bottoms of small cavities where the cavity wall defines a fairly narrow acceptance cone. The disadvantage of using lunar samples compared to man-made spacecraft is that, due to lunar rotation, information about one dimension of the direction of travel is lost. Fortunately, the moon's equator is very near the ecliptic plane (less than 2° difference) so the critical ecliptic latitude dimension is retained. The advantage of using lunar samples is their long integration times--exceeding 10^5 years.

As IS particles come into the solar system in hyperbolic orbits, some of their aspects will be considered. For a given set of conditions --position of the particle detector, sun, and initial velocity (far from the sun) of IS particles--there are two types of hyperbolic impact paths: "direct" and "around". These are illustrated in Fig. 1 wherein the detector is near the earth, on the moon for example. The around paths can be ignored only if their perihelion distances are so small that they are interfered with by the sun, but this is not usually the case. The importance of the around path particles is realized when one considers the example of limiting the problem to gravitational influences alone. In this case, the flux per unit solid angle (specific flux) is the same for both the around and the direct path particles! Furthermore, if the initial IS particle direction distribution is isotropic, then the detected distribution will be isotropic when the direct and around path particles are combined. These results are not changed by considering radiation pressure effects, but they are somewhat modified when the earth's orbital motion is taken into account.

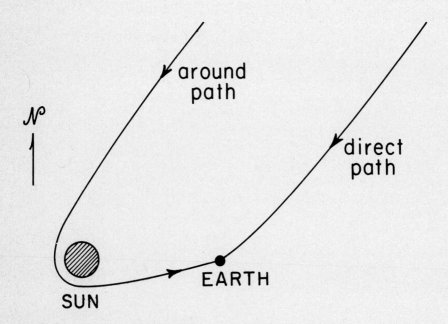

Figure 1. The two types of impact paths: "direct" and "around"

A computer model was developed to calculate the distribution in impact directions from an initial direction of travel of IS particles. The initial direction was assumed to be that due to the sun's local motion in the galaxy as measured with respect to nearby O and B stars. Fortunately, this direction is well away from the ecliptic plane: toward 53° north ecliptic latitude at 20 km/s. Recent measurements of local hydrogen L_α and He I and II emission indicate that IS gas is coming toward the solar system from a direction near the ecliptic plane (Fahr 1974). However, if the solar wind and solar UV flux are asymmetric, the apparent direction of the L_α maximum would be altered. Also the dust particles may not be coupled to the gas. The particles approach the earth between 30 and 60 km/s; at such velocities the earth's and moon's gravitational effects can be neglected. The calculations then include the effects of the sun's gravity, the orbital motion of the earth about the sun, and solar radiation pressure as given by Mie scattering theory.

Typical results of the calculations are given in Fig. 2. Only the direct path impact directions are shown. As the earth changes its position with respect to the sun and IS initial direction, the impact direction changes along the curve of Fig. 2. As mentioned, if a lunar sample is used as a detector, lunar rotation will spread the line into a band. The around path particles have impact directions 10-25° below the ecliptic equator. Thus, the around path particles may be too close to the ecliptic to be distinguished from IP dust. The direct path particles, however, should be discernible on lunar samples and, if enough particles are detected, on spacecrafts.

References:

Wolf, H., Rhee, J.W., and Berg, O.E. (1976). This Volume.

Fahr, H.J. (1974) Space Sci. Rev., 15, 483.

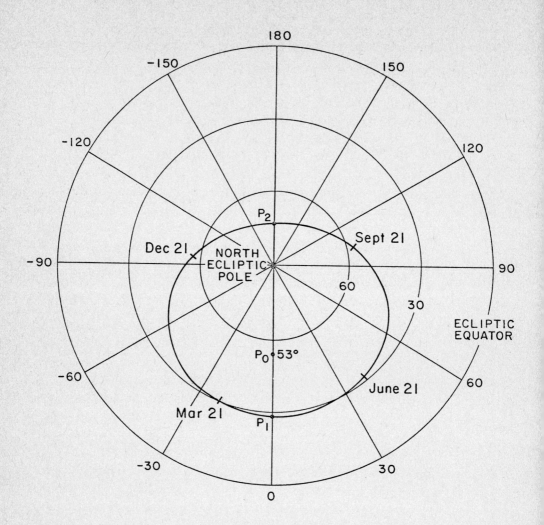

Figure 2. Polar plot of the annual change of the radiant (impact direction) of interstellar dust grains (direct path only). The coordinates, fixed with respect to the stars, are ecliptic longitude $(\lambda - \lambda_o)$ and latitude (β). P_o is the initial direction $(\lambda_o = -90°, \beta_o = 53°$ N). The initial velocity (V_∞) is 20 km/sec. Points P_1 and P_2 are the approximate minimum and maximum latitude points. For no radiation force these are 28° and 73°. For the radiation force equal to one-half the gravitational force these are 26° and 68°.

6 CONCLUDING SUMMARIES

6.1 THE ZODIACAL LIGHT

H. Elsässer
Max-Planck-Institut für Astronomie
D 69 Heidelberg-Königstuhl, F.R.G.

As one of the most important results of what we heard in these days I consider the density law of interplanetary dust derived from zodiacal light observations by the deep space probes going out to Jupiter and going in to 0.3 AU. The dependence on the distance to the sun R seems to be nearly as R^{-1}. This finding is in agreement with a new discussion of ground based observations which was reported by Dumont. The density law was one of the open questions for a long time; for me this represents a break-through.

In the case of number per unit volume, particle flux and size distribution we have now a delightful situation too, since different methods lead to the same picture. Whereas years ago we were facing large differences in fluxes determined by registrations in situ and based on zodiacal light, respectively, the agreement is excellent now and we are already looking for real variations due to the moon etc. May I remind you that the size distribution of the particles can be derived, too, by analyzing the scattered light in interplanetary space, that means in particular from the F-corona as forward scattered sunlight. This was done first by van de Hulst who found an exponent 2.6 for the power law spectrum of the particle radii. My own work 20 years ago led to an exponent 2.0; and this is very much the same as what we see nowadays from the other measurements in this range of particle dimensions.

No comparable progress was experienced in what we know about the particles' nature and their chemical composition. This depends in part on the scattering function, the other function which appears under the well known integrals for brightness and polarization of the zodiacal light. An unique solution of this integral equation is as out of sight as before. On the other hand, in my opinion evidence is accumulating that we have to take into account seriously deviations from the spherical Mie scatterers. But the final answer to the question of the particles nature, I do not expect from derivations of the scattering function.

New and important results on the spectrum and the color of the zodiacal light were reported which show that within a wide range of wavelengths no significant deviations from the solar spectral distribution curve can be observed. There seem to be pecularities in the UV; to verify and to look for such spectral features as Lillie and his colleagues have done is certainly one of the important tasks of the future. Here we could see specific properties of the interplanetary dust which might be on the other hand similar to properties of interstellar dust. These spectral features certainly could give essential hints as to the particles' nature and could inform us about links between interplanetary and interstellar dust.

Now I would like to add a few words about submicron particles (radii ≤ 0.1 μm) which were discussed by Dr. Hemenway with enthusiasm. From the color of the zodiacal light, in particular from the new observations presented here and the model calculations Dr. Giese reported, it is obvious that the appearance of the zodiacal light is not determined by submicron particles. The color would be bluer than what we really observe and from this consideration one can derive an upper limit to the number of submicron particles which could be present in interplanetary space. This upper limit seems to lie near the flux curve given for instance by Dr. Grün and his colleagues.

In addition I should point out that the evidence for submicron particles which was presented here, in my opinion rests at least in parts on an unsound basis. Let me pick out two points: The limb darkening of the sun certainly has nothing to do with particles in the solar atmosphere. It is due to a temperature stratification of the solar atmosphere which causes not only the limb darkening of the continuum but also a center to limb variation of the Fraunhofer lines. - The discrepancy in the coronal temperatures was solved some years ago by new studies of the ionization equilibrium in the solar corona. It became clear that the older investigations had overlooked one important physical process, namely dielectric recombination, and that the correct temperature lies between 1 and 2 million degrees. I could add further points, but do not like to go in more details here.

It is not at all my intention to deny the existence of submicron particles in interplanetary space. Mass loss of stars, in this case of the sun, is a modern concept of astrophysics. Circumstellar dust shells of late type stars which radiate in the IR seem to be one result of stellar mass loss. Also the amount of mass loss which would correspond to Dr. Hemenway's submicron population is not excessively high. But what we need is more reliable evidence and this will most certainly not come from zodiacal light studies.

In a final remark concerning future problems I would like to mention one aspect, the thermal emission of interplanetary dust. In my view it is one of the most important tasks for the future to measure this emission at wavelengths between about 5 and 20 µm and, if possible, to look for spectral features of the same type as they are indicated in the UV. By such observations we could get rid of the scattering function and the Mie scatterer problems and approach the question of the particles nature from a new direction because we observe the dust in a very different mode of behaviour from the visible and UV. Measurements of this kind are not easy. Even from the balloon they are very difficult; one of the problems there is our poor knowledge about airglow emission in that part of the spectrum. It may well be that these measurements need Spacelab.

IN SITU MEASUREMENTS OF DUST

O.E. Berg
Goddard Space Flight Center
Greenbelt, Maryland/USA

In this colloquium, my colleagues and I have reported on data from both the lunar surface and from interplanetary space. I am going to comment briefly and individually on each of these disciplines insofar as they pertain to cosmic dust, but first I would preface those comments with a personal philosophy for each of the two stated disciplines. The two philosophies, you will find, are directly opposite. Concerning the lunar surface, one must recognize that it is a relatively unexplored area for in situ measurements. Accordingly, one must be sufficiently flexible and open-minded to include and consider all available experimental evidence and data until a somewhat consistent model of the lunar surface is derived. Further, and in a similar vein, one's thinking must not be limited to what little we already know about the lunar surface nor to what is known about the Earth's surface and terrestrial particles -- assuming that lunar particles are similar to Earth particles. Our lunar experiment shows strong evidence for the electrostatic transport of fines, and yet it is difficult to accommodate that phenomenon on the lunar surface in view of associated parameters that have been derived for the lunar surface, and in terms of what we know about the electrostatic surface properties of terrestrial particulates. But the electrostatic surface properties have been determined in the laboratory on terrestrial particles which are known to have vastly different properties from lunar fines. Thus, it becomes important to compile and consider all available data during this early exploration.

Now in considering the interplanetary space particle, I advocate adopting a philosophy of greater rigidity. We now have a fair idea for the meteoroid model in our solar system, and we can afford to become selective toward the available experimental results and data that should be included in further refining the model. No longer should we carry along the very early and purely exploratory results that are so far out of line with the rapidly refining model we see emerging. In situ measurements of cosmic dust began in 1949 with Drs. Bohn and Nadig, of Temple University. Their results and those of several subsequent exploratory experiments, using acoustical sensors for cosmic dust fluxes, strongly indicated that cosmic dust constituted perhaps the major hazard to manned space travel. Consequently, a major effort in cosmic dust measurements was supported with the sole objective of evaluating the meteorite hazard in space. Today, as we all know, more sophisticated experiments and improved data analyses have erased the existence of a significant cosmic dust hazard and although the cosmic dust particle has emerged as an extremely interesting astronomical object, some of us adhere to the outmoded practices of the past. The prime example of this and the one I choose to espouse today is the flux curve which obviously needs total revamping:

1) There are data points on that curve which date back to experiments performed in 1955, and earlier, when the field was purely exploratory. These data should be reconsidered in terms of being obsolete, deleted from the curve, and permanently archived in favor of the more recent and more precise experimental results.
2) It is now obvious that no single curve can reasonably represent the flux of dust particles in our solar system because the flux varies markedly with different sets of conditions. There is an established difference between near-earth measurements and deep-space measurements of flux. There are more than two orders of magnitude variation as the sensor axes point toward and away from the sun.

Offhand, the flux curve does not appear to be an important issue to cosmic experimental results. In fact, however, it is generally the first criterion for judging the accuracy or success of an in situ measurement. Consider the case of a bright young experimenter who has been successful in flying a new type of impact device in a satellite or probe. His first question is concerned with how his measured flux agrees with previous results. If he finds his results do not agree precisely with the well established curve, he may make an adjustment within the extremes of experimental error in order to reasonably comply, even though his results may be very accurate for those specific sets of experimental conditions. That is not good. So I would recommend a near-earth flux curve; a deep-space flux curve; and probably an adopted correction factor to accommodate the exposure angle relative to the sun.

I have truly enjoyed this colloquium and I would like to commend the persons who have organized and directed it so efficiently and effectively. I am sure it has been helpful to everyone here to have, gathered under one roof, the investigators from other disciplines in the overall interplanetary dust and zodiacal light ensemble and to hear their results and comments. In this age of rapidly expanding volumes of information on space, it becomes increasingly difficult to keep up with the literature and progress even in our own specific limited field. A colloquium such as this one offers a summary for some of the more significant contributions from several areas concerned with the meteoroid complex. One cannot help but derive an improved overall portrait of the nature of our subject.

I am quite hesitant to select any particular papers as being outstanding lest through the prejudice of my own specific interests I slight other participants or contributors. Many outstanding papers have been presented, but one that comes immediately to my mind is the work presented by Brownlee, et al. I was involved in cosmic dust collections, many years ago, in attempts to recover evidence of craters from micrometeorites, and I recognize some of the difficulties associated with the work and the omnipresent terrestrial contamination. From the evidence that has been presented, I am quite convinced that their particles are indeed of extraterrestrial origin. Their results should be considered a major advance in the field of space science.

Another set of results presented at the colloquium that, in my opinion, warrants special recognition is the successful compositional analyses of cosmic dust as reported by the Max-Planck-Institut für Kernphysik, Heidelberg. Now that this technique has been proved possible and practical, I would assume that ion analyses of particles in space should become the major objective in cosmic dust studies over the next several years. Through this achievement, it becomes possible to distinguish between cometary, asteroidal, and interstellar origins for the particles.

And now a final note on "stardust" and speculation. Speculation is often very good and certainly scientifically acceptable, because it promotes complete explorations of the extremes of variables, and ties in associated fields of science. Speculation over a long period of time without supporting experimental evidence, however, should be discouraged because it tends to confuse the issue. One can assign speculative origins to almost any set of results, I suppose, and a natural consequence is that the speculative origins become validated or invalidated with time. In my opinion, time has run out for stardust.

6.3   CAN COMETS BE THE ONLY SOURCE OF INTERPLANETARY DUST?

A.H. Delsemme
Department of Physics and Astronomy
The University of Toledo
Toledo, Ohio/USA

Abstract: The steady-state of the meteoritic complex can be achieved by the present input rate of cometary dust, if the long-period and "new" comets release a sizeable fraction of their dust, in grains larger than 100 microns.

I want to limit my concluding remarks to the matter of the balance of the budget needed to maintain the meteoritic dust cloud to its present steady state. This is an important topic, since we cannot really guarantee that we understand anything about the origin or the fate of the meteoritic dust cloud, if there are problems balancing its budget.

The dissipating mechanisms require a source of 10 tons per second at steady state (Whipple, 1967). Contrarily to the previous belief, the production rate of dust of the short-period comets observed during this century can only explain 2 % of this amount; those observed during the previous century could explain 3 % of it (Delsemme, 1976). However, fluctuations from one century to another are large. It makes sense, because short-period comets decay quickly, and their set is replenished by comets captured by Jupiter; it happens that only one comet as bright as comet Halley is likely to be captured every century into a short-period orbit; one comet five magnitudes brighter (as bright as the Great Comet of 1577), every 20 millenia, if Vsekhsvyatskii's (1964) statistics have any sense. It is clear that wide statistical fluctuations are going to happen, mainly introduced by the brightest and rarest comets to be captured into short-period orbits.

In order to explain the present steady state, we can formulate several hypotheses:

FIRST HYPOTHESIS

A comet five magnitudes brighter than Comet Halley was captured less than 20,000 years ago into a short-period orbit. Comets decay fast, and we do not have any direct evidence of such a capture. However, the meteoroid stream associated with Comet Encke suggests that it has decayed for a long time, and could have been such a great comet a few millenia ago: this was Fred Whipple's (1967) idea, and it should not be rejected lightly. The probability of capture of such a bright comet is of the order of 0.005 per century, therefore, among the 100-odd short period comets that we know, the probability that at least one of them represents the remnants of a very bright comet is 0.5; this probability is large enough to feel comfortable with an identification with Comet Encke.

SECOND HYPOTHESIS

The set of long-period and "new" comets produces enough mass in particles large enough not to be ejected out of the solar system by the radiation pressure of the sun. I have shown previously (Delsemme, 1976) that the long-period comets produce an average of 20 tons of dust per second, if the cutoff of their brightness distribution is set at the (observed) absolute magnitude of -2. If the major portion of this mass were in particles of the millimeter size, so that the light pressure could be neglected, then symmetry arguments on the vector addition of the particle velocity to the comet's velocity show that half of this mass would be ejected on hyperbolic orbits, but 10 tons/sec would be kept on more or less elongated orbits. These orbits would then decay into the inner solar system, much as the larger bodies' orbits do.

There are possible variations of this hypothesis. For instance:

a) the cutoff in the brightness distribution of comets can be put elsewhere than at those comets that have been historically observed, namely at magnitude -2. If Vsekhsvyatskii's distribution is extrapolated up to magnitude -7, 99 % of the dust can be lost on hyperbolic orbits, and 10 tons/sec still are steadily captured by the inner solar system.

b) Large particles may represent a much larger fraction of the cometary dust than expected before. Of course, we cannot easily detect them in cometary tails because they do not reflect enough light per unit mass, but several circumstantial arguments point to their existence. As a matter of fact, one of the concluding remarks I would like to submit, is that large particles have become fashionable during this colloquium: we must use them to explain some of the cometary antitails; some of the meteor spectra; or the peculiar polarization of the zodiacal light (specular reflection on rather large crystal facets); we even collect them in the upper atmosphere. They obviously are also in the meteor streams even if they do not reflect enough light to be identified in space: we have learned here that 50 % of the mass of the new meteor streams lies in particles of the order of a gram.

c) All comets (short period, long period and "new") with perihelia up to Jupiter and probably Saturn, should possibly be included in the balance of the budget. The argument is that, if they cannot vaporize water, they could vaporize huge amounts of other gases more volatile than water. Some other faraway comets have been known to have comas or tails at distances where water could not vaporize. An excellent example is P/Schwassmann-Wachmann I. The total population of comets vary in approximate proportion to their perihelion distances; therefore, we multiply it by five if the radius of the vaporization sphere goes from 2 to 10 AU. For long-period or "new" comets we would therefore be allowed to loose 90 % of the dust on hyperbolic orbits, and still be left with the desired rate of capture of 10 tons/sec in the inner solar system. As far as short-period comets are concerned, Comet Schwassmann-Wachmann I has the advantage of being a giant comet on a quasi-circular orbit and could explain most of the dust needed if its outbursts represent a steady state that has lasted ten millenia.

d) The hypothesis of large particles has reminded us that we may be wrong to try to balance the budget with what is easily seen; we see the fine component only because it reflects much light. The total mass of the short-period comet set is not known even within two orders of magnitude. However, it is likely to be very much larger than the dust fraction that is dragged away in the cometary tails. If we accept that most of these comets are not going to die into big asteroids, but that a major mass fraction is going to decay into invisible large chunks of smaller and smaller size, as possibly evidenced by the meteoroids associated with Comet Encke; then we certainly may have enough mass to balance our budget. The spectacular phenomenon of the cometary dust tail, this "bagful of nothing" may have distracted our attention from something more important going on: the steady escape of the major mass into unobservable chunks that decay by steps into meteoroid streams.

SUMMARY

It is clear that the present production rate of dust by the short-period comets cannot provide more than 2 or 3 % of the mass needed to explain the steady state. In particular, my analysis has been independently confirmed by S. Röser (1976) during this colloquium.

Therefore, either a very bright comet was captured by Jupiter some five or ten thousand years ago. Comet Encke could be its remnants, and the zodiacal light is slowly decaying, waiting for the next bright comet to be captured;

or the major fraction of the dust of the short-period comets is <u>not</u> dragged away by the vaporization of water, but by a stuff more volatile than water, like methane or carbon monoxide or dioxide, in a much larger sphere around the sun. Giant Comet P/Schwassmann-Wachmann would then be a good candidate among the short-period comets of large perihelion distance, although its rare outbursts do not suggest a vaporization steady-state;

or the long-period comets contribute a large fraction of the dust captured. They indeed produce a <u>total</u> dust mass which is at least twice what we need; possibly twenty times if we include comets beyond Jupiter; possibly two hundred times if we include unobserved but historically predictable very bright comets. The real question becomes: what is the fraction of their dust lost on hyperbolic orbits? It could be 50 % only if the particles are in the centimeter size range; 90 % in the millimeter size, 99 % in the 100 micron size, so that the contribution of the long-period comets cannot be neglected if the particles dragged away are large, which is becoming an acceptable idea.

NSF Grant GP 39259 is gratefully acknowledged.

<u>REFERENCES</u>

Delsemme, A.H. (1976) IAU Colloquium No. 31, this volume.

Röser, S. (1976) IAU Colloquium No. 31, this volume.

Vsekhsvyatskii, S.K. (1964) Physical Characteristics of Comets, Translated from Russian, Israel Program of Scientific Translation, NASA-TT-F-80.

Whipple, F. (1967) p. 409 in "The Zodiacal Light and the Interplanetary Medium", edit. J.L. Weinberg, NASA-SP-150.

6.4

METEORS

Z. Ceplecha
Astronomical Institute
Czechoslovak Academy of Sciences
Ondrejov Observatory

After hearing so many technical descriptions of different high velocity particle experiments, one easily recognizes the paramount importance of sensors for gathering the experimental data on meteoric dust. In contrast the "classical" meteor astronomy and physics is using a less expensive "sensor", the Earth's atmosphere. The interaction of the meteor body with the air has to be recorded and thus our experimental data are just records of the natural phenomenon itself: integral light photographs, spectral photographs, recently also image intensifier photographs and videorecords. The only "active" method is the radar observation of the ionized trail. The close distance of the phenomenon to the observational sites enables one to determine the complete geometrical and dynamical data (heights, distances, velocities); the light recording gives the intensity of the emitted light at individual points of the trajectory, the radar recording gives the ionization intensity. A suitable physical theory is necessary to convert the basic observational data into meteor mass and other parameters. The drag equation yields the "dynamic" mass, the luminosity equation yields the "photometric" mass, and the ablation equation yields the "ablation" mass. For a single meteor, these masses are dependent on parameters defined by the drag coefficient, shape, bulk density, ablation rate and luminous efficiency. Differences in the resulting masses of a meteoroid computed from the different methods at the same trajectory point are good relative measures of the structural and compositional differences. The absolute calibration may be a problem, but laboratory and rocket measurements of the luminous efficiency, calibrations by the Lost City and Pribram fireballs and several other direct and indirect methods can be used.

The most important result in respect to the structure and composition of the meteoroids yielding meteors from +3 to -22 absolute stellar magnitude (masses from 10^{-4} to 10^{8} g) is the detection of several widely different types of the material. Bigger bodies (giving fireballs) can be related to three basic structures (Ceplecha, McCrosky: in preparation for JGR): strong stones (ordinary chondrites), carbonaceous (and "precarbonaceous") chondrites, weak cometary material.

The cometary material has two distinctly different structures: the classical cometary material (with densities of the order of several tenths of g/cm^3) and the Draconid type material with densities approaching 0.1 g/cm^3. Thus for bodies of quite big dimensions (tens of centimeters to tens of meters), we have observational evidence of greatly different structure: the strongest body could be hardly destroyed by hitting it with a hammer, the weakest body would disappear and only make finger tips dirty after touching. Each of these three structural types of meteoroids participates approximately equally in the whole population of fireball bodies. The number of all fireballs and the corresponding incoming mass within the range of 10^5 to 10^9 grams is enough to explain the enhancement of the atmospheric and surface dust of meteoric origin.

The faint photographic meteors (photographed by Super-Schmidt cameras) contain only material of the cometary group and the carbonaceous group; the ordinary chondrite group is almost missing: few cases of assumed "stones" were used previously by some authors as calibration points and often called "asteroidal" meteors (this is not a good name, because there is no direct evidence for asteroidal origin in contrast to the direct association of individual comets with meteor showers). A huge difference in the penetration ability of two bodies of approximately the same dynamical and geometrical parameters of the trajectory was presented by Ceplecha (Table 2) during this colloquium.

When interpreting our observational results on meteors a decade ago, we started with the wrong, but simple assumption that we are searching for one average structural composition. Thus I would strongly recommend to consider the distribution of meteor particles of any range of sizes (meteoric dust, too) in terms of several different populations statistically superimposed in the experimental data. Nature is not so simple that she prepares just one statistical distribution for our comfort!

The groups of different structure and composition have also different orbits, but the question of their origin is not trivial. There is the direct evidence through showers that the cometary material originates from comets, but also the carbonaceous material seems to be present in some meteor showers with known cometary associations. The statistics of orbits of the carbonaceous material seems to be not much different from the distribution of the ordinary chondritic material and the possibility that the strongest meteoroid constituent is also of cometary origin is still open.

The problem of identification of a cometary orbit with a meteor shower is a well known and established procedure, because the bodies are relatively big, gravitation being the decisive force, and the dispersion time being sufficiently long to preserve the orbital elements until the collision with the Earth. But this does not apply to tiny picogram dust grains observed in space. Kresák (paper of this Coll.) investigated the gravitational, radiative, and destructive effects governing the rate of displacement, dispersion, and removal of interplanetary dust streams. He found drastic effects and he sees no way how a compact dust stream might be maintained over a number of revolutions. The proposed comet-micrometeoroid associations seem to be fictitious and the origin of the observed dust streams should not be connected with distant large bodies such as comets.

The interpretation of the spectral records of meteors would seem to be the most direct way to arrive at the chemical composition. However, the meteor spectra mostly reflect the impact velocity. In case of fireballs, the light is radiated mostly from the very surface of the luminous gas volume, which casts doubts on any abundances of elements determined from meteor spectra. The only exception is the case of the Draconid-type meteors: they start to emit light very high, where the free molecular flow enables us to see all the light produced. The abundances of elements computed for such meteors by Millman correspond roughly to carbonaceous abundances. The future quantitative study of spectra of much fainter meteors (around +3 stellar magnitude), not available today, may give better results on the chemical composition, because of much less importance of the selfabsorption phenomenon. Qualitatively the overwhelming majority of meteor spectra are similar to each other and are roughly independent of the meteoroid structure. A very small percentage (increasing with decreasing brightness) of meteors exhibit spectra without iron lines. Another extreme is formed by a few known spectra containing only iron lines. The best existing spectrum of this type corresponds to a typical cometary meteor (orbit and atmospheric trajectory as well).

Meteors observed by radar are closer to the range of sizes of meteoric dust particles. The observational data are not so accurate as the optical photographic data, but much fainter meteors can be studied independently of weather. The structural variety of these small bodies seems to be not so wide as for the bigger bodies and the "stony" population is missing. But the experimental data are derived from very short trajectories under several strongly selective conditions, which put more uncertainity into the results.

The future progress in meteor astronomy and physics depends on several promising space experiments. First, it would be very important to calibrate spectral observations of meteors by producing artificial meteors from a suitable orbiting station. The UV spectra of natural meteors (and artificial meteors as well) are not accessible from the ground due to the absorption in the lower part of the earth's atmosphere. Observations of UV spectra from an orbiting space station is one of the important experiments in the near future. The calibration of radar observations by artificial faint meteors fired with rather high velocity from an orbiting station would enable more reliable interpretations of natural meteors observed by radar. The recording technique should at least partly turn to electronic systems giving videorecords, which could be directly processed by a computer.

The main task of all such experiments and observations is the study of different meteoroid populations and their relative importance among all bodies within a given range of sizes. Inter-corporating all results into one frame, we will arrive at the complete picture of interlocking populations over a huge range of meteoroid masses.

6.5 FINAL REMARKS

F.L. Whipple
Center for Astrophysics
Harvard College Observatory and Smithsonian Astrophysical Observatory
Cambridge, Massachusetts/USA

My summary will certainly leave out much fine work but will stress a few of the points that impressed me.

First, I note remarkable progress since 1967 when I last worried about the interplanetary complex, particularly the new observations from space and of very small particles. I am delighted to see the ground based Zodiacal Light observers, like Weinberg and Dumont, converging on their observations both of polarization and of intensity. The larger particles are clearly the major contributors particularly away from the Sun. This forces the theoreticians and the laboratory people to find out how these larger particles produce the observed polarization including the negative polarization. Theoreticians always can prove the answer when it is known.

The law of intensity variation from the Sun has been summarized. There is a discrepancy with the Pioneer 10 particle measures. I think that observations of the Zodiacal Light in space will be of more value near the Sun than farther away. No north-south asymmetry remains and it is now completely proven, I think, that the Zodiacal particles are concentrated with respect to the fundamental plane of the solar system, by the Apollo 15, 16, 17 observations. The high reflectivity in the very far ultraviolet is of great importance and, as Dr. Elsässer mentioned, observations are needed in the infrared as well as in the ultraviolet.

Skylab indicates no short-term intensity variations. Levasseur and Blamont find evidence for near-Earth variations produced by meteor streams. My own personal opinion: MAYBE?

Observations from space have proven conclusively that the Gegenschein is an optical phenomena of back-scattering by particles. As for observed clouds of particles in the Lagrangian points of the Earth-Moon system I again say MAYBE.

Delsemme has given a fine account of the comets as a source of the interplanetary particles and I agree with his conclusion. There is a great deal of mass to be contributed by comets and Dohnanyi's evidence against the asteroids confirms my own opinion.

Next we have the exciting discovery of beta-meteoroids, beta representing a finite ratio of solar light pressure to gravity. Berg and his co-workers have proven their point. With the HEOS, Fechtig and his group have proven finally that there is a strong tendency for particles to break up near the Earth to produce little clusters and groups. With regard to the refractory Hemenway particles generated in the Sun, I think it is very important to look for them. Here I have my own opinion: MAYBE?

On the atmospheric collection of small particles I think Dr. Brownlee and his group well deserve the resounding congratulations already given them for capturing what are almost certainly cometary particles that we can hold in our hands and study. The structure is truly remarkable for these sub-micron clusters. They look like fish eggs. They will certainly be of extreme importance in determining the nature of the formation of comets.

The Moon crater work is really extremely impressive to me, that such detailed studies can be made of craters down to submicron size. The same is true of the other laboratory studies on lunar samples. I congratulate the workers and stress that in the laboratory some genius must find a method for accelerating low density, fragile particles or clusters of small particles to velocities of kilometers per second. Otherwise calibration of craters versus meteoroid velocity and mass remains obscure. But I do congratulate the group here on their very fine work on laboratory crater formation.

There appears to be a big question as to whether the rates of impacting debris on the Moon at the present time are greater than they were 50,000 years ago. That question was answered in the positive by one worker and in the negative by another. The question revolves around dating processes on the Moon by means of high-energy particles from solar flares. The problem demands resolution.

With regard to the interstellar dust observed in space, particularly by Pioneers 8 and 9, I must say I began looking for interstellar meteors in 1933 when I made some calculations identical in principle to those by Tomandl. I calculated the radiants of meteors coming from a great cloud around the star Sirius, following Öpik's proof in 1931 that a cloud of particles could be stable over billions of years against the passage of stars at great distances from a star. My calculations led nowhere. They were never published and I could never find any clear evidence of any hyperbolic particles. I hope he has better success than I.

The experiments on the Moon have proven there are mobile charged dust particles on the Moon. I do wish that Berg and Rhee could agree whether they were going from the light side plus-to-minus or from the dark side minus-to-plus but I am sure this will be resolved. Continued studies of this phenomena should yield very important results.

Over thirty years ago I asked the question "Does the Moon gain or lose mass by impact accretion?" The question is still not answered, at least by concensus. There seems to be some strong evidence on both sides. Laboratory studies show that the crater debris won't go out fast enough. Some lunar evidence, not discussed here at this meeting indicate that there must be some loss of mass.

Paddock and Rhee have at last produced some good theory and laboratory work on the spin up of small particles in space by solar radiation, mentioned so many years ago by Öpik. With the experimenters I feel that you can spin them fast but you can't break them up that way. If they are fragile enough to break they will not last long after they are released from a comet. But there is clear evidence that meteor dust is revolving rapidly: the Soberman experiment. I congratulate the group with HEOS 2 in finding dust from Comet Kohoutek. Their calculated rate of loss of matter from that comet is amazingly large. If finally confirmed the result will give us food for thought about the surfaces of new comets. Can we ever observe this phenomenon again in other comets?

It is very encouraging to see good theories for the antitails of comets fitting so well with observation. I think the resultant particle size distribution among meteoroids has been discussed adequately except for one point. I believe it is now firmly established that reduction in slope, or nearly a "stillstand", occurs on the upgoing curve of log (cumulative number) versus log (mass curve). This indicates a drop off in the normal particles of cometary origin around 10^{-8}-10^{-9} grams. Then there is a build-up of the smaller particles which may or may not extend out to 10^{-18} grams or further, probably due to fragmentation. I think this is enormous progress since 1967 when we had no knowledge of what went on much below about 10^{-7} grams.

The particle density clearly varies inversely and only statistically in some fashion with particle size. Is the explanation a) survival of the fittest in the collisional processes or b) that the interiors of great comets contain harder material with more coherence than material from the outer layers? The answer bears heavily on the processes of comet formation.

In meteor streams the time arrow is very clear. Meteor streams from comets start out very narrow and compact and continuously spread out, violently in conflict with the jet-stream concept of Alfvén and Trulsen. The fact that the densities of meteoroids also correlate with the ages, further shows that the jet-stream theory among cometary meteor streams is wrong. On the other hand, I must congratulate Trulsen that this theory provides such a fine relationship between the inclinations and the eccentricities among the asteroids in the early <u>dissipative</u> stages.

I am delighted to see the clear-cut relationship between meteors and the solar cycle finally established by Lindblad. I worried about such a relationship so many years ago and could never prove a correlation.

From the theories presented this morning, small conducting particles with an imaginary term in the index of refraction will definitely spiral away from the Sun, counter to the Poynting-Robertson effect. Dielectrics will spiral on into the Sun. It is a very interesting question whether indeed we have any pure silicates, dielectrics, in mixed material that must be the basis of a comet. There must not be any pure crystals of any sort, or are there? Iron must surely be reasonably abundant to produce opacity. The particles must be radiation damaged, and that, I understand, will produce some opacity, following Harwit's experiments. Finally it seems to me that close to the Sun, when the particle begins to warm up and the vapor pressure becomes significant, then the particle will begin to radiate and therefore to absorb. Thus it is difficult to believe that any particles will spiral into the Sun. Nevertheless, I think that we need more theory and more laboratory work on the problems of light pressure on tiny particles, particularly those with very irregular shapes.

Now to summarize some of the jobs to be done. We need both laboratory work and theory on the radiation problem just mentioned and desperately need them on the polarization and reflectivity problems of particles as functions of irregular shapes, varying sizes, phase angles, etc. Such data are most important for the larger particles in the millimeter range, which contribute most to the Zodiacal Light.

Also at last, electric charge is beginning to count with regard to these very tiny particles in space. In 1940 I first calculated the likely charges but they turned out to be only a few volts, unimportant for sizeable particles (unpublished). Now I think we must watch for charge effects, particularly on these particles that break up near the Earth.

Laboratory experiments on projecting low-density particles to very high velocities to study crater formation are especially important.

With regard to the Zodiacal Light there is a plea for more observations with respect to latitude, longitude, elongation from the Sun and, as Dr. Elsässer mentioned, in the far infrared. Lillie's very interesting results in the far ultraviolet below 2,000 angstroms demand more observations from space, again as a function of solar distance.

I see I have 20 seconds left. In this time I make a plea for space missions to comets and to asteroids. I am of the opinion that many of the atoms in our bodies come from comets, perhaps a major portion, and many from asteroids. Life on Earth may exist only because of comets in the early history of the solar system. Thus in situ studies of comets may be critical to studies of life in the universe.

Authors Index

Alexander, W.M.	458	Hoffmann, H.-J.	159, 334
Alvarez, J.M.	181, 182	Hofmann, W.	52
Andreev, V.V.	383	Howard, R.A.	66
Bandermann, L.W.	101	Jousselme, M.F.	443
Beeson, D.E.	29		
Belkovich, O.I.	383, 400	Kissel, J.	159, 164, 334
Berg, O.E.	165, 233, 478	Koomen, M.J.	66
Blamont, J.	58	Koutchmy, S.	343
Brownlee, D.E.	279	Kresák, L.	391, 396
Burnett, G.B.	53	Lamy, Ph.L.	343, 437, 443
Ceplecha, Z.	385, 485	Leinert, C.	19, 24, 120
Cruvellier, P.	74	Lemke, D.	52
		Lena, P.	67
Dalmann, B.-K.	164	Levasseur, A.C.	58
Delsemme, A.H.	314, 481	Lillie, Ch.F.	63
Dohnanyi, J.S.	170, 187	Lindblad, B.A.	373, 390
Donn, B.	345	Link, F.	107
Drapatz, S.	464	Link, H.	19, 24, 120
Dumont, R.	85, 115	Llebaria, A.	78
Eichhorn, G.	243	Matsumoto, T.	323
Elsässer, H.	475	Maucherat, M.	74
		Michel, K.W.	328, 464
Fechtig, H.	143, 159, 275, 290	Millman, P.M.	359
Frey, A.	52	Mori, K.	36
		Morrison, D.A.	227
Gammelin, P.	159	Mujica, A.	122
Giese, R.H.	135	Munro, R.H.	65
Grün, E.	135, 159, 164, 334		
		Nagel, K.	241, 275
Hahn, R.C.	45	Neukum, G.	275
Hall, D.	67	Newburn, R.L.	346
Hallgren, D.S.	270, 284	Nishimura, T.	328
Hanner, M.S.	29	Nock, K.T.	346
Hartung, J.B.	209		
Hayakawa, S.	323	Ono, T.	323
Hemenway, C.L.	251, 270, 284, 290		
Hodge, P.W.	279	Paddack, S.J.	453

Cont'd

Pittich, E.M.	396	Yeates, C.M.	346
Pitz, E.	19, 24, 120		
Porubčan, V.	379	Zerull, R.	130
Potapov, I.N.	400	Zinner, E.	227, 232
Poupeau, G.	232		

Rhee, J.W.   165, 233, 238, 448, 453
Richards, M.A.   458
Roach, J.R.   68
Robley, R.   121
Röser, S.   124, 319
Rosinski, J.   289
Ross, C.L.   64, 73

Salm, N.   19, 24
Sánchez, F.   122
Schneider, E.   241, 242, 275
Schwehm, G.   459
Sekanina, Z.   339, 434
Soberman, R.K.   182
Soufflot, A.   67
Sparrow, J.G.   29, 41, 45
Stähle, V.   241
Staude, H.J.   106

Tanabe, H.   36
Teptin, G.M.   389
Thum, C.   52
Tokhtas'ev, V.S.   383
Tomandl, D.   279,469
Trulsen, J.   416

Vanýsek, V.   299
Viala, Y.   67

Walker, R.M.   232
Weinberg, J.L.   3, 29, 41, 45, 182
Whipple, F.L.   403, 489
Wlochowicz, R.   284
Wolf, H.   165, 233
Wolstencroft, R.D. 101

SPRINGER TRACTS IN MODERN PHYSICS

Ergebnisse der exakten Naturwissenschaften

Editor: G. Höhler

Associate Editor:
E. A. Niekisch

Editorial Board:
S. Flügge, J. Hamilton,
F. Hund, H. Lehmann,
G. Leibfried, W. Paul

Volume 66
30 figures. III, 173 pages. 1973
ISBN 3-540-06189-4

Quantum Statistics
in Optics and Solid-State Physics

R. Graham: Statistical Theory of Instabilities in Stationary Nonequilibrium Systems with Applications to Lasers and Nonlinear Optics.
F. Haake: Statistical Treatment of Open Systems by Generalized Master Equations.

Volume 67
III, 69 pages. 1973
ISBN 3-540-06216-5

S. Ferrara, R. Gatto, A. F. Grillo:

Conformal Algebra in Space-Time
and Operator Product Expansion

Introduction to the Conformal Group in Space-Time. Broken Conformal Symmetry. Restrictions from Conformal Covariance on Equal-Time Commutators. Manifestly Conformal Covariant Structure of Space-Time. Conformal Invariant Vacuum Expectation Values. Operator Products and Conformal Invariance on the Light-Cone. Consequences of Exact Conformal Symmetry on Operator Product Expansions. Conclusions and Outlook.

Volume 68
77 figures. 48 tables. III, 205 pages. 1973
ISBN 3-540-06341-2

Solid-State Physics
D. Schmid: Nuclear Magnetic Double Resonance — Principles and Applications in Solid-State Physics.
D. Bäuerle: Vibrational Spectra of Electron and Hydrogen Centers in Ionic Crystals.
J. Behringer: Factor Group Analysis Revisited and Unified.

Volume 69
13 figures. III, 121 pages. 1973
ISBN 3-540-06376-5

Astrophysics
G. Börner: On the Properties of Matter in Neutron Stars.
J. Stewart, M. Walker: Black Holes: the Outside Story.

Volume 70
II, 135 pages. 1974
ISBN 3-540-06630-6

Quantum Optics
G. S. Agarwal: Quantum Statistical Theories of Spontaneous Emission and their Relation to Other Approaches.

Volume 71
116 figures. III, 245 pages. 1974
ISBN 3-540-06641-1

Nuclear Physics
H. Überall: Study of Nuclear Structure by Muon Capture.
P. Singer: Emission of Particles Following Muon Capture in Intermediate and Heavy Nuclei.
J. S. Levinger: The Two and Three Body Problem.

Volume 72
32 figures. II, 145 pages. 1974
ISBN 3-540-06742-6

D. Langbein:

Theory of Van der Waals Attraction
Introduction. Pair Interactions. Multiplet Interactions. Macroscopic Particles. Retardation. Retarded Dispersion Energy. Schrödinger Formalism. Electrons and Photons.

Volume 73
110 figures. VI, 303 pages. 1975
ISBN 3-540-06943-7

Excitons at High Density
Editors: H. Haken, S. Nikitine
Biexcitons. Electron-Hole Droplets. Biexcitons and Droplets. Special Optical Properties of Excitons at High Density. Laser Action of Excitons. Excitonic Polaritons at Higher Densities.

Volume 74
75 figures. III, 153 pages. 1974
ISBN 3-540-06946-1

Solid-State Physics
G. Bauer: Determination of Electron Temperatures and of Hot Electron Distribution Functions in Semiconductors.
G. Borstel, H. J. Falge, A. Otto: Surface and Bulk Phonon-Polaritons Observed by Attenuated Total Reflection.

Springer-Verlag
Berlin
Heidelberg
New York

Selected Issues from

Lecture Notes in Mathematics

Vol. 507: M. C. Reed, Abstract Non-Linear Wave Equations. VI, 128 pages. 1976.

Vol. 501: Spline Functions, Karlsruhe 1975. Proceedings. Edited by K. Böhmer, G. Meinardus, and W. Schempp. VI, 421 pages. 1976.

Vol. 495: A. Kerber, Representations of Permutation Groups II. V, 175 pages. 1975.

Vol. 490: The Geometry of Metric and Linear Spaces. Proceedings 1974. Edited by L. M. Kelly. X, 244 pages. 1975.

Vol. 489: J. Bair and R. Fourneau, Etude Géométrique des Espaces Vectoriels. Une Introduction. VII, 185 pages. 1975.

Vol. 485: J. Diestel, Geometry of Banach Spaces – Selected Topics. XI, 282 pages. 1975.

Vol. 484: Differential Topology and Geometry. Proceedings 1974. Edited by G. P. Joubert, R. P. Moussu, and R. H. Roussarie. IX, 287 pages. 1975.

Vol. 481: M. de Guzmán, Differentiation of Integrals in R^n. XII, 226 pages. 1975.

Vol. 480: X. M. Fernique, J. P. Conze et J. Gani, Ecole d'Eté de Probabilités de Saint-Flour IV–1974. Edité par P.-L. Hennequin. XI, 293 pages. 1975.

Vol. 477: Optimization and Optimal Control. Proceedings 1974. Edited by R. Bulirsch, W. Oettli, and J. Stoer. VII, 294 pages. 1975.

Vol. 474: Séminaire Pierre Lelong (Analyse) Année 1973/74. Edité par P. Lelong. VI, 182 pages. 1975.

Vol. 470: R. Bowen, Equilibrium States and the Ergodic Theory of Anosov Diffeomorphisms. III, 108 pages. 1975.

Vol. 468: Dynamical Systems – Warwick 1974. Proceedings 1973/74. Edited by A. Manning. X, 405 pages. 1975.

Vol. 464: C. Rockland, Hypoellipticity and Eigenvalue Asymptotics. III, 171 pages. 1975.

Vol. 463: H.-H. Kuo, Gaussian Measures in Banach Spaces. VI, 224 pages. 1975.

Vol. 461: Computational Mechanics. Proceedings 1974. Edited by J. T. Oden. VII, 328 pages. 1975.

Vol. 459: Fourier Integral Operators and Partial Differential Equations. Proceedings 1974. Edited by J. Chazarain. VI, 372 pages. 1975.

Vol. 458: P. Walters, Ergodic Theory – Introductory Lectures. VI, 198 pages. 1975.

Vol. 449: Hyperfunctions and Theoretical Physics. Proceedings 1973. Edited by F. Pham. IV, 218 pages. 1975.

Vol. 448: Spectral Theory and Differential Equations. Proceedings 1974. Edited by W. N. Everitt. XII, 321 pages. 1975.

Vol. 447: S. Toledo, Tableau Systems for First Order Number Theory and Certain Higher Order Theories. III, 339 pages. 1975.

Vol. 446: Partial Differential Equations and Related Topics. Proceedings 1974. Edited by J. A. Goldstein. IV, 389 pages. 1975.

Vol. 445: Model Theory and Topoi. Edited by F. W. Lawvere, C. Maurer, and G. C. Wraith. III, 354 pages. 1975.

Vol. 444: F. van Oystaeyen, Prime Spectra in Non-Commutative Algebra. V, 128 pages. 1975.

Vol. 443: M. Lazard, Commutative Formal Groups. II, 236 pages. 1975.

Vol. 442: C. H. Wilcox, Scattering Theory for the d'Alembert Equation in Exterior Domains. III, 184 pages. 1975.

Vol. 441: N. Jacobson, PI-Algebras. An Introduction. V, 115 pages. 1975.

Vol. 440: R. K. Getoor, Markov Processes: Ray Processes and Right Processes. V, 118 pages. 1975.

Vol. 439: K. Ueno, Classification Theory of Algebraic Varieties and Compact Complex Spaces. XIX, 278 pages. 1975.

Vol. 438: Geometric Topology. Proceedings 1974. Edited by L. C. Glaser and T. B. Rushing. X, 459 pages. 1975.

Vol. 437: D. W. Masser, Elliptic Functions and Transcendence. XIV, 143 pages. 1975.

Vol. 436: L. Auslander and R. Tolimieri, Abelian Harmonic Analysis, Theta Functions and Function Algebras on a Nilmanifold. V, 99 pages. 1975.

Vol. 435: C. F. Dunkl and D. E. Ramirez, Representations of Commutative Semitopological Semigroups. VI, 181 pages. 1975.

Vol. 434: P. Brenner, V. Thomée, and L. B. Wahlbin, Besov Spaces and Applications to Difference Methods for Initial Value Problems. II, 154 pages. 1975.

Vol. 433: W. G. Faris, Self-Adjoint Operators. VII, 115 pages. 1975.

Vol. 432: R. P. Pflug, Holomorphiegebiete, pseudokonvexe Gebiete und das Levi-Problem. VI, 210 Seiten. 1975.

Vol. 431: Séminaire Bourbaki – vol. 1973/74. Exposés 436–452. IV, 347 pages. 1975.

Vol. 430: Constructive and Computational Methods for Differential and Integral Equations. Proceedings 1974. Edited by D. L. Colton and R. P. Gilbert. VII, 476 pages. 1974.

Vol. 429: L. Cohn, Analytic Theory of the Harish-Chandra C-Function. III, 154 pages. 1974.

Vol. 428: Algebraic and Geometrical Methods in Topology, Proceedings 1973. Edited by L. F. McAuley. XI, 280 pages. 1974.

Vol. 427: H. Omori, Infinite Dimensional Lie Transformation Groups. XII, 149 pages. 1974.

Vol. 426: M. L. Silverstein, Symmetric Markov Processes. X, 287 pages. 1974.